Robert Adam and his brothers

New light on Britain's leading architectural family

Robert Adam and his brothers

New light on Britain's leading architectural family

Edited by Colin Thom

Historic England

Published by Historic England, The Engine House, Fire Fly Avenue, Swindon SN2 2EH
www.HistoricEngland.org.uk

Historic England is a Government service championing England's heritage and giving expert, constructive advice.

First published 2019

ISBN 978-1-84802-359-8

British Library Cataloguing in Publication data
A CIP catalogue record for this book is available from the British Library.

Brought to publication by Victoria Trainor, Publishing, Historic England.
Typeset in Georgia Pro 9.5/11.75pt
Edited by Merle Read
Indexed by Catherine Hookway and Colin Thom
Page layout by Pauline Hull
Proofread by Kathryn Glendenning
Printed in the UK by Gomer Press Ltd

Frontispiece
Robert Adam, Mountain Landscape with Two Castles and Smaller Towers beside a Loch*, 1780s.*

Contents

Notes on contributors

Adriano Aymonino is a Lecturer and Director of Undergraduate Programmes in the Department of History of Art at the University of Buckingham. His main interest is the reception of the Classical tradition in the Early Modern period, with a particular focus on 18th-century Britain. In 2013 he organised an exhibition on the Topham Collection at Eton College Library, and in 2015 curated the exhibition *Drawn from the Antique: Artists and the Classical Ideal*, held at Sir John Soane's Museum in London and the Teylers Museum in Haarlem. He is currently working on a book on the patronage of the 1st Duke and Duchess of Northumberland, which will be published by Yale University Press; on a critical edition of Robert and James Adam's grand tour letters (with Colin Thom); and (with Nick Penny and Eloisa Dodero) on a revised edition of Francis Haskell and Nicholas Penny's *Taste and the Antique*.

Eileen Harris is an independent architectural historian and author and a leading authority on the work of Robert Adam, which she has written about extensively since her first book, a groundbreaking study of *The Furniture of Robert Adam*, appeared in 1963. More recently her account of Adam's interior designs – *The Genius of Robert Adam: His Interiors* (Yale UP, 2001) – has proved to be a landmark in Adam studies. Eileen is also an acknowledged expert on the Adam revival, both in Britain and America, and is Honorary Librarian and Consultant to the Adam Project at Sir John Soane's Museum, London.

Miranda Hausberg is a PhD Candidate at the Department of Art History at the University of Pennsylvania in Philadelphia, Pennsylvania. She received her Bachelor's degree from Williams College in Williamstown, Massachusetts, and Master's degrees in art history from the Williams College Graduate Program in the History of Art and the University of Pennsylvania. Miranda is currently completing her thesis, which in part analyses and celebrates Robert Adam's picturesque and scenographic interiors. She specialises in the architecture of 18th-century Britain and 16th-century Italy, and has strong interests in Renaissance plays and theatres.

Jerzy J Kierkuć-Bieliński started his curatorial career at the Holburne Museum, Bath, and subsequently worked at Tate Modern and the Department of Prints and Drawings, the British Museum. His current role is as regional curator for the National Trust, looking after Sissinghurst, Chartwell and Scotney Castle, among other properties in Sussex and Kent. From 2015 to 2017 he was the curator of The Iveagh Bequest, Kenwood, where the *Portrait of James Adam* by Antonio Zucchi was put on public display for the first time since 1867. Previously, from 2007 to 2015, he was exhibitions curator at Sir John Soane's Museum, where he curated a number of exhibitions on the Adam office, Sir John Soane and G B Piranesi. He has published on Sir John Soane as a collector, as well as an architect, on the Adam office and on American 20th-century art. Having studied at the Courtauld Institute of Art, he continues to teach there as a visiting lecturer for their public programme.

David King is Emeritus Professor of Economics at the University of Stirling. He has a long-standing interest in architecture, particularly that of Robert and James Adam. His 1991 book *The Complete Works of Robert and James Adam* is a catalogue raisonné of their executed architectural and decorative work, and in preparing it he visited every surviving example. This book was re-issued in 2001 with a companion volume, *Unbuilt Adam*, which covers their unexecuted designs. He is currently analysing the geometry of Robert Adam's 650 or so ceiling designs.

Peter N Lindfield FSA is a Leverhulme Early Career Research Fellow at Manchester Metropolitan University. He has published widely on Georgian Gothic architecture and design, as well as heraldry, forgery and antiquarianism. His first monograph, *Georgian Gothic: Medievalist Architecture, Furniture, and Interiors 1730–1840*, was published in 2016, and his second book, *Unbuilt Strawberry Hill: Designs for Horace Walpole's Gothic Villa* will appear in 2019. Peter has held four research fellowships at Yale, as well as named fellowships at Oxford and Durham.

Conor Lucey is Assistant Professor in the School of Art History and Cultural Policy at University College Dublin. He has written widely on 18th-century domestic architecture and interior decoration, with a special focus on the contested relationship between design and production. His book, *Building Reputations: Architecture and the Artisan, 1750–1830*, was published by Manchester University Press in 2018.

Alistair Rowan is an Irish architectural historian and a leading Adam scholar. He was a lecturer in the Department of Fine Art at the University of Edinburgh, Professor of the History of Art in University College Dublin, Slade Professor of Fine Art in Oxford in 1988, and Principal of the Edinburgh College of Art from 1990. In 2000 he was invited to set up a new department of the History of Art at University College Cork. He retired in 2003 and lives in Dublin. He has published *Designs for Castles and Country Villas by Robert and James Adam* (1985); and *Robert Adam: Catalogue of Drawings in the Victoria and Albert Museum* (1988). He also organised two Adam exhibitions for Sir John Soane's Museum: *Bob the Roman* in 2003 and *Vaulting Ambition* in 2007.

Colin Thom is a Senior Historian with the *Survey of London*, the leading reference work on the history and architecture of England's capital city – now part of UCL's Bartlett School of Architecture, where Colin also teaches architectural history and theory. He edited the *Survey*'s landmark 50th volume, *Battersea Part 2: Houses and Housing* (Yale UP, 2013) and was co-editor of the recent *South-East Marylebone* volumes (Yale UP, 2017), which won the SAHGB's Colvin Prize for 2018, and for which Colin provided the first comprehensive account of the Adams' speculative London street development at Portland Place. He has written and spoken widely on the Adams, and is now embarked on a new research project (with Dr Adriano Aymonino) to produce a critical edition of Robert and James Adam's grand tour letters.

Marrikka Trotter is a Los-Angeles-based architectural historian and theorist whose research examines the historical intersections between geology, architecture and landscape in the 18th and 19th centuries. She is co-editor of the contemporary architectural theory collection *Architecture is All Over* (Columbia Books on Architecture and the City: 2017), and her writing has appeared in publications such as *Harvard Design Magazine*, *Log* and *AA Files*. She teaches at the Southern California Institute of Architecture (SCI-Arc), where she heads the history and theory department.

Jonathan Yarker is a scholar of British Art and the grand tour. He has published widely, contributing to a number of publications on the grand tour including: *Digging and Dealing in Eighteenth Century Rome* (Yale, 2010); *The English Prize, the capture of the Westmorland, an incident of the Grand Tour* (Yale, 2012) and the recent exhibition *Richard Wilson (1713–82): A European Master* (Yale 2014). His current research projects include work on the Thomas Gainsborough catalogue raisonné, which is to be published by the Paul Mellon Centre for Studies in British Art in 2019, as well as a book-length study of the British artistic community in Rome in the second half of the eighteenth century entitled *Savage Pilgrims: Rome and the Grand Tour 1750–1798*.

Acknowledgements

This book evolved from an international symposium on the Adam brothers, held by the Georgian Group at the RIBA in London in September 2015. Neither the symposium nor the present volume would have been possible without the continuing support of the Georgian Group and its Secretaries, past and present – Robert Bargery, David McKinstry and David Adshead – and I am grateful to the Group for providing a subvention towards image costs. The Howard de Walden Estate also contributed generously towards the symposium expenses, principally through its creative director, Jenny Hancock, as did my employers at UCL's Bartlett School of Architecture, from whom I also received a grant towards image acquisitions. Thanks are due also to the speakers at the symposium, who in addition to those represented in this volume included Stephen Astley, Dr Sue Berry, Dr Maria Celeste Cola, Simon Green, Richard Ireland, Dr Anthony Lewis, Katherine McHale, Dr David Pullins and Professor John Wilton-Ely. But above all my biggest debt is to the book's contributors: their knowledge, expertise and generosity of spirit have been an inspiration.

I have been fortunate in being able to draw upon the patience and skill of the publishing team at Historic England, in particular Clare Blick, John Hudson and Victoria Trainor, as well as two exceptionally talented individuals who have made this book far better than it otherwise would have been – copy editor Merle Read and designer Pauline Hull. My Survey of London colleague Helen Jones very kindly took the time to provide the excellent line drawings; and I also owe a debt of thanks to Nick Haynes, who supplied many fine photographs of Adam buildings in Scotland, and to Chris Redgrave, who took new photographs in London and improved a few others – he is a true artist and a good friend. Other individuals who contributed to the book's production include Catherine Hookway for providing the index, Kathryn Glendenning for proofing the final copy, and Paolo Gattavari, Annie Pritchard and Marion and Robin Page for translating certain Italian, French and German sections of the text.

Several friends and scholars kindly read parts of the text and offered advice and encouragement. Principal among these were Dr Geoffrey Tyack, my co-organiser at the Adam symposium, and Harriet Richardson, my Survey of London colleague – I am much indebted to them both. Others who helped in various ways include Sydney Ayers, Dr Adriano Aymonino, Dr Iain Gordon Brown, Eileen Harris, Dr Olivia Horsfall Turner, Professor David King, the late Treve Rosoman (who is sadly missed) and Professor Alistair Rowan.

Finally, I must thank the owners and custodians of original Adam material. Keith Adam of Blair Adam was unfailingly generous in offering regular access to the Blair Adam Muniments and allowing me to quote from letters and reproduce images in his ownership. Sir Robert Clerk of Penicuik kindly showed the same degree of generosity towards the Adam letters and drawings in his possession. Both men have been enormously encouraging and this book would not have been possible without their support. I also relied heavily on the munificence of the staff at Sir John Soane's Museum in London: Stephen Astley, Helen Dorey and Sue Palmer, but most of all Dr Frances Sands, the museum's Curator of Drawings and Books, who has been endlessly patient and accommodating, not only in providing items from the Adam office drawings but also in imparting freely her knowledge of that collection.

Colin Thom

Foreword

Keith Adam of Blair Adam

Such was Robert Adam's achievement and reputation as an architect that in 1792 he was buried in Westminster Abbey with dukes and earls as pallbearers. In its obituary, the *Gentleman's Magazine* observed: 'His talents extended beyond the line of his own profession.' Nevertheless, his nephew William, son of John, the eldest of the Adam brothers, from whom I am directly descended, shortly afterwards declared of his uncles: 'Their arduous attempts at striking great strokes in their line, testified by the Adelphi and many other buildings in the metropolis, kept them incessantly in difficulty. They died bachelors, so here their story ends'.

In my youth, on the shelves of our library at Blair Adam I was aware of the books about Adam architecture by John Swarbrick, A T Bolton and James Lees-Milne. Yet it seemed that during the 20th century little attention had been given to the brothers until John Fleming's engaging biography, focused on Robert's grand tour in Italy, was published in 1962. With that study reaching its conclusion soon after Robert's return to London from his survey of the Emperor Diocletian's palace in Split, there remained, I felt, much still to discover about the talents of my forebears.

Happily, from the 1970s, excellent works by a range of eminent architectural historians, architects, librarians and archivists have followed one after another, deepening year by year our understanding of the Adam practice. It has been a personal pleasure to come to know many of these writers. If I named them, I might miss someone out, but suffice to say there are about 20. Let that be testimony to John's son, William: the brothers' story did not end in the 18th century as he had predicted.

Our library at Blair Adam was diminished by an unavoidable sale in the 1920s, but now it is freshly enhanced by these various books, as well as articles and exhibition catalogues which illuminate the Adam style, along with copies of Vitruvius and Robert Adam's own *Ruins of the Palace of the Emperor Diocletian*, which I have fortunately been able to repurchase. Such was the extent and range of this authorship that I personally did not consider much more could be written about the Adam architects.

Thus I was intrigued when Colin Thom, while undertaking some research at Blair Adam for the *Survey of London* series a few years ago, told me of his plans for a two-day Adam brothers symposium in London, followed by a well-illustrated book based on a selection of the conference papers. This has now been achieved most notably, as you will perceive on turning the pages. In this publication we are treated to an interlocking series of essays which enable us to gain further insight into the spectrum of Adam architectural endeavours. I have greatly enjoyed adding an extra dimension to my understanding of their achievements. I have also welcomed the contributors' clarity of expression in drawing upon such rich material.

This standard is well established in Colin's introduction, which lays before us Robert Adam's personality, his frustrations and imperfections, as well as his talents, his stylistic development towards neoclassicism, his acute regard for the Italian craftsmen who enabled so much personal self-development; the significance of his brother James's role and of the necessary networking as the practice gathered momentum; the vicissitudes of the brothers' commercial speculations and consequent business downfall; and eventually the change of direction to Scotland with a 'remarkable style full of picturesque drama', entailing fusion of building and landscape, while yet able to include expressions of neoclassicism. Alistair Rowan develops the story in the first chapter, highlighting John Adam's personal architectural achievement, and drawing us into the interactions between the brothers, their businesses in Scotland and their eventual separation. In Alistair's inimitable style, he helps us to understand John's personal feelings as well as his architectural philosophy of 'rational simplicity'. The meticulous notes which John kept attest to his diligence.

From such a foundation, the subsequent papers develop and extend our understanding of the brothers' practice. Whether learning about their relationship with Roman dealers, agents, financiers and merchants, as they combine commercial ambition with the collection of antiquities to help shape their 'highly personal classical language of ornament', or whether exploring the provenance of Zucchi's portrait of James Adam, further levels of insight are reached as material unfolds in this most enterprising collection of papers on the Adam brothers.

May the reader delight, as much as I have, in this evolving story, which so evidently is far from reaching its end.

Introduction

'Some promising young men': Robert Adam and his brothers

Colin Thom

> John [is] a good Lad & has wrote me a long
> Letter
> Jamies a Sloathful Son of an indulgent
> Mother,
> Will is a never do well Banker.

So wrote Robert Adam from Rome in January 1756 to his mother, Mary, in Edinburgh. As in most of his letters home to Scotland during his grand tour of the 1750s, Adam's irrepressible tongue-in-cheek humour is laced with strands of truth.[1]

The timing of the letter is important. Robert's elder brother, John, mainstay of the family architectural practice in Edinburgh, had been expecting Robert on his return to rejoin him and their younger brother James in Scotland as partners – it was anticipated that Robert's fresh knowledge and understanding of the antique would add a valuable new dimension to the business the three siblings (Figs 0.1–0.3) had inherited from their father, William Adam senior (1689–1748). The 'long Letter' Robert refers to was Johnnie's magnanimous response to discovering unexpectedly from Jamie that his two younger brethren had been planning secretly to establish a separate Adam practice in London on Robert's return. The good feeling that then still existed between the two older Adam brothers resonates in Robert's words – as indeed does his irritation at Jamie for letting the cat out of the bag. William (or 'Willie'), the youngest of the

Fig 0.1 (opposite)
'Bob'. Robert Adam (1728–92), oil portrait attributed to George Willison, c 1770–4.

Fig 0.2 (far left)
'Johnnie'. John Adam (1721–92), oil portrait attribute to Francis Cotes, c 1760.

Fig 0.3 (left)
'Jamie'. James Adam (1732–94), oil portrait by Allan Ramsay, c 1754–8

four, was the only one not to have trained as an architect but as a banker and merchant, and was at the time flitting between various speculative projects, unsure of his true vocation.

The letters written by Robert and James in Italy to their family, now preserved in the Clerk of Penicuik papers at the National Records of Scotland in Edinburgh, constitute one of the fundamental resources for anyone wishing to understand the Adam brothers as people and artists.[2] That regular exchange of ideas and observations between Scotland and Italy highlights the two great stimuli on Robert's life and work. Rome was the catalytic experience that transformed him into an artist of international significance, but much of his work also drew upon his early family life and background in Scotland. It is there that we begin.

The 'Brotherhood': partnership and practice

So meteoric was Robert Adam's rise to prominence in England in the early 1760s that his earlier grounding in his family's Scottish architectural practice and building firm is often overlooked. The brothers' father, William, was Scotland's leading architect when he died in 1748 and was also the head of a vast building materials and construction empire. He left behind a thriving business with an income bolstered by new contracts for repairing and building government forts in the Scottish Highlands, in the wake of the Jacobite rebellion of 1745.[3]

Around the time of the '45, young Robert left Edinburgh University with his studies unfinished to join his father's office, never to return. With his father's death and the continuing increase in workload – especially the building of a new coastal fortification, Fort George at Ardersier, east of Inverness – Robert was required at the age of 20 to become his brother Johnnie's permanent partner and assistant in the Adam practice. Any hopes he may have nursed of a career as a landscape painter (a boyhood passion) now evaporated, though landscape drawing and painting were to remain precious to him, both as an escape and as an outlet for his creativity.[4]

Though he later claimed to have reinvented himself artistically during his years of study in Italy in 1755–7, Robert's early practical education under his father and elder brother never left him. The commissions pursued by the Adam Edinburgh practice ranged widely: small villas and town dwellings; country houses – occasionally, as at Hopetoun House, on an extensive scale; the design of public buildings, such as the Royal Exchange in Edinburgh (Fig 0.4); the inspection and repair of numerous Ordnance fortifications and barracks in the Highlands; and the construction of the monumental Fort George, where Robert and Jamie spent many a summer from around 1750 overseeing the works. In this way the younger brothers learned to build and manage complicated contracts and, just as importantly, to cultivate professional friendships with aristocratic clients. Robert's later close relations in England with patrons like William Murray (1st Earl of Mansfield), Sir Nathaniel Curzon (1st Baron Scarsdale) and Hugh Percy (1st Duke of Northumberland) show in him the same ability to attract 'uncommon friendship and attachment from men of the highest rank' that was noted in his father, who did much in this way to raise the status of his profession.[5] As well as the practical experience, Bob and Jamie also benefited from the intellectual milieu of the Adam household in Edinburgh's Cowgate, where regular visitors and family friends included leading thinkers of the day. And there was the architectural inspiration to be had from the cosmopolitan library assembled by their father at his country seat of Blair Adam in Fife.[6]

Harder to gauge is Robert's contribution to the work of the Edinburgh practice, known by 1750 as 'John & Robert Adam's Architects'.[7] As Alistair Rowan explains in the opening chapter of this book, whatever creative influence Robert exerted, it was his elder brother, John, seven years his senior, who was in control of the firm's design work in the years before Robert left for his grand tour in 1754. Their patrons in Scotland valued continuity, reliability and prudence, and would have expected no less. Putting to one side his early Gothic fancies (discussed by Peter Lindfield in Chapter 7), Robert's surviving architectural designs of this period sometimes exhibit the Vanbrughian Classicism favoured by his father, or alternatively follow the cleaner, Italianate-Palladian style promoted by the Edinburgh practice under John. Unfortunately, we have no record of Robert's first successful private commission – a house built at Nairn in 1753 for an Edinburgh doctor, Joshua Mackenzie (father of the writer Henry Mackenzie) – by which to judge his development.[8] But his obvious pride in his prominent role in the family's commission at Dumfries House in Ayrshire that year, where he had special responsibilities –

providing many of the office drawings and attending the founding ceremony – suggests a collaborative design with John, and the house is perhaps our best guide to Robert's architectural progress on the eve of his departure for Rome (Fig 0.5).[9]

As head of the practice, John Adam was careful to foster a strong sense of brotherly purpose and familial cohesion. Correspondence of the mid-1750s between 'Messrs John Robt & James Adam, Architects, in Edinburgh', as it had become by *c* 1754, and Innes & Clerk, their financial agents in London, often includes expressions of concern for 'our R A' or 'our brother Robert' and his whereabouts. Also, much of the practice's design output intended for reproduction in *Vitruvius Scoticus*, the publication originally begun by their father, is credited collectively to 'Adams architects'.[10] How hurt must Johnnie have been, then, when the precocious Robert, after his epiphany in Rome, defected to establish his own office in London, taking with him his two younger brothers (and two of his sisters, Jenny and Betty, as housekeepers), leaving John alone in Edinburgh. John's was no facile, selfish hurt – indeed his first response when he heard of Bob and Jamie's London plans was to state that he would never 'stand in the way of people so nearly connected by blood & friendship doing whatever shall appear most for their advantage'. Later in life, when early London successes had evanesced, leaving mostly bad feeling and financial turmoil, John still gave all his fortune in aid of his wayward younger siblings, 'without leaving a penny to my own family'.[11] Any grieving of John's was for the dispersal of a 'wonderfully loving family' and the loss to the Edinburgh practice of Robert's blossoming, exceptional talents, rather than for personal financial reasons.[12]

Robert's decision to establish himself in London at the end of his grand tour in 1758 was not a sudden one. His letters suggest the idea came to him in Rome, but it is possible that he and Jamie had been hatching plans in London in 1754, at the start of his journey – or maybe even earlier, during those long summer evenings together at Fort George. Perhaps Johnnie's control of the Edinburgh firm's design work encouraged Robert to branch out on his own.

Fig 0.4
City Chambers (formerly the Royal Exchange), High Street, Edinburgh. Built c *1753–61 to designs by John Adam, probably assisted by his brother Robert.*

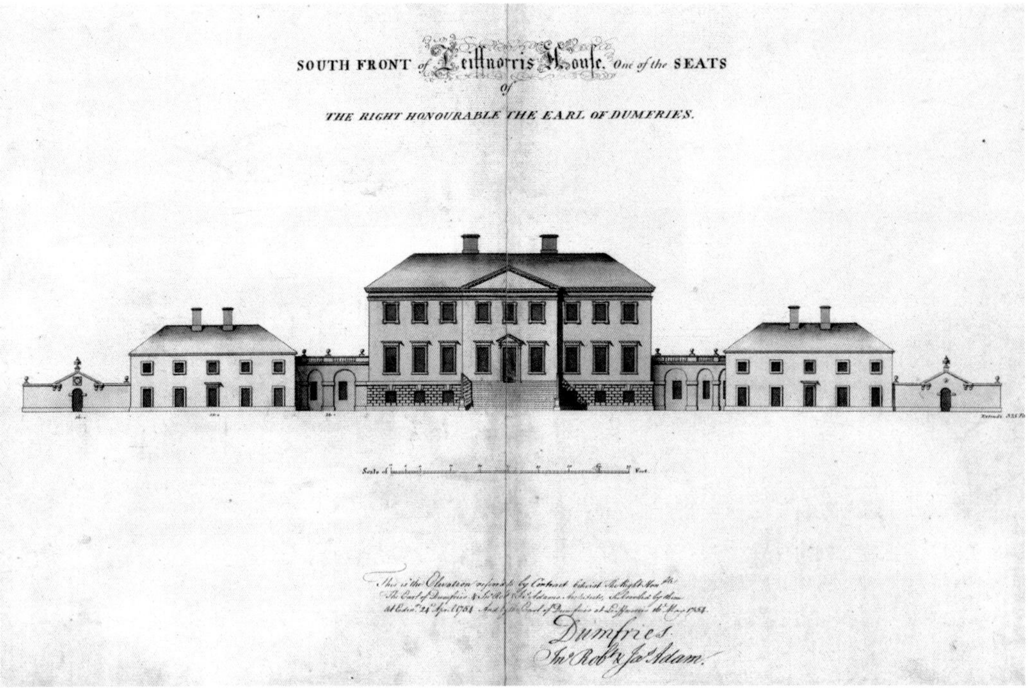

Fig 0.5
Dumfries House, Ayrshire, contract drawing of the south elevation, April 1754, signed by John, Robert and James Adam, and Lord Dumfries.

It is suggestive that Jamie, who accompanied Robert and his travelling companion Charles Hope (the Earl of Hopetoun's younger brother) for the first leg of their trip, as far as Lille, was to remain in London from December 1754 until March 1755 before returning to Edinburgh. What was he doing there all that time? He may have been, inter alia, surveying the London architectural scene as his brother's scout. Robert wrote to Willie from Florence early in February 1755, eager for news of Jamie's activities: 'I rejoice at James Negociations. I long for more account of his London affairs.'[13]

Certainly by April 1755, only a few months into his tour, Bob was discussing his London plans with Jamie in considerable detail. Always a sharp and perceptive judge of character and situations, Bob already understood the threat to English success posed by William Chambers and, to a lesser extent, Matthew Brettingham. He had also by then conceived of the idea of having his sisters keep house and his youngest brother, Willie, help with the accounts. Although there were elements of good fortune to Robert's advancement in Rome, with his London establishment he left little to chance. Throughout 1755–6 his letters home contain detailed discussions of his schemes, and in Rome he also sounded out close acquaintances for advice, such as the painter Allan Ramsay and the grand tourist Sir William Stanhope (1702–72), brother of Philip, 4th Earl of Chesterfield.[14]

In Italy the need for a cadre of talented draughtsmen to work on Robert's recording and publishing projects soon became apparent. The system he initiated there – employing specialist artists like Laurent-Benoît Dewez (his 'plan man & Line drawer') and Agostino Brunias (for 'all my monuments and all my figures') alongside other 'drudges' for the more everyday copying duties – was to be repeated in the Adam drawing office in Mayfair on his return. Bob and Jamie's reliance on such a team of 'beagles' or 'Elves', as they often called them, to interpret and develop the antique motifs of the emerging Adam style was immense. These London assistants were a cosmopolitan bunch, many of them Italian or French, trained in or familiar with the Italian academies and with first-hand experience of the grand manner of Italian drawing aspired to by the brothers. Robert brought Dewez and Brunias back with him from Rome and they were instrumental in the early years of the office in laying the foundations of the Adam style. Robert's letters are full of justifications for this expense: 'At London there is not one that knows my manner of Drawing, nor would learn it in 2 years ... These two I have actually bred ... I realy woud not have the Courage to Settle in London without them and that is Saying much'; and later: 'But what can I do, without them Boys I coud absolutely do nothing, But work like a Slave to little Purpose.'[15]

When Robert returned to Britain in January 1758 and set up his first London office in St

James's Place, close to Piccadilly, James hurried south to join him for a time, to be followed soon after by Willie. Jenny and Betty came later, probably in the summer, by which time Robert had procured larger premises in Lower Grosvenor Street, thanks to a hefty loan from John. By May, James was back in Edinburgh, his own projected grand tour, which had been greatly urged by Robert, postponed for another year at John's advice. Jamie then returned to his regular routine of the annual contractual visit to Fort George, where he worked through the summer in order to 'lay up cash' for his foreign trip. He seems to have rejoined Robert in London only briefly for a month or so before embarking for the Continent in the spring of 1760. With England and France still at war, he adopted the disguise of a Jacobite officer.[16]

So, from the outset in January 1758 until Jamie's return late in 1763, Robert was mostly in practice alone, though with the able assistance of Dewez (till the end of 1758) and Brunias, and also Jenny and Betty helping with the household accounts – and perhaps those of the practice, too. There was thus a clear demarcation between Robert's architectural business and that of the Adam family in Scotland, which now bore the name 'John & James Adam, Edinburgh'. Young William's role at this stage is less certain. During the late 1750s and early 1760s his energies were taken up mostly by his partnership in a London merchant's business, Bayne & Adam (the Bayne family being acquaintances of the Adams), with offices in York Street, St James's, very close to his brother Robert's architectural practice.[17]

If Robert had hoped to set London afire with his new antique-derived style, he must have been sorely disappointed at how few genuine opportunities to express it came his way during his first two years of independent practice in 1758–9. The much-hoped-for commissions from the *milordi Inglesi* he had cultivated in Italy failed to materialise, and his letters of the period bristle with frustration. He spent 'long London mornings' holding forth, made visits and paid 'his respects to the great' – some of whom made encouraging noises about his drawings – but all to little avail. One job was a commission from Archibald Stewart (1697–1780), a successful wine merchant and former Lord Provost of Edinburgh, who owned a villa and country estate at Mitcham, Surrey, just south of London, to design wooden bridges to cross streams in his grounds. 'On such pityfull Objects is my attention bent,' Bob complained to Jamie.[18]

Robert found himself in a 'pinch for money' (a favourite Adam phrase), forced to borrow from both John and Jamie.[19] Family experience taught him the value of holding a sinecure and the security of a regular income, and much of 1758 was taken up in the search for an official 'position'. Here some unpleasant facets of Robert's personality rose to the fore alongside his drive and ambition – chiefly a casual callousness and deviousness in order to get his own way.

The family's Ordnance commissions had brought them good contacts in London, including the surveyors George Mercer, father and son, and there was an expectation that Bob would exploit these connections as a route to such employment (the Mercers were among the very first people he visited with Jamie when he arrived in the capital in October 1754 before embarking on his grand tour). In 1758 Bob told Jamie that the Duke of Argyll, whom the Adams had worked for at Inveraray, and Colonel Skinner, the chief engineer at Fort George, might use their influence to force Mercer senior into accepting him as a partner in an Ordnance post at Milford Haven – 'Skinner can frighten Mercer most damnably, To throw him out altogether unless I am instantly agreed to. I'd make Skinner a present of all my Zucarellis at once if he'd do something Clever for me.' Days later Robert went to the Ordnance office in person and harangued 'that Idiot Young Mercer' about another possible appointment, as a successor in some capacity to the military engineer J P Desmaretz, who Adam told Mercer was now too old and infirm to be capable of duty. Nothing came of this, either. (Later the London-based Adams negotiated several important contracts with the Board of Ordnance in England, which helped keep them afloat when the financial tides turned against them.)[20]

Another possible sinecure that Adam hit upon that summer was Master of the King's Works in Scotland, a post held by George, 23rd Laird of Dundas (d 1762), an elderly, failing man. This position Robert reckoned worth £300 a year for doing nothing. Jamie in Scotland was to enquire if the Laird were likely to die; if not – if he might 'linger out any Years' – then Jamie was to arrange for him to learn via a third party that an advance of two or three years' salary might come his way if he were willing to resign in Bob's favour.[21] By September, Bob was also angling for a post in the Board of Works, which as part of the royal household was funded through the civil list. His good friend Gilbert Elliot of Minto (1722–77), a lord of the Admiralty and a protégé of the Duke of Argyll, spoke in Adam's favour to the Duke of Newcastle, 1st Lord of the Treasury – as indeed did Argyll himself and also the 9th Earl of Kinnoul (a former Paymaster of the Forces), and the influential Earl of Bute, who had come to appreciate Adam's potential. But despite this support, and Adam's own importuning at one of Newcastle's famous 'Levees', no position was forthcoming. It was not until 1761 that Bute finally solved Adam's dilemma by having him appointed to a new post of architect to the king, jointly with William Chambers, at a salary of £300 pa each.[22]

For those fond of omens, Bob's first audience with George III at St James's Palace, in November 1761, perhaps did not augur well. Apparently he was 'agitated' and stumbled, dropping several of the rolls of drawings he had brought to show the king – though the latter quickly dispelled any awkwardness by chatting freely with Adam and helping him pick them up. Also, there is a suggestion in one of Jamie's letters home to his sister Nell that George III mistook Bob for Jamie, who was shortly to be the king's agent in the acquisition of the Albani collection of drawings in Rome: 'I don't think it was necessary for him [Bob] to explain any mistake. I am of opinion they are not much in the wrong who take one of us Brothers for the other.' Bob had to resign his honorary post in 1768 when he was elected MP for Kinross-shire, with Jamie taking his place. But by then Bob had secured another position, in 1765, as clerk of works to Chelsea Hospital, the gift of Charles Townshend, the new Paymaster General, who was married to a daughter of the Duke of Argyll. Adam retained this post until his death in 1792.[23]

By the early 1760s important commissions were beginning to come in, including a series of grand town and country houses where Adam's new and sophisticated antique manner reached its first heights. He established good relations with patrons who were to stick with him in the years to come – such as Lord Bute at Lansdowne House, and Lord and Lady Northumberland at Syon Park (*see also* the section on *Robert Adam's patrons*, p 20). During the three years that Jamie was away on his grand tour, between 1760 and 1763, Robert established a busy, successful London office and developed a style of interior decoration that put him at the forefront of the architectural profession in Britain for more than a decade.

Jamie's departure in 1760 marked the final severing of the two Adam practices in Edinburgh and London. Johnnie wrote to their agents Innes & Clerk that summer: 'My Brother James being now gone abroad, He & I have dissolved our Copartnership during his absence, which is the reason of writing in my own name.'[24] A further shift, or realignment, came in 1764–5, after Jamie's return to join Robert in London. Firstly, in 1764 Johnnie was forced into administration, having been dragged into financial distress when his banker and schoolboy friend Adam Fairholm (or Fairholme), with whom he had speculated greatly in government stocks, went bust and disappeared (*see also* p 58). Bob and Jamie immediately offered Johnnie a share of their London practice but he declined, 'on second thoughts', preferring not to leave Scotland. Instead, all four brothers decided to push ahead with a joint business as building supplies merchants and contractors in London, something they had apparently been planning for some time. This was a natural progression for them, and entirely in keeping with the commercial speculations established by their father in Scotland to augment his architecture business.[25]

Named William Adam & Company after the youngest of the four brothers, who was its managing partner, the Adam firm embraced a wide range of activities, including a timber business and composition factory at Thames Bank in Pimlico, a lime wharf at Rotherhithe, brickfields at Tottenham Court Road and in Essex, and granite quarries in Guernsey and Scotland to supply a London paving business (Adams, Campbell & Company). It was hoped that Bob and Jamie, through their connections and 'influence with the Artificiers', could promote the trade in building materials. Johnnie was to help procure funding, and an unexpected windfall in the form of compensation for his losses in the Fairholm debacle was ploughed into the family business. This was just as well, as from the beginning the Adam company was beset by financial problems. Willie had lost heavily in Bayne & Adam's speculative venture dispatching a ship to Havana (perhaps an attempt to join in the clamour at the time among London agents for captured Spanish bullion), and the debt and its interest stayed on the Adam company's books for many years to come. Also, parts of the business, such as the Aberdeen quarries, did poorly, exacerbated by the loss of ships laden with some 300 tons of granite.[26]

The 1770s are generally considered a 'watershed' in the affairs of the Adam brothers, so dramatic were the changes in their fortunes and to their modus operandi. But the decade began well enough, with work proceeding satisfactorily on their London speculations in Marylebone and at the Adelphi, and in 1772 Robert secured the contract for the new Register House in Edinburgh, one of his key public commissions. This decade also saw his abilities as a domestic designer reach new heights in a quartet of exceptional aristocratic London town houses: Chandos House, Marylebone (1769–72), Wynn House, 20 St James's Square (1771–4), Derby House, Grosvenor Square (1773–4, now demolished), and Home House, 20 Portman Square (1775–7; Fig 0.6). Architectural historian Sir John Summerson thought these represented 'perhaps, the highest point of imagination and artistry in the handling of the London house'.[27]

Fig 0.6
Robert Adam's spectacular imperial staircase of c 1775–6 at Home House, 20 Portman Square, London, in 1995.

The Adams' eventual fall was the result of overreaching – specifically the London-based brothers' attempts to take on two large and potentially lucrative developments, at the Adelphi and Portland Place, at one and the same time. The banking crash and credit shortage of 1772 that brought things to a head simply tipped them over a precipice that they were already fast approaching. The more level-headed John, in Scotland, had seen it coming. In a letter written to Jamie more than a year before the crash, John tried to discourage his younger brother from entering into yet another uncertain business venture:

> We are stretched to the uttermost farthing here ... considering that every Nerve is strained to find supplies for our present Engagements it strikes me as highly imprudent to enter upon new ones till the old are cleared up ... It is my opinion that we have hitherto attempted too many things, and if we had confined ourselves to fewer Branches we should have done better.[28]

Johnnie concluded that he would never feel at ease until the Adam businesses were put on a more certain footing, 'so as to contract rather than enlarge our dealings'. This was, in effect, what was forced upon them. The public nature of their financial crash, the sale by lottery of much of the Adelphi and some of their private art collections, and the embarrassment of houses standing unfinished and ground lying waste in Portland Place, year after year, all left a stain on the Adams' reputation, particularly in England, that they were never able to remove.

The Adelphi (Fig 0.7), with streets named after each of the brothers, was to have been a symbol of the Adams' potency as artists and builders, and a new centre for their activities. In 1772 Bob and Jamie moved from Grosvenor Street to live at No 4 in the Adelphi's Royal Terrace, with an office at 13 John Street. Willie lived at 6 Adam Street.[29] But in terms of their architecture, from this time on there was a gradual diminution in the number of sizeable English commissions that the brothers were able to attract. The great age of country house building in the south was now over, and the escalation of the revolutionary wars in America brought a financial nervousness to the London money markets, hindering the Adams from selling any remaining spec-built houses profitably or securing credit for new adventures.

But there were still loyal patrons. In April 1780 Jamie wrote to one of them, Lord Findlater – 'the most difficult letter I ever wrote in my life' – begging for money to invest in yet another doubtful speculation. Jamie explained:

Fig 0.7
The Adelphi, London, and Thames riverside, looking east. Engraving of c *1770 by Benedetto Pastorini, later reproduced in the third (posthumous) volume of the Adam brothers'* The Works in Architecture, *1822.*

For some time past the very particular state of affairs in this country, has prevented the expencive Art of Architecture from being cultivated as it was formerly. so that both my Brother & I have for many months found ourselves far from being fully employed.[30]

Letters to clients at this time emphasise the brothers' 'pinch for money', their inability to 'lay out' cash to promote building in advance of payment or mortgage, and a new desire 'not to venture beyond our depth'.[31] By the mid-1780s the practice had sunk so low that Robert was complaining that he was unable even to pay his clerks' wages. Further heavy losses from investments in risky business enterprises, such as Liardet's cement and the Battersea & Sand End Company (*see* p 209), brought the brothers to the brink of bankruptcy, and their personal relations suffered.[32] John, in Scotland, became increasingly disillusioned with his brothers' behaviour and ungratefulness. It was around this time that Jamie withdrew from architecture altogether, preferring instead to spend more time with his mistress on his farm at Shenley, Hertfordshire. By then John too had been retired from active design work for some time, though he still had income from a few remaining Ordnance contracts in Scotland, and from his shares in the Carron Company, a leading ironworks near Falkirk, which produced items for the mass market (often to designs by his brothers). Robert was therefore able to enlarge his operations in his homeland, which was to be his focus for the remainder of his life.[33]

This change in balance between Robert Adam's English and Scottish commissions in the latter part of his career was remarkable, with around 30 in England against 5 in Scotland in the late 1770s, compared to only 1 in England and 29 in Scotland by the 1790s.[34] Such a late flurry of activity in his home country marked something of a renaissance for Adam and included designs for important public buildings in Edinburgh and Glasgow, as well as private villas and castle-style houses. Despite having suffered so many setbacks, his energy and abilities were undiminished. Still based in London (from the late 1780s at 13 Albemarle Street), Robert relied heavily on clerks of works on-site in Scotland, who were employed to keep an eye on things and to implement his instructions, but he also began to travel north annually to check on progress and discuss new designs with clients. On these occasions he usually lodged with a sister in Edinburgh, and by around 1790 the house of his widowed sister Mary Drysdale in Nicolson Street had become his makeshift office. Drawings often travelled with him in his post-chaise, or were sent separately by coach; larger items such as moulds for plasterwork details went by sea.[35]

Though he had been suffering for some time from a stomach complaint and the stresses of overwork and financial uncertainty, Robert Adam's death at Albemarle Street in March 1792 was nonetheless unexpected. Big brother Johnnie's death followed only a few months later, in June of the same year. As a result, Jamie was flushed out of early retirement by Willie and other family members and briefed rapidly on Robert's unfinished building work in Scotland, such was his unfamiliarity with the world of Adam architecture.[36] He and Willie helped to further the construction of Charlotte Square in Edinburgh (Fig 0.8) and Fitzroy Square in London, Robert Adam's last two great contributions to British town planning, as well as oversee several public and private commissions in Scotland.

When Jamie died in 1794, Willie was left alone to shore up the tottering Adam construction business, manage the remaining holdings at the Adelphi, and complete any outstanding projects. Without Jamie he also had the added responsibility of dealing with the architectural side of company affairs, which he generally did by contracting work to architects like George

Fig 0.8
Charlotte Square, Edinburgh, north side, designed by Robert Adam in 1791.

Young of George Street, Edinburgh (to whom Willie still owed over £500 in unpaid fees in 1817). With such a heavy workload, friends and acquaintances suggested he engage a clerk, but by this date company debts had risen again to around £35,000 and he was 'continually harassed with money matters', and struggling to pay his workmen.[37] With bankruptcy looming, Willie decided around 1800 to enter into a new partnership with two young relatives, the builders and developers Daniel and Alexander Robertson. Together they won a large contract that year, to supply the bricks for and construct the warehouses and dock walls at the new West India Docks at Poplar, in east London – an improvement scheme unprecedented in its size and complexity, but one which Willie characteristically thought he could 'get through ... with great ease' with the Robertsons' assistance. But the usual Adam traits of overreaching and bad luck conspired against him; he was soon bankrupt, and forced to give up the contract. The fallout was wider still, as his nephew, Commissioner William Adam of Blair Adam, had stood as security for loans that Willie needed to help pay for brickmaking in advance of work beginning at the docks, and so was liable for some of the debts.[38]

In recognition of their good faith, the dock company in 1802–3 gave Adam and the Robertsons new building contracts, enabling them to remain in business, but Willie was unable to turn things around. Like Johnnie and Jamie, he had developed an interest in agriculture, and around 1790 had acquired an extensive farm at Mount Nod, Streatham, where he spent an estimated £8,000 on new ranges of buildings and modern machinery to feed 600 head of cattle on grain. This farm was described in 1795 as 'conducted with great judgment, without considering expence', and Willie's free spending on farming ventures became an additional strain on his already overstretched resources.[39] To add to his woes, the Robertsons took to drawing large sums of money from him, and racking up debts of their own which they expected Willie to honour. Another bankruptcy was inevitable, and came in 1817. After the Adelphi crisis of the 1770s and his failure of 1801, this third financial disaster, coming at the age of 79, took a heavy toll. Willie had been living with his sisters in London, but after Peggy's death in 1820 seems to have moved to a rented property in Welbeck Street, Marylebone, with his young cousin, Susan Clerk, where they lived in relative poverty. It was there that the inspirational story of the daring and talented Adam brothers came to its sad end, with the suicide in 1822 of a broken, disappointed old man, the last of the *Adelphoi*.[40]

The 1760s and the Adam revolution: the birth of a new style

If Robert Adam's London practice took time to flourish after his return to Britain in January 1758, so too did the formulation of his new style in the light and delicate manner that we now associate so closely with him. Adam had set great store by his collection of art and antiquities, and his great trunk of drawings, new and historic, that he showed off to prospective clients in London. These symbolised his anticipated status as the foremost modern interpreter of the glories of Rome – yet his earliest works exhibit only limited signs of a new direction under their influence. Take, for example, the 'great room' he designed for an extension wing at Thistleworth (now Gordon) House in Isleworth in 1758 for General Humphry Bland, a former governor of Edinburgh Castle – Adam's first interior after his return from Rome (Fig 0.9). There is certainly evidence of a new desire to incorporate Classical motifs, such as anthemia, urns, griffins and rinceaux, but overall the decorative style is heavy, and more in keeping with the work of the Adam practice in Edinburgh before Robert's grand tour, as at Arniston House in Midlothian, or Hopetoun House (*see* Fig 1.4).[41] A similar tendency towards thicker detail and conservative architectural features is also evident at Hatchlands Park, where Adam designed interiors in 1758–9 for a new house recently built for Admiral Boscawen, a son of the 1st Viscount Falmouth. These early designs suggest a disinclination on Adam's part to shake off his earlier, Scottish style, and a need for more time to absorb and process his recent experience of the antique before he could express it fully in the delicate, refined manner he desired through his architecture. Even in later work the looser, rococo spirit of his Scottish apprenticeship sometimes resurfaces alongside his new, more refined and linear neoclassical style, as in the dining room at Osterley Park.

What changed? It was perhaps the shift in 1760 from heavy plasterwork as the central element of his interiors to schemes of delicate painted decoration at Kedleston and Syon that first allowed the full beauty of Adam's new language of antique decoration to shine through (*see* Figs 4.2–4.4 and 4.15–4.16). Here the influence of Agostino Brunias, the draughtsman who worked for Adam during the decisive period in 1758–64 when the early Adam style was forged, is conspicuous. Brunias was able to develop Adam's blossoming ideas fully and was also personally responsible for copying many of the studies of antique interior decorations made by Francesco Bartoli, held in the Topham collection at Eton, which were fundamental to the new style's formulation. All of this is discussed and analysed in greater detail by Adriano Aymonino in Chapter 4.

Another important factor was Adam's first known professional encounters at Kedleston and Croome Court in 1760 with the plasterers Joseph Rose, senior and junior, uncle and nephew. Beforehand, when discussing his Thistleworth House commission, Adam had expressed frustration at the inability of English plasterers to translate into that medium the antique motifs and effects he desired – to leave behind 'their Angly Stiff Sharp Manner'.[42] But the skilful Roses seem to have understood intuitively his aspirations and so successful was the partnership that Adam stuck to them like glue ever after. Joseph senior was probably his main plasterer in the early 1760s, when the nephew was still an apprentice. Later, following a trip to Rome in 1769–70, Joseph junior appears to have taken charge of the family firm, though his uncle remained active until his death in 1780 – for example designing at least one ceiling in Portland Place (*see* Fig 9.15, bottom). Like Brunias, the Roses were more than craftsmen, and many of the Adam-style ceilings at Portland Place were designed as well as manufactured by them. Joseph junior particularly was aware of the latest developments in architectural style and theory, and understood the intellectual background to the neoclassical revival of the late 18th century.[43]

As the Roses gained in experience and became more attuned to Adam's vision, they were able to recreate in shallow-relief plasterwork the free-flowing grotesques and figurative plaques and panels that had been such an integral part of Brunias's painted rooms. The improvement is especially noticeable in the panels of *grottesche* that featured prominently in Adam interiors, and which over time became more elegant as Adam's designs and the Roses' abilities were refined. It was the integration of these stylistic elements in plaster with varied room shapes, screens of columns, niches and painted Classical scenes or capriccios by the likes of Michael Angelo Pergolesi and (after 1766) Antonio Zucchi that defined the intricate early Adam style as seen at its height in the mid- to late 1760s at Syon House, Harewood House (Fig 0.10), Lansdowne House, Nostell Priory, Kenwood House, Newby Hall and many other locations (*see, for example,* Figs 2.9, 5.2, 5.3, 5.7 and 5.8).[44]

Fig 0.9
Adam design for the drawing-room ceiling at Thistleworth House, 1758–9, largely as executed, inscribed Cieling *[sic]* for General Bland at Isleworth.

Once refined to this degree, Adam's interior style changed little for the remainder of his life. There were variations of course, and from the late 1760s and 1770s the treatment of his decorative motifs becomes ever more delicate and linear, compared to the thicker forms of his earlier work. But what was recognisably the same neoclassical mode of interior decoration was still being employed by Adam in the middle and late years of his career and throughout his *oeuvre*, in public buildings like the Register House in Edinburgh, and even in his later castle-style houses, such as Mellerstain and Culzean. The Adam style also reached a wider clientele through the brothers' town-planning speculations, as at the Adelphi and Portland Place, where even the smaller terraced houses had simplified or reduced versions of characteristic Adam ceilings and chimney pieces.

Two stylistic developments are, however, worthy of further note here. One was Robert Adam's increasing use of strong colour from the mid- to late 1760s, possibly as a result of the presence in

Fig 0.10
Harewood House, Yorkshire. The music room, c 1765–71, in 2007.

his drawing office of the Roman artist Giuseppe Manocchi (d 1782), who is thought to have been first employed by Jamie while on his grand tour to record antique decorations. Manocchi, like Brunias before him, was sent by Robert to Eton to make copies of the brightly coloured drawings there after the antique by Francesco Bartoli. But Manocchi's tastes went beyond these antique colour schemes, and he seems to have introduced Adam to the use of stronger, darker tones.[45]

The other notable advance was what Adam referred to as his Etruscan style, a new mode of interior design that he developed in the 1770s, and which he typically claimed as unique – different 'from anything hitherto practised in Europe'.[46] Most of its decorative elements – sphinxes, medallions, swags and so forth – were characteristic of Adam's work of the period; what was new was the restricted colour scheme of terracottas and blacks set on a pale ground and the increased delicacy of its forms. His inspiration was the decoration found on Classical Greek urns and vases – a form of antique ornament that Piranesi had suggested in his *Diverse maniere* (1769) could be applied to modern interiors. Indeed Piranesi's work and his publication were major sources for Adam's new style. Another stimulus was the antique vase collection of Sir William Hamilton, the British ambassador to Naples, illustrated in a series of folio volumes in 1766–76, which had such an impact on Josiah Wedgwood's pottery. Hamilton was of the opinion that these Classical vase designs were derived from ancient wall paintings – one of the antique sources that fascinated Robert Adam. Also, Wedgwood's own black basalt tablets exerted a further influence, and were sometimes incorporated by Adam into his designs.[47] A bedroom for the Countess of Derby at Derby House in Grosvenor Square, London, of around 1773 (now demolished), was Adam's first Etruscan interior, described and illustrated in detail in the second volume of the brothers' *The Works in Architecture*. Another lost specimen, the Countess of Home's bedchamber at Home House, of 1775–7, was reinstated in spirit as part of a modern restoration there, incorporating the room's original spectacular white marble chimney piece, inlaid with scagliola in Etruscan colours. The best surviving example of this style is the roughly contemporary Etruscan Dressing Room at Osterley Park, with its exquisitely painted wall decorations on a pale blue-grey ground (executed by Pietro Maria Borgnis; Fig 0.11).[48]

Fig 0.11
Osterley Park, Isleworth, Middlesex. The Etruscan Dressing Room, designed c 1775–9.

Architectural influences

When it came to architectural inspiration, surprisingly for two young neoclassical architects with an eye for harmony and proportion, Bob and Jamie claimed to have had little regard for the masters of the Classical tradition, such as Palladio or Inigo Jones. The manuscript 'short retrospective view' of British architecture, written by their brother-in-law John Clerk of Eldin, intended to precede his unfinished life of Robert Adam, suggests that Jones's work was overvalued and open to criticism. Like much of that document, this probably reflects Robert Adam's own opinions. In a letter of 1756, sent from Rome to his sister Nell, Robert asks if Jamie knew 'how I despise our famous Inigo, he would be astonish'd and that even Palladio Was much to be criticized. He would marvell.' A later letter about a planned publication on Roman baths outlines his intention 'to Attack Vitruvius Paladio & all those Black guards of Antient & Modern Architecture Sword in Hand'.[49] In both cases Adam is writing to amuse his family, but as always his flippancy masks an element of truth. For all that the Adam brothers' architecture was rooted in the past, they liked to see themselves as iconoclasts. They revelled in throwing brickbats at the stolid Palladianism of their eldest brother and its ultimate sources, without apparently recognising that the style underpinned so much of their own work. Whatever Jamie's thoughts in 1756, a few years later on his own grand tour he was toeing his brother's line. A visit to Vicenza in October 1760 offered an opportunity to appreciate Palladio at first hand but Jamie was not impressed (*see also* p 61, note 27). Even the great Basilica he found 'abominably maigre in every respect'. Only the Villa Capra or Rotunda raised a favourable reaction, with its 'pretty' plan: 'There is somewhat to make a good thing of,' he noted in his travel journal, 'which is more than can be said for most of Palladio's buildings.'[50]

The works of the Venetian Mannerist Michele Sanmicheli (d 1559) in Verona were more to Jamie's taste, though again he found fault: 'He seems to have had rather more genius than Palladio in the ornamental part of architecture, and is sometimes bold and male in his decorations, but is often very impure and very trifling.'[51] In Robert and Jamie's writings and in the biographical notes compiled by Clerk of Eldin, the names of architects recur whom the brothers evidently held in high regard. Their admiration of the 'movement' found in the works of Sir John Vanbrugh is well known from *The Works*, this being the quality in architecture they valued above all else. But the Adams were also influenced in their approach to design by Italian artists of the cinquecento, whose works they would have encountered on their grand tours and through publications and prints. These included two giants of the period, Michelangelo and Raphael. But perhaps more significant were several now lesser-known figures, most of them painters as well as architects, working at the cusp of high Renaissance Classicism as it began to transmute into Mannerism; many were connected with the circle of Raphael or were members of his large workshop. They included Raphael's former assistant Giulio Romano (*c* 1499–1546); the Sienese Baldassare Peruzzi (1481–1536); Giovanni da Udine (1487–1564), whose specialities were frescoes and stucco grotesques; Polidoro da Caravaggio (*c* 1499–1543), best known for his paintings on the façades of Roman houses; the versatile Pirro Ligorio (*c* 1513–83), who attempted to assemble a 50-volume encyclopaedia of antiquity for publication; the painter-architect (and historian) Giorgio Vasari (1511–74); and the fresco painter and draughtsman Federico Zuccari, or Zuccaro (*c* 1542–1609).[52]

Most of these individuals appealed to the brothers as reinterpreters of Classical antiquity. They came from an era when much more ancient fabric in and around Rome was still intact and available for study – particularly the stuccoes and highly decorated interiors that intrigued Robert Adam, and which provided Peruzzi, especially, and after him Raphael and Pirro Ligorio, with the inspiration for the grotesques, gems and other motifs with which they decorated the façades as well as the interiors of their buildings. James Adam said of these artists in his unfinished manuscript on architecture, in a section called 'Of Exteriors or outside decoration': 'They have render'd morcells otherwise indifferent realy precious to us & have imitated in no inconsiderable degree the beauty & elegance of the Ancients from whom they have Visibly borrow'd their first Ideas.'[53] In fact, Robert Adam's creative reuse of such sources is comparable to the way that Renaissance artists like Giuliano da Sangallo and Baldassare Peruzzi themselves drew on newly discovered antique remains (such as the Domus Aurea) for their imaginative reconstructions of ancient wall and ceiling designs.[54]

It is also likely that, through their friendship and working relationship with Giovanni Battista

Piranesi, the Adam brothers would have been made aware of the spatial and scenographic qualities of the more experimental baroque architects of the seicento, such as Francesco Borromini (1599–1667) in Rome and Baldassare Longhena (1598–1682) in Venice. Another artist of this period commonly mentioned by the Adams was the Roman sculptor Alessandro Algardi (1595–1654). This may seem surprising, given the high baroque style of Algardi's statues and monumental marble bas-reliefs, but they probably had in mind only his interior decorations at the Villa Pamphilj (now Villa Doria-Pamphilj) in Rome – especially the stucco ceilings, based on antique sources, which influenced some of Robert Adam's earliest English ceiling designs, at Hatchlands (of 1759–61).[55]

But Rome itself, above all else, was the single great, galvanising experience of Robert Adam's life; it was there that he gained a new understanding and self-belief, which enabled him to grow into an artist of true genius. This metamorphosis was something Adam himself understood at the time, as it was beginning. At first he wandered the city for several days in a haze, awestruck by the majesty of its ancient remains. Then came a realisation that here was the opportunity for him to become 'the Man I may be', and develop 'a Taste much Superiour to what I ever thought of before I saw Rome'.[56] This was no mere boasting. Adam was blessed with a rare shrewdness, a capacity to assess candidly his own abilities and worth. He also had the great good fortune to meet and come under the artistic guidance of Charles-Louis Clérisseau and Piranesi – two leading lights of the emerging neoclassical movement and, like Adam, mavericks who had either shunned or been shunned by the academic establishment. Thrust into their company, alongside Laurent Pécheux, Jean-Baptiste Lallemand, Laurent-Benoît Dewez, his fellow Scotsmen Gavin Hamilton and Allan Ramsay, and many other artists in the magical atmosphere of Rome, Adam was quick to realise his deficiencies as an architect, draughtsman and interpreter of antiquity – deficiencies he then worked supremely hard to overcome. Also, his travelling relationship with Charles Hope of Hopetoun, as well as a quickly established close friendship with Abbé Peter Grant of Blairfindy, the Scottish Catholic agent and cicerone, provided an invaluable entrée into the opaquely choreographed social world of the British aristocracy in Rome, which Adam hoped might help pave the way to future commissions. His letters home record honestly his campaign of self-improvement on both fronts – artistic and social.

The immediate impact can be seen in the many inventive sketches and capriccios that Adam made during his time in Rome. Recurring architectural features, such as coffered barrel vaults, domes, column screens and ancient sarcophagi, all show how much he had fallen under the spell of Roman monuments such as the Baths of Diocletian and Caracalla, the Pantheon, Hadrian's Villa at Tivoli, the many triumphal arches and columns – and how keen he was to absorb their lessons (Fig 0.12).[57] Then there are the more elaborate, grandiose fantasy palaces that Adam drew towards the end of his stay in Rome, showing off his new learning and attempting the mode of presentation associated with the architectural competitions of the Italian academies – such as the *Concorso Clementino*, sponsored by the Accademia di San Luca in Rome. Such drawings also reveal the extent to which he followed Clérisseau's lead in terms of his new artistic style, and Piranesi's imaginative, picturesque approach to interpreting the remains of antiquity.[58]

Of the two, the dynamic Piranesi was the greater genius. For a while he was close to Adam, whose abilities he genuinely admired and whom he was happy to assist as a sort of protégé. Piranesi even dedicated his *Il Campo Marzio* (1762), a fantastical recreation in map form of the Campus Martius area of ancient Rome, to Adam, whose name also features on its title page (though apparently he had to put his hand in his pocket for the privilege). Adam was struck by the invention that Piranesi could apply to his very personal reinterpretations of the ancient city, but was never quite able to match the liveliness and drama of Piranesi's work until much later, in the 1780s and 1790s, when his picturesque landscape drawings and paintings could be said to possess a comparable potency. Piranesi's designs and theories were to remain an inspiration for Adam throughout his career.[59] However, it was Clérisseau, Robert Adam's daily companion and drawing instructor in Italy, who was the bigger immediate influence, as can be seen in the many Roman sketches by Adam that consciously ape his teacher's style. The architectural details and imaginative capriccios of Classical ruins that Clérisseau drew were exactly the kind of source material that Adam craved, and he strove to copy him. As he told Jamie: 'He rais'd my Ideas, He created emulation and fire in my Breast. I wish'd above all to learn his manner, to have him with me at Rome, to Study close with him ... What I wish'd for I obtain'd.'[60] Their relationship was close, if odd, given the tension of master and pupil also being employee and patron.

Once back in England, Adam's emerging style owed so much to Clérisseau that Adam was determined to keep the Frenchman away for as long as possible, lest his presence in England should make the debt more obvious and dull Adam's growing reputation. Only once that reputation was secure would there be any consideration of Clérisseau joining him, and even then, as Bob told Jamie, he should still be brought 'secretly like a thief in the night & no one regard him'.[61] The large panels of painted views of Classical ruins and landscapes that became such an important feature of Adam interiors of the 1760s – for example those by Zucchi in the dining room at Osterley Park and the music room at Harewood House (*see* Fig 0.10) – are obvious echoes of Clérisseau's Roman capriccios, but in truth his influence was everywhere.

The last months of Robert Adam's grand tour brought an added dimension to his understanding of Roman architecture and his stock of ancient motifs via a detailed survey of the remains of the Emperor Diocletian's fortified palace at Spalato (now Split, Croatia). It gave Adam the subject for his first publication, which helped establish his reputation internationally, but more than that it gave him the opportunity to position himself as the leading interpreter of ancient Roman domestic buildings. Its influence can be seen throughout Adam's architecture: in

Fig 0.12
A typical Adam grand-tour capriccio of Roman remains, c 1756–7, in pen and pencil with colour washes.

the vaulted construction and riverside elevation of the Adelphi, a modern urban reinterpretation of the palace's dramatic waterfront setting; in his so-called 'Spalatro' order of columns that feature in many of Adam's designs, such as Luton Hoo, Bowood, St James's Square and elsewhere, based on the distinctive pilaster capitals in the palace's peristyle; and in the effective entrance sequence of columned rectangular hall and domed saloon, as found at Kedleston, repeating the form of the palace's atrium and vestibulum.[62]

In his modern-day recreations of antique interiors, Adam made use not only of his own vast collection of drawings, casts and antiquities, but also the increasing array of 18th-century publications that catalogued and illustrated the latest archaeological finds, such as Robert Wood's *Ruins of Palmyra* (1753) and the *Antichità di Ercolano esposte* (1757–92). As has been said elsewhere, Adam made use of details from Classical antiquity 'on a scale and with a consistency unknown to his predecessors and contemporaries'. His and his artists' favourite sources for motifs were the many drawings after the antique by the Bartolis, either those published in the *Admiranda Romanarum antiquitatum* (1693) or originals held in the Topham collection at Eton. Their impact is seen throughout Adam's interiors.[63]

Rome was always Adam's prime motivation, but French architecture and design also exerted an influence. Though he had little to say on this subject in his grand-tour letters, it would have been common and essential knowledge for a partner in a leading 18th-century Scottish practice. His admiration for French planning comes across strongly in his London house-plans of the mid-1760s to early 1770s, and in other writings. For example, he interrupts his explanation of the plans for Syon House in *The Works* to comment: 'A proper arrangement and relief of apartments are branches of architecture in which the French have excelled all other nations.' Designs for Lansdowne House in Berkeley Square, for General Clerk's house in Marylebone and for Lord Holland's house at The Albany in Piccadilly relied heavily on the courtyard arrangement of the Parisian *hôtel*. An unbuilt Marylebone town palace for the Duke of Portland took this form to a new level, with a colonnaded circular courtyard extending some 150ft (46m) in front of the large Palladian-style mansion. General Clerk's house included a sequence of rectangular private rooms in a classic French enfilade, while the floor plan of Derby House in Grosvenor Square had a succession of three interconnected first-floor reception rooms, which Adam himself described as 'an attempt to arrange the apartments in the French style, ... best calculated for the convenience and elegance of life'. French neoclassical interior design, though closely connected to Adam's style, does not seem to have interested him to the same degree as French planning. For instance he made little attempt to adapt his Italian-derived manner of interior decoration to suit rooms which his clients specifically intended for the display of French tapestries (such as the Tapestry Room at Osterley Park).[64]

Other influences emerged or re-emerged late in Adam's career, such as the inherent Palladianism of his middle-class villas or the romantic and picturesque elements of his Scottish castles; these subjects are dealt with later in this chapter (pp 27 and 28).

Adam architectural theory: 'the true, the Simple & Grand architecture'

The brothers' buildings are the definitive guide to their architectural philosophy, but there are many unexecuted designs and sketches, as well as publications and other writings, to help flesh out our understanding of the ideas that lay behind the Adam style. Any student of this subject is immediately confronted by a conundrum. Though Robert was the principal and often sole designer – and originator of the Adam style – the accepted interpretation is that he had no interest in architectural theory, which is considered instead as the domain of his languid, less gifted brother James.

Such a view is reinforced by the striking similarity of manuscript notes made by James in Rome and London in 1762–72 on the subject of architecture with passages in the first volume of *The Works in Architecture*, in particular the key sections dealing with 'movement' – the 'rise and fall, the advance and recess' of the different parts of a building – which the Adams thought contributed so much to their style's picturesque qualities. Robert himself encouraged this interpretation by noting, in a well-known letter of March 1763 to the Enlightenment thinker and writer Henry Home, Lord Kames (1696–1782), that Jamie wrote 'with that love and enthusiasm of architecture, which no one could feel that has not formed very extensive ideas of it'. In the same letter, Robert also said that he himself had 'but too few moments to dedicate to theory and

Fig 0.13
Kedleston Hall, Derbyshire. The south or garden front, designed in 1760.

speculation' – though he followed that statement with a particularly erudite and convincing assessment of the best forms and uses of the Classical orders. On another occasion Robert wrote to Sir James Clerk of Penicuik with considerable authority on the subject of country house planning, and his preferred room arrangements.[65]

In weighing up the brothers' contributions in such matters, we should also bear in mind Jamie's vivid description of his brother, in the early days of his practice, 'holding forth' to potential clients for 'long London mornings ... at the expence of much time & many words'.[66] What could Bob have been 'holding forth' on, if not his opinions on architecture? Also, during the formative years of the early 1760s, while James was abroad, Robert would have had to explain his planning and decorative schemes to well-informed connoisseur-clients like Lords Bute, Mansfield, Northumberland and Shelburne – an impossibility for a man who was uncomfortable expressing architectural ideas (and Robert Adam was known to be a 'quite high converser').[67] Furthermore, several passages in *The Works* are reminiscent of Robert's writings rather than Jamie's – for example his letter to Lord Kames and his unpublished draft introduction to the *Ruins* folio – and there is sometimes a verve and focus to the text that is often missing from Jamie's rambling notes on architecture, not to mention his disappointingly dull Italian travel journal. Of course, this ability of Robert's to discuss the practice of architecture with clients may not have extended to the finer, more philosophical points of architectural theory. Nevertheless, it seems certain that he was more actively engaged in theory than has hitherto been recognised, and that even if Jamie *were* the official 'theorist' of the partnership, the ideas he was expressing were probably mostly those of his older brother.

What were those theories? Judging by what Robert Adam first put into practice on his return from the Continent, his principal goal was to introduce to British architecture something of the spirit of heroic grandeur that he had witnessed among the remains of ancient Rome – what he referred to in his letters as 'the true the Simple & Grand architecture'. The garden and entrance fronts at Kedleston, marked by a monolithic portico on one side and a giant triumphal arch on the other, were early manifestations of this approach, as too was the 'transparent' portico of giant Ionic columns at Osterley Park – at one and the same time contrasting with and integrated within the existing Jacobean house, which as a result took on an entirely new character (Figs 0.13 and 0.14). But perhaps the finest early expression of Robert Adam's quest to bring 'the Antique, the Noble & Stupendous' to these shores is the sparkling entrance sequence of marble hall and domed saloon at Kedleston Hall – the former especially with its monumental columns of pink alabaster and Classical statues seems to recreate the spirit of ancient Rome (Fig 0.15; *see also* Fig 5.11). Other examples of this grand Roman manner of the early 1760s include the entrance hall and anteroom at Syon House (*see* Fig 2.9), the Admiralty screen at Whitehall

Fig 0.14
Osterley Park. The east front, with Robert Adam's portico, designed 1760s.

and an extraordinary neoclassical design for a circular chapel for the London Lock Hospital, topped by a Pantheon-style dome.[68]

Though this modern reinvention of ancient Rome remained an essential aspect of Adam's work, by the late 1760s and early 1770s he had begun to develop a more personal style, especially in his interior designs, based on a complex (and often random) combination of delicate, linear Classical motifs and low-relief stucco decorations. The influence of Piranesi's evolving theories on architecture is discernible here. In his writings Piranesi stressed the importance of diversity, decorative detail and an imaginative, contemporary reinterpretation of antiquity, rather than mere slavish imitation.[69] 'Variety', 'novelty', 'movement' and 'surprise' became the watchwords of the Adam style, and as such recur frequently throughout the text of *The Works in Architecture*. Always one of the most painterly of architects, especially in his approach to elevational design, Adam espoused theories in *The Works* that place him firmly within the emerging aesthetic movements of the picturesque and sublime of the 1780s onwards – an association that becomes increasingly evident in the latter part of his career in his landscape drawings and paintings, and castle-style houses.

Fig 0.15
Kedleston Hall. Adam office sectional drawing through the marble hall and saloon, 1760.

Robert Adam's patrons

When Robert Adam threw himself into a London career in 1758, he was assisted in his pursuit of new clients by several influential expatriate Scots, many of them old acquaintances.[70] His family's work at Inveraray for the 3rd Duke of Argyll, Scotland's leading politician, stood him in good stead: initially it enabled him to bring part of his art collection into the country duty-free, but Argyll also spoke well of Adam to potential employers. More immediately useful in terms of commissions were the Adam family's old friend Gilbert Elliott of Minto and Jean Campbell, Lady Lindores, a relative by marriage of George Drummond, Lord Provost of Edinburgh. Elliott was a lord of the Admiralty and it was probably he who secured two significant early contracts: alterations at Hatchlands Park, Surrey, for Admiral Edward Boscawen, and the Admiralty screen in Whitehall, Adam's first public work in London. Lady Lindores's regular salon at her house in Half Moon Street was a magnet for fellow Scots aristocrats and their English friends, and was thus a perfect vehicle for spreading news of Adam's arrival on the London scene. It was through her that he met Edwin Lascelles (later 1st Baron Harewood), who was building a new country house at Harewood, West Yorkshire, where Adam was hired in the summer of 1758 to improve on John Carr's work there. Also, the Adam family's long-standing connection with the Hopes of Hopetoun brought dividends, as it was Charles, Lord Hope (d 1766), the 2nd Earl's eldest son, who drew Robert to the attention of Lord Mansfield, his patron at Kenwood House. Another useful acquaintance was Lord Charles Hay, younger brother of the 4th Marquess of Tweeddale, who late in 1758 introduced Adam to Sir Nathaniel Curzon (later Baron Scarsdale) of Kedleston Hall.

But the relationship Adam coveted above all others was with John Stuart, 3rd Earl of Bute, special adviser to the Prince of Wales (from 1760 George III). That introduction probably came via Gilbert Elliott or Adam's good friend John Home, who had become Bute's confidential secretary. Though their first meeting went badly, in time Bute and Adam developed a close relationship. In 1762–3 Bute, then at the height of his political ascendancy, brought Adam one of his key early commissions at Bute (later Lansdowne) House in London's Berkeley Square; later came improvements to Bute's house in South Audley Street, as well as further work at his country seats at Luton Hoo, Hertfordshire (from 1766), and Highcliffe Castle, Hampshire (1773). As Alan Tait has pointed out, Adam publicly acknowledged his debt to Bute's patronage in *The Works in Architecture* at a time when the former prime minister was in disgrace and deeply unpopular.[71]

Adam valued a connection with Bute for the direct route it offered to George III and the renewed interest in public architecture that was anticipated during his reign. But though Adam was appointed royal architect in 1761 (jointly with William Chambers), he never won the trust of the new king and queen or the big public contracts that Chambers, their favourite, was able to secure. The fifth part of the first volume of *The Works in Architecture* was devoted to the Adam brothers' royal commissions, but it was rather a hotchpotch of unexecuted designs and relatively minor alterations – for example a chimney piece and ceiling for the Queen's House, and a sedan chair – fleshed out by the addition of a harpsichord design for Catherine the Great of Russia. Adam's temporary illuminations, created for George III's birthday celebrations at Buckingham House in June 1763, were the most spectacular and successful of these works (Fig 0.16); their brilliance and ephemerality are somehow symbolic of his fleeting engagement as a royal architect. Apparently the queen thought him too expensive, to which Adam countered: 'Her majesty of All people has least reason to complain,

for I have done many things for her & refused all recompence.'[72]

Although unable to captivate the royal couple, Robert enjoyed the loyal patronage of many influential aristocratic clients in addition to Earl Bute. Foremost of these were probably Hugh Smithson Percy, Earl (later 1st Duke) of Northumberland, and his wife, Lady Elizabeth Seymour Percy – two knowledgeable, sophisticated aesthetes who strove to restore the glory of the Percy name through rebuilding and improvement. They had, as Adam himself wrote of the duke, 'not only wealth to execute a great design, but skill to judge of its merit'.[73] Following the success of his new suite of Roman-style state rooms for Syon House (*c* 1760–8), the couple's suburban retreat, Adam was put in charge of alterations at two further Northumberland properties: their northern seat at Alnwick Castle, Northumberland, and the family's 17th-century London townhouse, Northumberland House in the Strand. (Adam also designed the duchess's monument in Westminster Abbey.)[74]

Another such figure was George William, 6th Earl of Coventry, who in 1760 employed Adam to create new interiors at his country seat, Croome Court, Worcestershire, and also at his London residence, Coventry House in Piccadilly. Also, though they later fell out, William Petty Fitzmaurice, 2nd Earl Shelburne, later 1st Marquess of Lansdowne, gave Adam the opportunity to finish the house in Berkeley Square (Lansdowne House) that he had begun for Lord Bute – the first big house that Adam was able to design and build from scratch in London.[75] While it has been suggested that other, lesser architects had a more elite client base, the evidence suggests quite the reverse. Adam's pallbearers included a duke and three earls, confirming the high status of his 'numerous circle of friends' and clients at his death, and the strong affection they still felt for him.[76]

For the most part Adam's private patrons were charmed by his genius. Even when, occasionally, he could be vexing – underestimating costs or being overly ambitious in his designs – most still held him in esteem. When his designs for a new South Bridge scheme in Edinburgh failed in 1785 to win the approval of the trustees there, Adam took advantage of their presence in London on parliamentary business the following April and arranged a meeting, at which he again presented his case. A powerful body of Adam's loyal clients and friends also attended this meeting, for no reason other than to support his cause; these included the Earl of Hopetoun, Lord Mansfield, the Duke of Buccleuch, Lord Frederick Campbell, Viscount Maitland, Sir Adam Fergusson and George Dempster.[77]

Perhaps the best example of the loyalty and good feeling that Robert Adam could engender is given by Dr John Turton, a society physician who numbered Adam, Lord Mansfield and George III among his patients. The first person

Fig 0.16
Design for the illuminations at Buckingham House for the king's birthday, 1763, by Robert Adam.

to purchase a house on the Adelphi's Royal Terrace in the 1770s, Dr Turton later commissioned Adam to design him a villa, Brasted Place, for a country estate at Brasted, near Sevenoaks in Kent, which he had purchased from Lord Frederick Campbell in 1784. Turton was 'cruelly disapointed' when Adam, who had agreed to design a house for a budget of under £5,000, presented plans that Dr Turton estimated would cost at least £11,000 to build. Dr Turton wrote to Adam expressing his dismay, and took pains to point out possible savings, but in the end was unable to prevent his regard for the architect from shining through. His letter concluded:

> I have ever admired you as an ingenious, I have ever esteemed you as an honest man. You have been woefully mistaken in your calculations. You have led me into difficulties, but it never hath, or I trust will shake my good opinion of you. I shall ever follow you with my good Wishes, & rejoice in every good that may happen to you.[78]

Adam publications

The Adams understood the significance of publishing in their quest to become the country's leading architects, in terms of asserting their status and disseminating their architectural theories and works. Such activities extended to the preparation of prints of new designs for sale as well as the issuing of folio volumes.[79] But it was not a pursuit that the brothers found easy. Robert especially was troubled by anxieties when it came to publishing his own designs, such was his fear of plagiarism, and he dreaded the reproduction of his work in prints or journals by lesser artists, whom he thought likely to be sloppy or inaccurate. Better by far, he believed, to let your renown spread 'in a more polite way', via the approbation of men of taste.[80] This caution may explain why only two of the many publishing schemes planned or begun by Bob and Jamie were to reach fruition. Neither was John Adam an active publisher, though his two younger, fashion-conscious brothers no doubt discouraged him. *Vitruvius Scoticus*, largely a celebration of his father's architecture, begun by William Adam in the 1720s, to which John had added some designs from the brothers' Edinburgh practice, was not published until 1812. By this time it was of little relevance and all the Adam architects were long dead.

At first Robert's reluctance prevented him only from committing his own designs to print; he was eager enough to engage in the archaeological–antiquarian vein of publishing then in fashion, such as Wood's *Palmyra* (1753). One of Adam's first visits on crossing the Channel in 1754 was to the engraver John Ingram in Paris, for advice on whom the brothers could engage for help with 'our books'. His principal aim then was that of self-promotion: a major scholarly work would introduce him to the London architectural world on his return with a 'puff' and signify his pre-eminence as a leading interpreter of antiquity. The end result was *Ruins of the Palace of the Emperor Diocletian at Spalatro in Dalmatia*, not published until 1764 but based on surveys undertaken in July–August 1757, near the end of his grand tour. However, the choice of site was eleventh-hour and opportunistic. Several other publishing projects had already been begun and pursued with vigour in Rome by Adam and his draughtsmen but had failed to reach completion. The first was a revised and improved edition of Desgodetz's *Edifices antiques de Rome* (1682, 1695), followed soon after by two further projects on Hadrian's Villa at Tivoli and the Baths of Diocletian and Caracalla. By July 1757 Adam and his 'myrmidons of art' were also busy measuring the remains at Pola. But soon the enormity of these tasks became apparent, especially for a freely eclectic and experimental artist like Robert Adam, who was temperamentally unsuited to the exactitudes required by the new, pseudoscientific approach to archaeological studies like Wood's, or Stuart and Revett's *The Antiquities of Athens* (from 1762). Diocletian's seafront palace at Spalato offered a nearby and relatively unknown Classical site, which had the added bonus of having been a private palace – one of the aims of the brothers' tours abroad being to track down examples of domestic antique decoration that could enrich their new style.

Robert Adam's approach to the book was typical of his working methods in general. He took a directorial role, overseeing a team of gifted artists, illustrators and engravers, and commissioned the main text from his cousin, the Edinburgh historian William Robertson. In many ways the *Ruins* folio as eventually published was a characteristic archaeological work of the period, mixing picturesque prints with measured drawings, text and descriptions of the plates. What made it special was the number and quality of the prints (and indeed the emphasis on illustrative material rather than text as the principal medium for conveying information), and also the degree to which it focused on the romantic, picturesque qualities of the site, rather than attempting an accurate record of the archaeological remains. Both Adam and Clérisseau, his principal collaborator on the *Ruins*, were artists who, like their associate Piranesi, looked to antiquity for its emotive properties, to fire their creative imaginations.

By the time the *Ruins* appeared, Adam had already been issuing prints for sale to promote his work. This began in 1760–1 with the preparation and publication of a plan, elevation and perspective view (engraved by Domenico Cunego) of the Admiralty screen in Whitehall. In the ensuing years he was to issue further prints of executed works for sale – for example a perspective of the riverfront elevation of the Adelphi (1770, *see* Fig 0.7), and also the elaborate temporary pavilion he designed for the *fête champêtre* arranged by General John Burgoyne at the Earl of Derby's house, The Oaks in Surrey (1774), to celebrate the coming wedding of his nephew Lord Stanley to Lady Betty Hamilton, daughter of the Duke of Hamilton (*see* Figs 5.9–5.10).

The logical extension of this activity was for the brothers to publish a volume of selected engravings of their best or most significant designs, with an accompanying text and notes. This was *The Works in Architecture of Robert and James Adam*, the first volume of which appeared in 1773–8 in a series of five parts, to be followed by a second volume, also of five parts, published in 1779. (A third book appeared posthumously in 1822 but had little connection with the brothers, comprising a collection of engravings probably rejected by them from the first two volumes.)

Robert Adam had been toying with such an idea for some time, even as early as 1760 when working at Kedleston (*see* p 118, note 4). There were several possible stimuli for publication of *The Works* in the 1770s. One was the loss to the Adams' reputation and the financial disarray brought about by the commercial failure of their two extensive London speculations at the Adelphi and Portland Place, and the calamity of the Scottish banking crash of 1772 that precipitated the sale of much of their property by lottery. Another was the increasing challenge to their status from younger neoclassical architects like James Wyatt and Henry Holland. It was time perhaps for a reassessment and a reminder to the public of the Adams' past successes and their continuing influence. Also, Robert had lent some of his drawings to John Woolfe and James Gandon for use in the fifth and last volume of *Vitruvius Britannicus* (published 1771), and it is possible that the crude reproductions there finally persuaded him to consider publishing his designs himself.[81]

The style of presentation in *The Works* was innovative. As with the *Ruins*, the focus was on illustration. Traditional elevations and plans were sometimes but not always included: the Adams' objective was to offer the reader a fragmentary, impressionistic mix of perspective views, details and assembled plates of fittings and decorations (Fig 0.17). This method of presentation aptly mirrored the visitor's experience of an Adam house, where the architectural qualities were gradually unfolded before his or her eyes, and it also conveyed visually the 'novelty' and 'variety' which the accompanying text stressed were the salient features of the Adam style – something that contrasted with the strictness of the revival in Greek architecture then coming into fashion.

The prefaces which introduced each section of *The Works* varied greatly in content and style, often repeating phrases from earlier writings by both brothers, loosely sewn together.[82] Critics disliked the declamatory nature of much of the text, which seemed intent on justifying the Adams' self-appointed position as the nation's chief arbiters of neoclassical taste. All the same, the first volume included some fine passages, especially those describing the ethos of the Adam style – such as that on 'movement' and the 'beautiful spirit of antiquity' that the brothers had succeeded in fusing into their modern architecture. By the time of the second volume, the commentary had started to dwindle, and other than Part I (which was devoted to designs for Derby House in Grosvenor Square), all that was provided in terms of text were extended image captions.

With the appearance of volume 2 of *The Works* in 1779 the Adams' publishing activities as architects came to an end. It failed to sell as well as its predecessor and there is a common theory that both brothers had become disenchanted with the business of book production by the late 1770s. But James Adam's fondness for writing and theory stayed with him. During the practice's barren years of the 1780s, when he turned his back on architecture and the Adam office to immerse himself in a new-found love of farming at his estate in Hertfordshire, he penned

Entablature & Capital of Column in the Eating-room.
Entablement et Chapiteau de Colonne dans la Salle à manger.

Trophies on the side of the Porters Hall.
Trophées sur le côté du Tinel du Portier.

Cornice & Freize of the Porters Hall.
Corniche et Frise du Tinel du Portier.

Impost in the Porters Hall.
Imposte dans le Tinel du Portier.

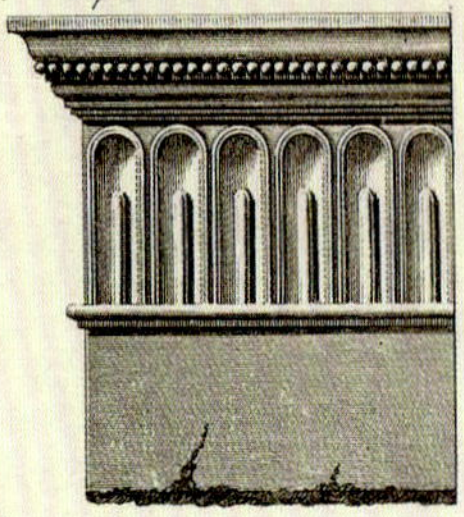

Center part of the Porters Hall Cieling. Centre du Plafond du Tinel du Portier.

Base & part of Column of the Eating room.
Base et partie de Colonne dans la Salle à manger.

Trophies on the side of the Porters Hall. Trophées sur le côté du Tinel du Portier.

Surbase & Base of the Porters Hall.
Surbase et Base du Tinel du Portier.

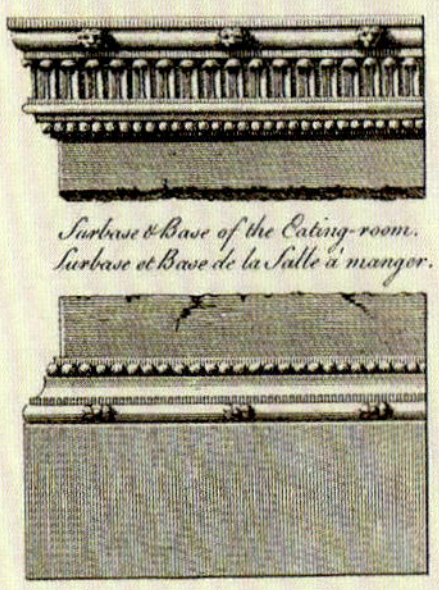

Surbase & Base of the Eating-room.
Surbase et Base de la Salle à manger.

Scale of 12 6 0 1 2 3 feet

R. Adam Architect 1773 Published as the Act directs 1777. J. Zucchi Sculp. London a two-volume study on the subject, *Practical Essays on Agriculture* (1789). Big brother Johnnie in Scotland could barely hide his surprise: 'Can this be our James?', he wrote to his son William in London. 'If so, in the name of all that is good, where has he got this Knowledge?'[83]

There is a further postscript. Alistair Rowan identified a distinctive series of mostly small-scale elevations and plans of the 1780s and 1790s among the Adam office drawings in the Soane Museum, which may have been prepared for an intended late publication of designs of neoclassical villas and castle-style houses. This hypothesis is supported by a reference by Robert Adam's Scottish clerk John Paterson, in a letter of 1790, to taking on a draughtsman 'to assist me in forwarding the Book'.[84] But if such a volume were envisaged, it would have been radically different in character from the other Adam publications, without the eye-catching perspective views or groups of details which were such a prominent feature of the *Ruins* and *The Works* folios.

Business speculation and town planning

An urge to speculate in building ran high in the Adam blood, as did a like desire to dabble in and profit from often risky commercial ventures beyond the realm of architecture. The brothers' father William Adam (whom they dubbed 'Old Stone and Lime') provided the model for such an approach. As well as his successful architectural and contracting firm, William Adam owned a string of other businesses – brick and tile works, barley kilns, timber mills, salt pans, coal fields, marble works and more – and so it was natural for the boys to do likewise. Hence the creation of their building supplies and contracting firm William Adam & Company in 1764–5 (described on p 6).[85]

Robert Adam had shares in the Scottish east-coast whaling industry, though characteristically seems to have had little grasp of what these were worth, or even if he still owned them – 'By Jove,' he wrote to Jamie, 'I know nothing of the matter.'[86] Older brother John was always the steadiest of the four but nonetheless also had a speculating streak. One of his commercial ventures in February 1755 was an attempt to trade by sea with London in a type of Dutch pearl barley that had been introduced to Scotland by his father. His plan was to undercut the popular London trade in imported barley from Holland, but the Adam barley was inferior and John misjudged the Dutch duties, and by 1757 had given up, unable to compete.[87] His bankruptcy in 1764 was also partly the result of business speculations, though his practical reaction to and recovery from this financial disaster (described on p 58) emphasises how much he differed in character from his younger brothers. In many ways Bob and Jamie's grand tours of the 1750s–60s, largely financed by John from the profits of the family firm, were also speculations – the potential return for the investment being the new insight the younger boys could bring to the work of the Edinburgh practice, as well as the forging of new links with prospective clients and patrons abroad. There were also the valuable collections of books, prints, drawings and artefacts that Bob and Jamie amassed and brought back with them; Jamie, for instance, is thought to have returned with more than 500 old masters.[88] (Jonny Yarker gives an insight into the brothers' activities as dealers and collectors in the Roman art market in Chapter 2.)

Unsuccessful forays into stucco production (with Liardet's cement) and the arms trade (via the Adams' own Battersea & Sand End Company) are well known, but the more one digs into the family correspondence, the more such ventures rise to the surface. Young Willie's partnership in London with John Bayne from *c* 1758, and his involvement in trade with Havana, has already been mentioned. Theirs appears to have been one of the leading firms engaged in sending ships to Cuba and may have acted as representatives for other London merchants, as it was to Bayne & Adam that Edward Sedgwick and Lovell Stanhope, secretaries to Lord Halifax, Secretary of State for the Southern Department, wrote in 1764 with updates about Spain's attitude to British merchants and their goods.[89] By 1761 Willie was also engaged in the wine trade, and his wares received the approval of several members of the aristocracy. He travelled to Bordeaux in 1763, perhaps in connection with his wine interests, and one of the objectives of Jamie's abandoned trip to Greece was to establish new commercial links that Willie could utilise to import Cretan wines and Ionian honey.[90]

When it came to architecture, a key strand in the Adam brothers' wide-ranging practice was their commitment to speculative town planning and street improvements. Robert Adam enjoyed architecture on a grand scale and understood implicitly the need for an architect of his ambition and aspirations to be engaged in public

Fig 0.17
A typical page from The Works in Architecture of Robert and James Adam *(Adam and Adam 1773–1822), showing a selection of details from the entrance hall and eating room at Wynn House, 20 St James's Square, London, in a carefully arranged composition. For other extracts from* The Works, *see for example Figs 5.1, 5.4, 5.5, 5.9, 11.4 (bottom) and 11.13 (far right).*

Fig 0.18
Robert Adam's first designs of 1785 for terraces flanking a new south approach to Edinburgh (the South Bridge scheme). West side (top), with, to the left, a new entrance screen for Edinburgh University and assembly rooms; below are the backs of buildings on the east side. Not executed.

works of national importance. The section in the brothers' *The Works in Architecture* on this subject began with the statement: 'Public buildings are the most splendid monuments of a great and opulent people.'[91] For Adam to succeed in this field he needed generous public or royal patronage but, despite his position as architect to the king, very few such commissions of any consequence came his way. Instead he turned to large speculative housebuilding developments to allow his architectural imagination free rein and to make his and Jamie's mark publicly.

In the 1760s and 1770s, London was the only British city advanced enough in terms of its capacity for large-scale street improvements for the Adams to realise these ambitions fully. The Adelphi and Portland Place were private speculations on a monumental scale, unmatched in Georgian London, and as such they transformed those areas of the capital and set new standards in their approach to civic planning. They were also intended to make the Adam brothers' fortunes, but because of their size and poor financial management were commercial failures. Frederick's Place, a cul-de-sac of eight houses built near the Guildhall in the City of London in the late 1770s, was a third such scheme, though very modest by comparison, and Fitzroy Square, though never completed on all four sides as intended by Robert Adam, showed just how expert he had become by the 1790s at handling the massing and detailing of continuous terrace façades around a square.[92]

Though personal ambition and financial gain were strong motives for schemes like the Adelphi and Portland Place, there was also an altruistic side to Robert Adam's work in this field. He had a genuine desire to improve British cities and understood intuitively that imaginative planning on a grand scale, if handled well, could reap public benefit as well as secure his architectural reputation. This goes some way to explaining the audacious, often overly ambitious scope of many of his designs. His unexecuted plans for his

native Edinburgh of the 1780s show Adam's genius for urban planning at its height. The decision by the local authorities to build a new South Bridge to connect the city and its New Town more effectively with the southern suburbs offered an opportunity for a major improvement. Adam's plan for a sweep of neoclassical terraces, raised up on arches, creating a new ceremonial approach into the city, was visionary (Fig 0.18). But the lord provost and trustees balked at the cost and took objection to the involvement and expense 'of an Overseer from London', preferring instead to leave the work to be done piecemeal by local builders. Adam realised the terrible loss to himself and his home town and kicked up a fuss, writing candidly to the lord provost of the scheme that had replaced his: 'It may become as ugly and deformed a piece of Work as ever disgraced a great City.'[93] Adam's architectural presence is strong in the Scottish capital, with Register House, Edinburgh University's Old College building, and Charlotte Square – his last great urban development (*see* Fig 0.8). But had his South Bridge scheme gone ahead, alongside other plans for more scintillating terraces on Leith Walk and further rows of houses connecting Princes Street with Calton Hill, it would have made a fitting culmination to the New Town at its eastern end and a neoclassical cityscape equal to any in Europe.[94]

Late villas

The shift in emphasis in the work of the Adam brothers' practice after the late 1770s from mostly English commissions to predominantly Scottish ones was accompanied by a corresponding change in the type of projects they undertook. Public buildings and town-planning schemes remained essential components of their *oeuvre*, but increasingly rare were the big aristocratic country-house commissions that had helped make Robert Adam's name in the 1760s. In their place came two diverse types of architecture: Classical-style suburban villas for the mercantile and professional classes, and houses in a synthetic, picturesque 'castle style' of startling originality.

As with his earlier neoclassical interior designs, Adam's late villas show a rare ability to devise endless variations on a theme, utilising the same or similar elements and motifs – pyramidal roofs, Venetian windows, pediments, square pavilions and so on – in myriad combinations (Fig 0.19).[95] Also, their planning is frequently dependent upon Adam's facility with complex interlocking forms (Robert Adam's genius for planning is discussed by David King in Chapter 8). The style is fundamentally Palladian, informed and influenced by the understated manner of his brother John and the mid-century English architects and theorists like Roger Morris and Robert Morris whom John admired. As with many of Robert Adam's castle-style houses (discussed separately on pp 28–31), the designs are sometimes severely rational, stripped of decorative detail and reliant upon geometry for effect. Occasionally they suggest his moving towards a more progressive, restrained Italianate style, anticipating the work of Soane or Nash in the following century.

The survival rate of Adam's late villas is poor: there was little understanding of their value or their vital role in the work of the Adam practice in its later years until Alistair Rowan's scholarly

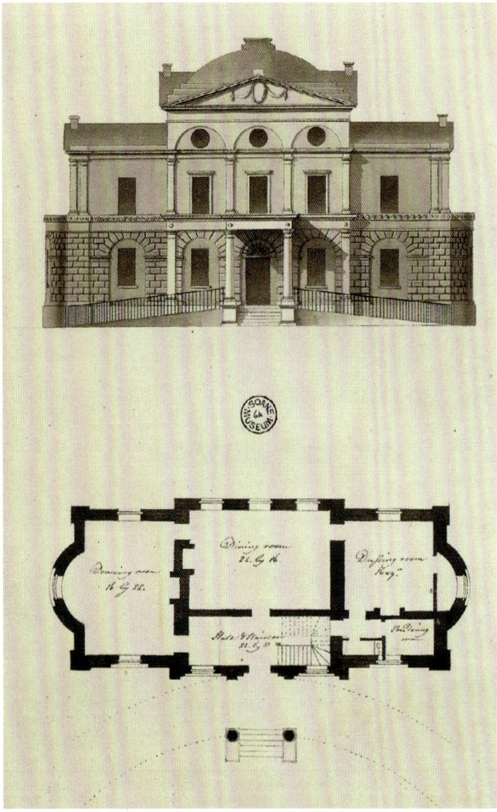

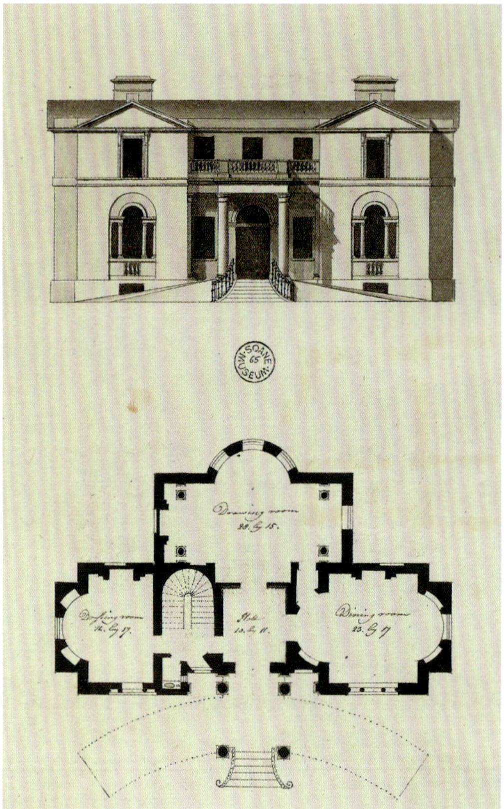

Fig 0.19
Late Adam villas. Plan and elevation of c *1790 for an unknown villa (left) and the final design for a villa called Sunnyside, Liberton (near Edinburgh), largely as built* c *1790–1 (right).*

study of the subject in the 1980s. Had the proposed Adam brothers' publication of villa and castle designs been completed during their lifetimes, the story might well have been different, and more of their fine villa buildings, like Jerviston House, near Motherwell (demolished in the 1960s), might have been spared.[96]

Robert Adam's late picturesque vision: the birth of another new style?

As noted already, for most of his life Robert Adam found solace and inspiration in drawing and painting picturesque views. His favourite themes were the brooding landscapes and castles of his native Scotland, or the ruins of ancient Rome as experienced on his grand tour. At times these compositions were faithful representations of topographical views, at others imaginative fantasies – or sometimes an amalgam of the two. Few surviving examples seem to date from the 1760s and early 1770s, when his star was in the ascendant and his abilities as a designer were in great demand. But with the Adelphi calamity and the financial crises of the 1770s, and the subsequent 'falling off' in his architecture business, particularly in England, Adam returned to this form of creative expression in the 1780s with a vengeance. And it was Scotland, on which he had turned his back so roundly in 1758, that was to be its focus.

The many drawings and watercolours Adam produced during these last years of his life generally followed a similar format, with a castle (or castles) as the principal object, usually set within a dramatic, wild landscape and viewed through a dark foreground, often with figures to provide scale and movement; other common motifs are waterfalls and ruined bridges (*see* Frontispiece, Fig 0.20 and Chapter 10). But such compositions are more than simply a sum of their parts. The sombre atmosphere they evoke is that of the Scottish Highlands, the architecture they contain is a hybrid, a seemingly timeless union of the Gothic and Classical, and the overall emphasis is a picturesque one, endlessly questioning and examining a building's relationship to its landscape setting. As always with Adam there were many possible sources to inform this side of his *oeuvre*. Aspects of the work of Paul Sandby, whom Adam knew and sketched with in Scotland, are evident, as are the techniques developed by William Gilpin and Robert's friend, relation and sometime sketching partner John

Clerk of Eldin. But in the end Adam's picturesque paintings are entirely personal, with a unique emotive power and enigmatic fusion of buildings and landscape.

It is perhaps easier to understand this body of work by considering it alongside Adam's late castle-style buildings in Scotland, which were a manifestation of the same interests in the sublime and picturesque, and which at their best recreate in three dimensions the bold massing and striking outlines of the castles in his pictures. Though the style reached its height in the 1780s it was not entirely a late development. Adam had been designing castellated houses since the 1760s, and he and his brothers had gained valuable early experience of 'castle' building in mid-18th-century Scotland, and not just in their roles as contractors to the Board of Ordnance fortifications. They had assisted their father in constructing a new castle at Inveraray in the 1740s–50s, designed by Roger Morris for the 3rd Duke of Argyll – an important exemplar for Robert, perhaps, as an entirely modern residence encased in a castellated exterior, expressing in masonry the Clan Campbell's status and ancient pedigree. Also John Adam, possibly with Jamie's assistance, had in the 1750s drawn up plans for Douglas Castle, Lanarkshire, for the Duke of Douglas, a work never completed. As designed, both buildings were entirely symmetrical in façade and plan, with only some Gothic detailing, battlements and corner towers to convey a castle identity. Robert Adam's earliest forays into the genre, as at Ugbrooke (1763), Mellerstain (1770, Fig 0.21), Cauldwell House (1771) and Wedderburn (1771), have a similar, almost Palladian 'boxiness' and regularity to them. But by the late 1770s, when

Fig 0.20
Robert Adam, Romantic Landscape with Castle, *1780s. Like many of Adam's landscapes, this composition appears to combine multiple perspectival vantages in one scene.*

Fig 0.21
Mellerstain House, near Kelso. The main (south) front, by Robert Adam, c 1770–8.

Fig 0.22
Culzean Castle, of 1777–92, on its clifftop setting on the Ayrshire coast.

he began rebuilding Culzean Castle for the 10th Earl of Cassilis in a spectacular setting on the Ayrshire coast (Fig 0.22), Adam had developed his ideas into a remarkable new style full of picturesque drama, perhaps with the intention of re-establishing the Adam brothers' status in British architecture.[97]

This mature castle style, seen at its height at Seton, Dalquharran and Culzean (*see* Figs 10.3, 10.4 and 10.7), was an eclectic synthesis of Adam's many influences and experiences, and should be considered as an expression of neo-classicism and the Picturesque. The pronounced massing of forms for effect rather than architectural detail recalls Vanbrugh; the circular towers and massive wall-heads suggest Roman military architecture or the great papal and Renaissance fortresses that Adam saw in Italy; and of course there was the inspiration of the ruins of the Emperor Diocletian's fortified seafront palace at Spalato, as well as the castles and tower-houses of his homeland, including his own ruined castle of Dowhill, on the family's Blair Adam estate. Buildings such as Culzean, Dalquharran and Seton were well suited to a Scottish environment, and it was a country where the castle form still held an appeal and meaning for Adam's clients.

Adam's late picturesque vision and his new, hybrid building type were revolutionary – more so than the style of interior decoration he had introduced to Britain so successfully in the 1760s on his return from Rome. This may help explain why so few architects were to follow his lead.[98] Often associated mistakenly with the Gothic Revival, Adam's castle-style houses and landscapes have been best understood and explained by writers like Alan Tait and Alistair Rowan (and Marrikka Trotter in Chapter 10 in this volume) who grasp the scale of what he was attempting: an architecture of continuity embedded in a changing landscape setting, and so pared down and fundamental in style that it encompasses both Gothic and Classical, yet defies easy classification. Nor was it Adam's intention, like other artists of the period, for this body of work to suggest a romantic view of a disappearing past. The castle-houses were up to date, planned for convenient living, sometimes (as at Culzean) with elegant neoclassical Adam interiors. Such buildings signify the modern perpetuation of an age-old building tradition – and in Adam's drawings and sketches, and sometimes even in his picturesque watercolours, that point is emphasised occasionally by the inclusion of a clock on the castle walls.[99]

After Robert Adam: death, disrepute, revival

Robert Adam's funeral at Westminster Abbey in 1792 was a private affair, albeit with an array of aristocratic friends in attendance and as pall-bearers; there was none of the public spectacle or pomp that marked the passing of his contemporary Sir Joshua Reynolds, with whom Adam was compared in his obituary in the *Gentleman's Magazine*.[100]

In architecture the early years of the 19th century were dominated by the Greek revival and expectations of strict archaeological correctness – an unfavourable context for a true appreciation of an imaginative and versatile artist like Adam. The degree to which he had dominated British architecture of the 1760s–80s was now forgotten and his work was viewed as flawed, trivial, over-exuberant and excessively 'feminine'. Even Sir John Soane, an Adam admirer, who believed the nation was 'indebted' to his taste, thought that at times Adam's 'flights of fancy ... descended to trifles'. His progressive late castle style was also condemned: J C Loudon accused Adam of deviating from 'good taste' by melding together Gothic and Classical in a way that 'may justly be thought so ridiculous as never to have been attempted'. The most callous attack, though, came in the 1840s from Joseph Gwilt, whose *Encyclopaedia of Architecture* paid little heed to Adam's legacy, castigating him as a 'mere tyro', his taste as 'vicious', his compositions as 'depraved'.[101]

It was not till the 1850s–70s, with shifts in architectural taste towards lighter, more nationalistic historic styles, such as Queen Anne, that the Adam manner experienced a revival, most notably in interior design and furniture.[102] A renewed interest in Wedgwood and Chippendale, with whom Adam was often associated, also contributed. Successful exhibits of Adam-style furniture by the cabinetmakers and decorators Wright & Mansfield of Great Portland Street at the 1862 International Exhibition in London and the 1867 Paris Universelle Exposition marked the onset of the Adam revival. The firm also undertook several Adam-style schemes of interior decoration for wealthy clients, notably Brook House, Mayfair, and Guisichan, Inverness-shire, for Sir Dudley Coutts Marjoribanks (1st Baron Tweedmouth), and Haddo House, Aberdeenshire, for Lord and Lady Aberdeen. Other firms followed suit, such as Gillows of Lancaster and London, whose Oxford Street showrooms displayed an entire octagonal Adam-style boudoir that had been exhibited at the Paris Exhibition of 1880.

High-class practitioners like Wright & Mansfield and Gillows could produce exceptionally fine items, and the best Adam revival work is today sometimes indistinguishable from the real thing. But many of the reproduction interior designs and fittings that were popular in the later 19th century were misinterpretations or parodies of Adam's true style, and the increasing emphasis on his significance as an interior decorator diminished the true scale of his achievements as an architect and artist. As in Adam's time, greater knowledge and understanding of his *oeuvre* came through publication. In 1880 the London publishers B T Batsford responded to the popularity of the style by issuing a selection of engravings – nearly all interiors – from the brothers' *The Works in Architecture*; editions also appeared in Boston and New York. Batsford's example was followed a few years later by a book of Adam designs edited and published by R Charles, and Batsford published a revised second edition of their folio in 1900. Shortly afterwards the first full reprint of all three volumes of *The Works* was produced by a publisher in France, E Thézard Fils, indicating the popularity of the Adam manner there.[103] But the biggest advance in Adam scholarship was Arthur Bolton's two-volume study of the brothers' works, published in 1922, where the preface drew attention to how frequently modern opinion of the architect was based on altered or incomplete remains, or 'possibly the Coffee Room of some hotel "in the Adam Style"'. Bolton's much broader and more accurate interpretation was also the first to make use of the Adam office drawings at Sir John Soane's Museum, where he was curator.[104]

The Adam (or frequently, though incorrectly, 'Adams') style became increasingly prevalent during the first half of the 20th century, especially in the interwar period, and not only in Britain but also in the United States, as Eileen Harris explains in the final chapter to this book. Its delicacy and refinement caught the mood of the period and the taste for a Georgian aesthetic; its adaptability made it suitable for public spaces such as hotel lounges and lobbies, restaurants, cinemas, theatres and transatlantic liners, where a setting of de luxe elegance was expected, as well as for domestic interiors.[105] An attendant development of this new expertise was the sympathetic recreation of 'Adam style' interiors in genuine Adam buildings, such as the improvements of the 1870s at Culzean for the 3rd Marquess of Ailsa (by the Edinburgh architects Wardrop & Reid), and A F Balfour Paul's de-Victorianising of the central houses on the north side of Charlotte Square in the early 1900s for the 4th Marquess of Bute.

Modern taste for minimalist interiors and off-white or neutral colours leaves little scope for the Adam style today, but in a sense the Adam revival has never ended. Original Adam fixtures and fittings remain in high demand. In 2008 a Roman-style giltwood sofa, one of a suite of four made in 1765 to Adam's designs by Thomas Chippendale for Sir Lawrence Dundas's drawing room at 19 Arlington Street, sold at Christie's for £2.17m, twice its estimated value. Or an Adam room can still be created to order. In 1997 the well-known London dealers Crowther of Syon Lodge installed an entire reproduction Adam library in an apartment on New York's Fifth Avenue. For those with less to spend there are plentiful Georgian pieces inspired by Adam and the revival work of the 19th century to bid for. And now Sir John Soane's Museum is licensing select commercial partners to produce decorative items and furniture based on Adam designs in its collection – chimney pieces, mirrors, jewellery and so on.[106] What would the man himself make of this easy availability of Adam-style pieces (you can get a timber Adam fireplace on eBay for £35)?

Notes

1 The quote in the title comes from William Adam's obituary in the *Caledonian Mercury*, in 1748, referring to the sons who would succeed him – *see* page 36; also Fleming 1962, 65. The 1756 letter quoted is in NRS, GD18/4797. Throughout this book the Adam brothers are referred to by their family nicknames as well as by their proper names. *See* Chapter 1, p 36, for an explanation of the diminutives used by the family.

2 NRS, GD18, esp GD18/4741–5000. The other great source is the collection of some 9,000 Adam office drawings in Sir John Soane's Museum in London, now being made easily accessible with informative notes and commentary as part of the museum's online catalogue – *see* http://collections.soane.org/drawings.

3 For this early period of Adam's career and his place in the family's Edinburgh practice, *see* Fleming 1962, 76–99; Simpson 1980, 4–5; Macaulay 1987, ch 5; Sanderson 1992, 19–37, and 1993; Kay 1993; and Green 2014, chs 1 and 2.

4 Early sketches and copies after old masters show already an awareness of the picturesque potential of depicting buildings in a romantic landscape setting that was to be the hallmark of Adam's later style. *See* for example SM, Adam vol 56/12, 14, 17, 23, 152. Some are reproduced in Tait 2008, 27–9, 33 (figs 11–13, 17).

5 Fleming 1962, 6ff; Kay 1993, 24–5; Sanderson 1993, 68; Gifford 1989, 77–83. William Adam's standing in Edinburgh society is indicated by the presence of his patron, James, 5th Duke of Hamilton, as a witness at Jamie's baptism in 1732 (Fleming 1962, 323, n3). On the development of the architectural profession in Scotland generally, *see* Colvin 1986; also Nenadic 2012.

6 For the Adam family circle, which included Adam Smith, John Home, David Hume, Adam Fergusson and the geologist James Hutton, *see* Fleming 1962, 4–6; NRS, GD18/4981, 4982. For William Adam's library, *see* Rowan 1990 and Mowat 1979.

7 Hopetoun MSS, bundle 639 (quoted in Macaulay 1987, 135).

8 NRS, GD18/4982. Some of Robert's pre-Roman architectural drawings are reproduced in Fleming 1962, 91 (fig 6), plate 27, and Tait 1993, 10, 96 (figs 3, 76). His plans for Dr Mackenzie's house and a written explanation were sold at Sotheby's in London in 1930, but their present whereabouts are unknown. *See* Sotheby's Sale Catalogue, 30 July 1930, item 59; also Fleming 1962, 140, 340n. *See also* Thompson, H W 1927 *The Anecdotes and Egotisms of Henry Mackenzie, 1745–1831*. Oxford: Oxford University Press / H Milford.

9 *See* Chapter 1 of this volume; also Green 2014, 12–25, 29–33.

10 LMA, MS CLC/B/227/MS03070, *passim*; Kay 1993, 26. *Vitruvius Scoticus* was not published until 1812.

11 NRS, GD18/4796; BAM, 1454/4/215/3. By 1760 Jenny and Betty seem to have been joined in London by their sisters Peggy and Nelly. *See, for example*, NRS, GD18/4857, GD18/4859, GD18/4862–3, GD18/4865.

12 Carlyle, A 1860 *Autobiography*, 2 edn. Edinburgh: Blackwood, 361 (also quoted in Sanderson 1992, 31); *see also* Fleming 1962, 159–61: Kondo 2012, 31.

13 NRS, GD18/4749, GD18/4762; LMA, MS CLC/B/227/MS03070, fols 6a, 7, 8. *See also* Fleming 1962, 159.

14 NRS, GD18/4793, GD18/4800, GD18/4820.

15 NRS, GD18/4333 (quoted in Fleming 1962, 217), GD18/4817. The Adam drawing office has been discussed in detail in Tait 1993, 71–102, and Rowan 1993; but *see also* Stillman 1966, 41–5, and Harris 2001, 2–3.

16 Fleming 1962, 245–6, 252–6, 262, 266; LMA, MS CLC/B/227/MS03070, fols 34–8.

17 One of Willie's London letters to Innes & Clerk, of 1759, bears the Bayne & Adam address in York Street (LMA, MS CLC/B/227/MS03070, fol 38). His mercantile partner was John Bayne, a younger son of Alexander Bayne, first professor of Scots law at Edinburgh University. John Bayne's older brother, Captain William Bayne (d 1782), a naval officer, was for a time Robert Adam's rival for the position of MP for the county of Kinross in the 1760s before changing sides and becoming his supporter. The Baynes were related to the family of the architect William Bruce, and Anne Ramsay (d 1743), the first wife of Robert Adam's friend the portraitist Allan Ramsay, was also a Bayne. *See* History of Parliament online, www.historyofparliamentonline.org; *also* National Portrait Gallery, Early Georgian Portraits Catalogue: Ramsay.

18 NRS, GD18/4847, GD18/4851, GD18/4946.

19 NRS, GD18/4848.

20 NRS, GD18/4852, GD18/4853. During the first quarter of 1788, the Adam company was paid over £6,000 by the Ordnance Board for works at Sheerness, Gravesend, Chatham and Woolwich (BAM, box 4, General Correspondence, 1786–8, pt 1).

21 NRS, GD18/4851. John Fleming mistakenly names the Master of the King's Works at the time as Robert Dundas, Lord Arniston (d 1787) – *see* Fleming 1962, 256.

22 NRS, GD18/4852, GD18/4982/1.

23 NRS, GD18/4982/1, GD18/4929; *Survey of London*, vol XI, *The Parish of Chelsea – Part IV*, 1927, 59; Fleming 1962, 248.

24 LMA, MS CLC/B/227/MS03070, fol 39.

25 BAM, 1454/4/1/6, copy letter from William Adam to John Adam, 12 April–17 May 1787. The Marquess of Annandale once said of William Adam: 'He has so many real, and so many imaginary projects, that he minds no body nor no thing to purpose' (quoted in Simpson 1980, 2).

26 BAM, 1454/4/16/18; 1454/4/18/3; 1454/4/22/2–5.

27 Rowan 1974, 661; Sanderson 1992, 74–5; Summerson, J 1970 *Georgian London* (rev edn). London: Barrie & Jenkins, 143. *See also* King 1991, ch 9.

28 BAM, 1454/4/190/6.

29 Robert later moved in 1778 to 3 Robert Street, and Willie to No 6. *See Survey of London*, vol XVIII, *St. Martin-in-The-Fields II: The Strand*, 1937, 105, 110, 118; Rowan 1974, 665.

30 NRS, GD248/3395/1.

31 Archives Nationales, Paris, T451/4–5, letters of 1774 from R & Jas Adam to Lord Kerry. For this correspondence and the Adams' designs for a house for Kerry in Portland Place, *see* Pilkington 2015.

32 BAM, box 3, General Correspondence, 1784–5. For a discussion of the decline in the Adam brothers' activities, *see especially* King 2001, appendix B.

33 Rowan 2007, 37–8; Sanderson 1992, 74–5. For the Carron Company, *see* Carron Company 1959 *The Story of Carron Company: Two Hundred Years of Service*. Falkirk: Carron Company, and Watters, B 1998 *Where Iron Runs Like Water: A New History of Carron Iron Works, 1759–1982*. Edinburgh: John Donald.

34 Macaulay 1987, 143, 144; King 2001, appendix B; Sanderson 1992, 82.

35 Sanderson 1992, 82, 83, 85, 119.

36 Ibid, 128–9, 132; NRS, GD18/4961/40, GD18/4973A.

37 TNA, B 3/64; Rowan 2007, 39; BAM, 1454/4/1/33, 44–5.

38 Willie's bankruptcy was announced in the *Morning Chronicle* in October 1801. For William Adam & Co and the West India Docks, *see Survey of London*, vols 43–44, *Poplar, Blackwall and the Isle of Dogs: The Parish of All Saints*, 1994, 254; NRS, GD18/4984; Rowan 2007, 39–40.

39 *Annals of Agriculture, and Other Useful Arts*, vol XXIII, 1795, 52, 54–6; Donaldson, J 1796 *Modern Agriculture, or, The Present State of Husbandry in Great Britain*. Edinburgh, 164–5.
40 Rowan 2007, 39–40; NRS, GD18/4995–6, 4999.
41 For designs of the Thistleworth ceiling, which is still *in situ*, *see* SM, Adam vol 11/2 and vol 54/1/17. Other designs for Thistleworth (now Gordon) House are in SM, Adam vol 9/177, 181, and vol 32/59–66.
42 NRS, GD18/4853.
43 For the careers of the two Joseph Roses, *see* Beard 1967, 1975 and 1981, *passim* but especially 279–83; *see also* Drury 1984 and Murray 2012. Rose designs for ceilings at Portland Place are in the Rose sketchbook now in the collection of the Earl of Harewood at Harewood House.
44 Only a brief overview of this important aspect of Robert's career is possible here; readers are instead referred to Eileen Harris's magisterial study of Adam interiors (Harris 2001).
45 Stillman 1966, 42–3.
46 Adam and Adam 1773–1822, vol II, no I, i.
47 Piranesi, G B 1769 *Diverse maniere d'adornare i cammini*. Rome: The Author, 9; Wilton-Ely 2006, 235; D'Hancarville, P F H (ed) 1766–76 *Collection of Etruscan, Greek and Roman Antiquities from the Cabinet of the Hon W Hamilton* (4 vols). Naples.
48 *See* Harris 2001, 11, 177–9, 290–3; Stillman 1967; also Stillman 1966, 94–5, plates 111, 112.
49 NRS, GD18/4983, GD18/4823, GD18/4834, but *see also* BAM, 1454/4/10/1: 'I have found Palladio most faulty in many things, & very unjust in his Measurements', R Adam to John Adam, Rome, 26 November 1756. Clerk of Eldin's 'retrospective view' has been published with a commentary by John Fleming – *see* Fleming 1968c.
50 Adam 1831, 167, 168.
51 Ibid, 169.
52 References by the Adams to these artists can be found in the preface to the first volume of their *The Works in Architecture* (*see* Adam and Adam 1773–1822, vol I, no II, 3); in Robert Adam's letters (for example NRS, GD18/4817); also in James Adam's notes on Italian architecture and architectural theory of 1762 and 1772, his travel journal and the list of books he acquired in Italy (NRS, GD18/4954, GD18/4920; Adam 1831). For John Clerk of Eldin's comments, *see* NRS, GD18/4981/1.
53 NRS, GD18/4954.
54 For the Italian Renaissance response to the antique, *see* especially the work of Hetty Joyce, for example: Joyce, H E 2004 'Studies in the Renaissance reception of ancient vault decoration', *Journal of the Warburg and Courtauld Institutes* **67**, 193–231; *also* Joyce, H 1992 'Grasping at shadows: Ancient paintings in Renaissance and baroque Rome', *Art Bulletin* **74**, 219–46. And *see also* Dacos, N 1986 *Le Logge di Raffaello: Maestro e bottega di fronte all'antico*, 2 edn. Rome: Istituto Poligrafico e Zecca dello Stato.
55 Wilton-Ely 2006, 220; Stillman 1966, 96–7, plates 117–19.
56 NRS, GD18/4766, GD18/4770.
57 In addition to the Adam drawings in Sir John Soane's Museum and the Clerk of Penicuik papers at Register House, Edinburgh (NRS, GD18), the chief sources for Adam's time in Rome are: Fleming 1955, 1956 and 1962; Rowan 2003; Tait 1984 and 2008; Wilton-Ely 2011.
58 For more on these drawings, *see* Rowan 2003 and Tait 2008. For the Italian architectural competitions of the period and British architects, *see* Stillman 1973.
59 For Piranesi and his influence on Adam, *see* Stillman 1967; also Tait 1984; Wilton-Ely 1995 and 2006.
60 NRS, GD18/4764 (quoted in McCormick 1990, 24). For examples of Adam and Clérisseau sketching together, *see* McCormick 1990, 32–42, figs 24, 25, 27, 28, 32–6, and Rowan 2003, 17, 18, 33.
61 NRS, GD18/4811; *see also* McCormick 1990, 51–2, and Fleming 1962, 159, 161–70.
62 For Adam and Diocletian's Palace, *see* Fleming 1958c; Brown 1992a; Salter 2013; and esp Belmarić and Šverko 2017. For the most recent accounts of the influence of Diocletian's Palace on Robert Adam's work, *see* Pinto 2017, Sands 2017 and Thom 2017.
63 The quote is from Aymonino *et al* 2013, 16. For more on this subject, *see* Adriano Aymonino in Chapter 4 of this book; *also* Aymonino 2010b.
64 Robert Adam's relationship with French art and architecture, and French neoclassicism in particular, is a subject of considerable potential interest but one that has yet to be explored fully. But *see* Stillman 1966, 38–40, and Harris 1963, 9–13. For the influence of French *hôtel* planning and fireproof construction on General Clerk's house, *see* Thom 2015, 131–8.
65 Jamie's notes on architectural theory are in his diary at NRS, GD18/4954; the section on 'the Elevation & its movement' is transcribed in Fleming 1962, 315. *See also* Kondo 2012, ch 3 ('"Movement": The picturesque in architecture'). Robert's letter to Lord Kames is transcribed in Tyler, A F 1814 *Memoirs of the Life and Writings of Lord Kames*, vol II, appendix 2, 184–90, and also in *The Builder*, 5 March 1881, 272–3 ('Some unprofessional memories of the Brothers Adam'); his letter to Sir James Clerk about the replanning of Penicuik House is in NRS, GD18/1758b, and is quoted at some length in Macaulay 1987, 170.
66 NRS, GD18/4946.
67 Adam was described as a 'high converser' by Fanny Burney (quoted in Sanderson 1992, 78).
68 NRS, GD18/4824; GD18/4764; for the Osterley portico, *see* Harris 1958.
69 For these theories *see* Piranesi's *Parere su l'architettura* (1765) and *Diverse maniere* (1769); *also* Stillman 1967, esp 200–1.
70 On Adam's patrons, and the Scottish connection in particular, *see* Fleming 1962, 247–56; Sanderson 1992, 53–62; and Harris 2001, 3.
71 *See* Tait 2004.
72 NRS, GD18/4982/1.
73 Adam and Adam 1773–1822, vol I, no 1, 7.
74 For the Duchess of Northumberland's monument, *see* Aymonino 2010a.
75 For the Northumberlands as patrons, *see* Aymonino forthcoming. For Lords Lansdowne and Coventry, *see* Parker 1968.
76 A recent account of the work of Adam's rival George Steuart, the painter-turned-architect who made

various designs for the Dukes of Atholl, Richmond and Montagu (among others), suggests that Steuart worked for a 'much more elite client base than Robert Adam ever achieved'. *See* Goodsir, 2010, 103. Adam's pallbearers at his funeral in Westminster Abbey in March 1792 were Henry Scott, 3rd Duke of Buccleuch; George Coventry, 6th Earl of Coventry; James Maitland, 8th Earl of Lauderdale; David Murray, 7th Viscount Stormont (later 2nd Earl of Mansfield); Lord Frederick Campbell (son of the 4th Duke of Argyll); and Sir William Pulteney, 5th Baronet. For Adam's funeral, *see Gentleman's Magazine*, March 1792, 282–3; also, info kindly provided by Sydney Ayers (*see also* note 100).

77 Fraser 1989, 76–7.

78 LMA, MS CLC/B/227/MS03070, fol 45.

79 The two best analyses of the Adam brothers as book publishers are Harris and Savage 1990, 71–94, and Middleton *et al* 1998, 3–18. Also, *see* Rowan 1985, 9–10, and Bolton 1917. For *The Works in Architecture, see also* Tait 1993, 103–34; for the *Ruins, see* Belmarić and Šverko 2017; also Fleming 1958c and Brown 1992a, 1992b and 1998. For the possible later publication on villas, *see* Rowan 1983 and Rowan 1985; and for *Vitruvius Scoticus, see* Simpson 1980, Harris and Savage 1990, 94–104, and Gifford 1989.

80 NRS, GD18/4825 (this letter is quoted at length in Fleming 1962, 363); BAM, 1454/4/10/1.

81 Bolton 1917, 257; Rowan 1985, 10.

82 For example, the preface to part 2, of 1774, which discusses the Doric order in Classical architecture, was based largely on Robert Adam's letter to Lord Kames (*see* note 65).

83 Quoted in Sanderson 1992, 101.

84 Rowan 1985, 13.

85 For William Adam senior's entrepreneurial activities, *see* Simpson 1980, 1–2; Gifford 1989, 109.

86 NRS, GD18/4851.

87 LMA, MS CLC/B/227/MS03070, fols 7, 8, 10, 20a, 21, 25a, 26, 28, 28a.

88 Fleming 1962, 296.

89 NRS, GD18/4853, GD18/4950, GD18/4952; *Calendar of Home Office Papers of the Reign of George III*, vol I, *1760–1765*, 408 (item 1303), 423 (item 1366).

90 NRS, GD18/4882, GD18/4935, GD18/4959, GD18/4955; Fleming 1962, 271.

91 Adam and Adam 1773–1822, vol I, no IV.

92 For Frederick's Place, *see* Imray, J 1991 *The Mercers' Hall*. London: London Topographical Society, ch 13, and King 1991, 96–7. For Fitzroy Square, *see* Rowan 2007, 32–5, and King 1991, 97–9; also *Survey of London*, vol XXI, *Tottenham Court Road and Neighbourhood (St Pancras Part 3)*, 1949, 52–63.

93 NRS, GD18/5838/5.

94 For Adam's Edinburgh designs, *see esp* Fraser 1989; also Bolton 1922, vol II, chs 31–2; Sanderson 1992, 103–5; MacInnes 1997; Rowan 1997 and 2004.

95 For Adam's late villas, *see* Rowan 1985 and 1983. Although predominantly a Scottish phenomenon, an important example of Adam's villa style survives in England at Brighton (Marlborough House). For this, *see* Miele 1998.

96 Rowan 1985.

97 For discussions of Adam's late castle style, *see* Rowan 1974; Tait 1971 and 1981; King 1991, 14–15; Sanderson 1992, 86–91; Astley 2000.

98 Those architects who later worked in an Adam-influenced castle style tended to have connections with Adam or had worked for him: for example John Paterson, Richard Crichton, Thomas Hamilton and David Hamilton. *See* Macaulay 1975, ch 11.

99 For the reference to clocks, *see* Astley 2000, 11, 35. Adam drawings of castle-style buildings with clocks include SM, Adam vol 2/27, 28, 33, 34, 41, 68; vol 10/58, 111, 127, 154, 207 (information kindly provided by Dr Frances Sands).

100 *Gentleman's Magazine*, March 1792, 282–3. This paragraph is also informed by 'The death of an architect: The obituary of Robert Adam in 1792', a paper given by Sydney Ayers at the conference 'Reading Architecture Across the Arts and Humanities', University of Stirling, 5–6 December 2015 (transcript kindly supplied by Sydney Ayers).

101 Watkin, D 1996 *Sir John Soane, Enlightenment Thought and the Royal Academy Lectures*. Cambridge: Cambridge University Press, 642, 650; Loudon, J C 1806 *A Treatise on Forming, Improving, and Managing Country Residences*, vol I, book I, part III, 115–18; Gwilt, J 1842, *An Encyclopaedia of Architecture, Historical, Theoretical, and Practical*. London: Longman, Brown, Green, & Longmans, 224.

102 The leading authority on Adam's posthumous reputation and the Adam revival, both at home and abroad, is Eileen Harris. *See* Harris 2001, 16–17; 1993; 1963, ch 4. But *see also* Parissien 1992, ch 7.

103 1880 *The Architecture, Decoration and Furniture of R & J Adam*. London: B T Batsford (the American editions were published by George H Polley of Boston and William Helburn of New York); 1900 *The Decorative Work of Robert and James Adam, Being a Reproduction of the Plates illustrating ... Their Works in Architecture Published 1778–1812*. London: B T Batsford; Charles, R (ed 1883), *The First Issue of Decorative Designs by Robert Adams Architect*. London: R Charles *The Works in Architecture of Robert and James Adam Esquires*, 3 vol repr Dourdan, France: E Thézard Fils, 1900–2.

104 Bolton 1922, vol I, viii.

105 For an account of neo-Georgian interior design in interwar Britain, *see* Taylor, C 2016 '"Modern Swedish rococo": The neo-Georgian interior in Britain, *c* 1920–*c* 1945', *in* Holder, J and McKellar, E (eds) *Neo-Georgian Architecture, 1880–1970: A Reappraisal*. Swindon: Historic England, 151–66.

106 *See* www.christies.com/lotfinder/Lot/a-george-iii-giltwood-sofa-designed-by-5092522-details.aspx [accessed 24 March 2018]; Moonan, W 1997 'Antiques: The mind of a genius at work', *New York Times*, 12 December 1997. For licensed goods inspired by the Adam drawings and designs in Sir John Soane's Museum, *see* 'Inspired by Soane: Brand licensing', www.soane.org/about/inspired-soane-brand-licensing [accessed 24 March 2018].

1

Johnnie, the eldest Adam brother

Alistair Rowan

For all of their lives, the members of the Adam family in Edinburgh called each other by familiar names. Thus the girls were Nelly, Betty, Jenny, Peggy and Susy. Only Mary Adam, the wife of the Revd John Drysdale, minister of the Tron church, who when left a widow was to look after Bob's architectural office in Scotland, escaped this penchant for diminutives. The boys always referred to each other as Bob, Jamie or Willie, and the eldest Adam brother was, for each of his siblings, simply Johnnie.

As we know from the parable of the prodigal son, it is not easy to be an eldest brother. Johnnie from an early age had to shoulder the burden of having others' expectations cast upon him. According to Archibald Campbell, the Duke of Argyll's agent at Inveraray, who had to provide a supervisor for the construction of Inveraray Castle in 1747, the boys' father, William Adam, was recognised 'as the ablest man we have in Scotland for carrying out so great a design'. Campbell therefore suggested that, at Inveraray, Adam or his son might visit the works monthly from March till November.[1] Thus Johnnie – and later Bob – came to work for the Duke of Argyll, and to enjoy the patronage of a brilliant politician who ran Scotland for the house of Hanover. When a year later, on 24 June 1748, William Adam died at the age of 59 of prostate cancer, *The Caledonian Mercury* added, in a note to its obituary, ''Tis fortunate he has left behind some promising young men to carry on what he has so happily begun.'[2] At that time Johnnie was 27 years old, Bob, who had been working in the family businesses for 4 years, was just 20 and Jamie was 16.

The Adam brothers in Scotland

In less than a month after William Adam's death, a testimonial of John's competence to take over from his father was written by the 2nd Earl of Hopetoun, a man of 44 who knew the Adam family well and at that time enjoyed, among his peers, something of a reputation as an expert 'in stone and mortar'. The Duke of Montrose had just embarked on some improvements planned by William Adam to Buchanan House, near Drymen in Stirlingshire. He wondered whether it would be prudent to give the conduct of the work to Johnnie and turned to Lord Hopetoun for advice. Hopetoun's letter, sent to the duke on 20 July 1748, gives an attractive picture of the eldest Adam brother and is unequivocal in its recommendation of him:

> Poor William's death is truly a general loss and the more so that there is nobody in this country in the smallest degree equal to him. His son John, I'm sure, is much preferable to any other, being a lad of very good parts and sense, unwearied application and perfectly sober; of knowledge in business far beyond his years. I cannot say I ever thought him quite so quick as his Father, indeed very few are, but I'm persuaded his modesty helped to make him appear less so, than he really is; and upon the whole I can venture to assure your Grace that he is abundantly qualified to execute all you have in view, and to undertake that he will serve you honestly. I think it no bad proof of his capacity that Morris with whom he was some months at Inverara [*sic*] last year, has wrote to him that he has satisfied the Duke of Agryle [*sic*] of it, and that he is to have the carrying on of that work which will make his visiting Buchanan the more convenient. He has, likewise, a brother, a very clever lad and well advanced, that will be a great assistance to him.[3]

William Adam's proposals for Buchanan House, recorded in *Vitruvius Scoticus*, were drawn up in 1741 for the duke's father shortly before his death the following year.[4] The new duke decided not to proceed with the scheme for a new house and instead began to collect materials to make a start on 'additions to the old house' in the spring of 1748, when the architect's death put a halt to the matter at least for that year. Though the duke wrote to John Adam to ask for his 'assistance and directions in executing what was designed by your father', including some 'sort of towers proposed at the corners' of the ends of the building, whatever may have been done was lost when Buchanan House burned down in 1848. Johnnie also drew up plans for a new parish church at Drymen, which was to contain a gallery for the duke's family, and he was asked to design a 'new court of offices' with an estimate of what that would cost if Adam's builders 'should undertake the whole'. Johnnie must have visited Buchanan shortly afterwards, and, though nothing came of his design for the church (which was not built until 1761), it seems likely that a plain, square courtyard of stone-built offices, entered through an archway with an eaves pediment – surviving on the estate – was built to his design.[5] These offices are possibly the earliest independent work undertaken by John Adam.

For the six and a quarter years between the death of their father and Robert's departure for Italy in October 1754, the two eldest Adam brothers worked together as architects in Scotland. James formed part of their team and continued to assist Johnnie for a further six years until he too left for Italy in May 1760. After that, John conducted the Adam practice in Scotland on his own. In this period of 12 years' cooperation between the brothers, there can be no question that John was the head of the business. He was 7 years older than Bob and 11 more than Jamie, so it is highly unlikely that William Adam's clients in Scotland would have chosen to put their business in the hands of the younger men rather than their elder brother. When William Adam died, Johnnie had already been working with his father for 11 years and was personally well known to a number of important patrons,[6] whereas Bob had been in the firm for just 4 years and was no more than the 'very clever lad' mentioned in Lord Hopetoun's letter.

On succeeding to his father's estate and businesses, John treated his siblings generously: he doubled the capital sum left to his brothers, each of whom thereby inherited £1,000, and he increased the portion for his six sisters from £300 to £500 each.[7] A year later he took his mother with an uncle and his brothers on a jaunt to London. The family needed the stimulus of a change of scene after the strain of William Adam's protracted illness; after their return Johnnie himself remained in London on business until March 1750.[8] At some stage and in a way that is not entirely clear, John entered into a partnership with Robert and James. Thus the accounts from the Adam firm for the work carried out by the brothers at their two great Scottish houses – Hopetoun and Dumfries House – are issued in the names of 'John, Robert and James Adam, architects'. This should not, however, be taken to mean that the three men shared equal responsibility for the design of these buildings.[9] Given the subsequent careers and celebrity of Robert and James, historians have tended to interpret whatever is found of quality or originality in the 1750s and early 1760s as the work of the younger men, rather than recognise that in this period it was Johnnie who headed up the firm and who largely provided its designs.[10] The partnership, in whatever form it may have existed, is mentioned in a letter of 18 April 1755 sent to James from Robert – recently embarked on his Italian travels and then in Naples – in which he speaks of the possibility of their continuing 'a Triumvirate Copertnary [*sic*]' on his return, even though he and James were already planning to settle in London and not to return to Scotland. 'It is not', he adds, 'that I think Johnie incapable to manage the whole Scotch business, on the contrary I think him more than Master for it, as [in a co-partnership] he would only have to reject all troublesome & triffling [*sic*] jobs & only retain such as would employ his time to purpose & gain money by advantageous contracts.'[11]

At this stage, with Johnnie providing much of the funding for Bob's and later Jamie's travels, something akin to parity of esteem existed between the brothers. Yet that was not to last throughout their lives: indeed from quite early on Bob and Jamie appear to have come together with secret schemes of their own from which Johnnie was excluded. Undoubtedly a part of the problem that emerged between the brothers who travelled and Johnnie, who stayed at home, was not only the fact that their eldest brother had inherited the bulk of their father's estate and controlled the firm's finances but also that, in their eyes, he lacked ambition. After Robert went abroad, in the scheme of things which the brothers had talked over, John was meant to follow him in visiting Rome, but in May 1757 that idea was quashed absolutely by one of the brothers' most significant patrons, Lord Dumfries, for whom they were then building a large new house and offices in Ayrshire. Though John had assured Lord Dumfries that by the time he went to Italy James would be perfectly capable of supervising any work that remained to be done at Dumfries House, his patron would have none of it. Johnnie wrote to Jamie following their discussion: 'Though he had a very great regard for you and thought extremely well of your abilities, yet it could not be expected your experience should exceed your years and that therefore he could by no means agree to my going away.'[12]

The clear preference shown by Lord Dumfries for John as the person to supervise his buildings rather than James is ample proof of the relative position in 1757 of the older and the younger brother within the firm. For a while Bob and Jamie still hoped to persuade Johnnie to make an Italian trip, until a letter which Jamie wrote from Edinburgh in June 1758 put the matter beyond doubt:

> I was accidentally talking to Johnnie lately about his going abroad & he then told me that he has spoke of his scheme to L'd M-l-n [Milton] & was resolv'd before he determin'd, to speak to L'd H-n [Hopetoun] concerning it & that if these two, after mature deliberation shoul'd say, that they thought it was quite imprudent to go & that the advantages did not balance the disadvantages, he would in that case lay aside all thoughts of going, which reference looks to me, as if his Heart had fail'd on the near approach.

At that time John was a married man of 37 and had a family of his own; with ties and responsibilities far beyond those of his bachelor brothers, it is no wonder he preferred to stay at home.[13]

The brothers' early work in Scotland

In 1753 the Adam brothers erected a monument in memory of their father in Greyfriars Kirkyard, Edinburgh. It was John who applied to the Town Council for permission to put it up and who reported on its completion in October that year. While the structure is simple, its scale is ambitious and more like a mausoleum than a monument. The design is of a square vaulted room sustained by three open arches and roofed by a shallow stone pyramid. While strong in form, the building eschews any flamboyance and has only a regular Doric frieze below the eaves cornice and a chaste Greek fret at impost level as decoration. The side walls are blocked to the impost level of the arch and have open Diocletian windows above them, while the back wall is solid and supports an aedicule containing a bust of William Adam, above a rococo sarcophagus. The taste of the Adam mausoleum – restrained, proportionate and uncluttered – is typical of the rational architecture which John developed, in contrast to the boisterous riot of different elements that his father had so often employed. In the drawings made by Johnnie to record the building, the front of the sarcophagus is left blank, and it is perhaps significant that, when the monument was finished, the panel

Fig 1.1
Hopetoun House, east front, designed by William Adam, with the end pavilions as redesigned in collaboration with John Adam. From Vitruvius Scoticus, *plate 17 (Adam 1812).*

was to be filled with a shallow perspective relief of one of the pavilions at Hopetoun House.

The east front of Hopetoun, heroic in scale and resourceful in its planning, is the design that launched William Adam on his career as the premier builder of country houses in Scotland from early in the 1720s (Fig 1.1). The work was to be protracted to such an extent that it was equally the completion of this same front which confirmed the position of Adam's eldest sons, John and Robert, as his successors.[14] Lord Hopetoun's good opinion of the brothers is proven not only by his letter to the Duke of Montrose (quoted on p 36) but also by the seamless way in which the completion of Hopetoun passed from William Adam to John, who had been working with his father on the house from the age of 16 and whom the earl knew well. It was during these first years of Johnnie's practical work as an architect that the wings at Hopetoun were built – the north pavilion from 1736, followed by a matching block on the south, which was finished in 1742 (Fig 1.2). So the wings and pavilions of the house, which Johnnie selected to represent his father's architecture on the family monument, were the most notable buildings by the firm in which Johnnie himself had had a hand.

Architectural historians have noted the extent to which the elevations of the pavilions are in a calmer and more reticent style than the front of the house, and, though they are clearly designs which were erected under William Adam, it is perhaps not unreasonable to attribute the change in character to discussions between the father and his eldest son.[15] The plates of the house in *Vitruvius Scoticus* illustrate the two proposals. The earliest is for a pair of identical L-shaped wings with two long elevations of 13 bays, fronting each other across a deep courtyard and balanced in the middle around an archway surmounted by a pediment in an all-too-obvious neo-Palladian way. The second design, which shows the wings as they were built, is an altogether more considered piece of work (*see* Fig 1.1). Here the pavilions take on an impressive architectural scale: there is more weight to the design and their bulk is sufficient to terminate the long sweep of the colonnades in a logical and satisfactory way. The length of the pavilion blocks is reduced to half that of the first proposal, and the courtyard and front elevations, looking out towards the Firth of Forth, balance each other exactly. There is a clarity and precision in the placing of the individual elements, which seem to recall a comment by William Adam on his son: 'John is not a little vain of the confidence reposed in him as to the fixing of proportions.'[16] That comment referred to the design of Garron Bridge at Inveraray, but Johnnie's 'vanity' in his command of proportions might equally have been extended to the final form of the Hopetoun pavilions.

Certainly it fell to John Adam to complete the front of this great house. His father had intended to place a showy, free-standing Corinthian portico, boldly projecting by the distance of an intermediate column, as the culmination of the façade, but that bold stroke, which would have been the first giant-order portico to be erected in Scotland, was never built. In its place Johnnie devised a scheme of sober magnificence in which a wide, straight flight of steps leads to a broad

stone platform, above which a centrepiece of four giant-order Corinthian fluted pilasters supports a flat entablature with an attic storey above (Fig 1.3). The restraint of the design, in which the centre three bays accord exactly with the width of the steps and are only marginally advanced from the line of the front, greatly enhances the sense of breadth and repose that are the distinguishing features of Hopetoun's entrance front. Behind this façade, completed in August 1751, and extending north from the entrance hall in the centre, Johnnie devised a sequence of rectangular state rooms – dining room, drawing room, state bedroom and dressing room – accommodated within the empty shell of the building, which his father had left unfinished (Fig 1.4). These rooms, fitted out with sumptuous plasterwork and refined joinery, were completed between 1752 and 1757. Long before they were finished, Bob had set out on his Italian tour, travelling with the

Fig 1.2 (right) Hopetoun House, the stable pavilion, completed 1742, from the south.

Fig 1.3 (below) Hopetoun House, the central block of the east front, as completed in 1751 by John Adam with his brothers.

Fig 1.4 (opposite) Hopetoun House, the Red Drawing Room.

Hon Charles Hope, Lord Hopetoun's younger brother, on the journey from Paris to Rome.

One building which is crucial to an understanding of John and Robert's joint activity as architects in Scotland is Dumfries House, at Cumnock in Ayrshire (Figs 1.5 and 1.6). This was the house where the patron was not prepared to part with John's services, effectively ensuring – or offering a convenient excuse for – his not travelling abroad. Two plates of the house, prepared for the continuation of *Vitruvius Scoticus* which John was planning in the 1760s, bear the inscription 'Adams Archt's', making it clear that Johnnie at least considered Dumfries House to have been designed by both brothers.[17] In any architectural practice where there is close collaboration, the constant exchange of ideas and consideration of different design solutions makes it impossible to disentangle retrospectively the separate contributions or lines of enquiry of the individuals concerned. Johnnie and Bob worked together to find a design that would satisfy Lord Dumfries, who was not always an easy client, and the resulting neo-Palladian house, handsome in its proportions and restrained in its detail, is a testament to their joint abilities. Nor does it seem correct, as has often been suggested, to attribute the creative side of the job exclusively to Robert.

Dumfries House has been the subject of a detailed and perceptive monograph investigating the origins of the building, its construction, furnishing and subsequent history.[18] Effectively two designs were prepared by the Adam brothers with no fewer than four estimates of the anticipated expense of construction as well. The first design, something of a freak, was for a tall, gawky house, three storeys high and nine windows wide, flat in effect and set awkwardly above a high semi-basement, with a complicated double ramp of external stairs. The central three windows were squashed closely together under a wide eaves pediment, while the window surrounds, different on each floor, were detailed with an untutored love of restless elaboration. Plain four-bay wings with tall chimney stacks and high-pitched roofs flanked the fussy central block. It is inconceivable that this first design – dating to 1751 when an estimate of its cost was prepared – could have been proposed voluntarily

Fig 1.5
Dumfries House from the south-west in 2010.

by either John or Robert Adam. Though the documentary proof is lacking, it seems most likely that the brothers, following their father's death, were asked by Lord Dumfries to work up an earlier proposal by William Adam, which they could hardly avoid doing.[19]

By April 1751 a 'New Design' had been prepared and endorsed by Lord Hopetoun, who wrote to Lord Dumfries to say that no less a person than Lord Burlington had approved the plans. Reading between the lines it would appear that John and Robert encountered some difficulty in weaning their new patron off the tall, old-fashioned house that had once been proposed for the site. In this regard a tip from Lord Burlington could well have been useful. Alterations and three further estimates aimed at reducing the cost were required before the contract for the building was signed in April–May 1754 (*see* Fig 0.5). That summer the foundation stone was laid on 18 July, with Bob in attendance to supervise the initial work.

In terms of its exterior, Dumfries House continues the type of restrained formal architecture which John Adam first explored in the pavilions at Hopetoun and in the unemphatic centre of its entrance front. Repose is the principal characteristic of the Ayrshire house, whose wide, low forecourt, astylar and without incident, seems almost to float in the parkland in which it is set. There is an innate sense of proportion and balance between the lower, two-storey service wings and the taller main block of the house, two storeys above a low basement and nine windows wide. Handsome is the adjective that might have been used in the 18th century, for the architecture is unquestionably aristocratic and at the same time reserved. The drawings of the north and south fronts of the 'New Design' are signed 'R A', which confirms Robert Adam's participation in the design and, no doubt, his pleasure in its august appearance. Yet in every aspect the sound good sense and know-how of Johnnie are evident as well.[20]

An undated sketch plan by John explores the idea of a wide projecting bay window at the centre of the north front. Although this notion was not carried forward, in every other respect – the plain flight of steps on the entrance front, the layout of the office pavilions connected by quadrant

Fig 1.6
Dumfries House entrance hall in 2007.

corridors, the service yards and the arrangement of the rooms in the house and pavilions – it is very close to what was built. Together John and Robert worked out a lucid, bilaterally symmetrical plan for the main house, remarkable for the rigour of its symmetry, with a wide corridor bisecting the building from east to west and ending in identical staircases lit by Venetian windows placed centrally in the side elevations. As Simon Green writes: 'The plan is well considered, the rooms are well proportioned, they connect with each other easily where required and there is a logical hierarchy in terms of the relative arrangement of the rooms according to their function.'[21] At Hopetoun, designed by Sir William Bruce, the brothers would have been familiar with the regular arrangement of the bedrooms in suites of three rooms – bedroom, dressing room and closet – set at the corners of the house within a rectangular plan. This consistent and generous arrangement is mirrored in their planning of Dumfries House.

John Adam's commonplace book and a book of architectural designs

Two manuscript volumes, each in use in the early 1750s, provide indisputable evidence of the extent of John Adam's professional involvement in the business of architecture and of his interests and tastes as a designer. The first is a large pocketbook which Johnnie began to keep, perhaps even before his father had died, and the second is a smaller architectural notebook with the plans and elevations of various buildings which he had designed or where he had been employed.[22] Some time in the 1740s John purchased the pocketbook for use in connection with his practice as an architect. The volume evidently accompanied him as he travelled about on business; it has a large flap that fits into a slit on the back cover, intended to keep the book closed when it was put into a pocket. Bound in card, the book is quite large, 114mm wide and 184mm tall. It contains 220 pages and has a printed alphabetical index, bound into the front as a separate 'signature'. The number of pages, the index and the manner in which it is bound all suggest that what Johnnie acquired was a purpose-made volume, regularly available in an 18th-century stationery shop, to be kept by its owner as a 'commonplace book'. Such a book could be used to make notes on a wide variety of topics, jotted down as occasion demanded, and located for future reference through entries made in the index pages.

Though commonplace books were used widely throughout the Georgian age, it is unusual to find such a volume put to a systematic use by a professional architect. In this regard, Johnnie's notebook is of special interest. Everything that caught his attention, and which he felt was worth recording, is contained within its pages, so that the reader today is given not only a vivid and immediate picture of the world that Adam knew but also more than a glimpse of the character of the man – meticulous, careful and courteous – as he makes practical notes of his discussions with tradesmen, transcribes long lists of prices, takes down a builder's recommended recipe or records the wisdom of an aristocratic client (Fig 1.7). There is nothing to indicate when John began his book; perhaps it would have occurred to him to make a compendium of this sort about the time of his father's death, when the burden of running the family architectural practice and overseeing the building yards at Leith devolved on him.

As a comparatively young man, John Adam must have found it useful to have a notebook in his pocket to back up, with examples based on experience, any advice or information on costs that clients might require. Though the book contains prices for items at several buildings where the Adam firm worked in the 1750s and 1760s – notably Dumfries House, Alva, Hopetoun, Arniston and Fort George – it is clear that these prices were transcribed by John as a record of comparative costs that he could refer to, in some cases long after the work had been completed. The entries on the prices charged by various tradesmen provide a salutary reminder that architects and builders throughout the 18th century lived in an age untroubled by inflation. For John, prices of work carried out 15 or 20 years earlier held good for current projects, and the know-how contained in his commonplace book remained relevant professional material. In places the material entered in the notebook dates back to a time when Johnnie was a child: the earliest reference is to a letter of 1724 by Walter Ferguson, steward to the Duke of Chandos, 'anent [concerning] the improvement of land by peat ashes', while another early note relates to the 'Expense of working marble at Leith in 1729'.

Entries referring to Johnnie's early years working with his father include Watson's Hospital in 1738; Tinwall House in 1739; 'Joisting for the Barrack proposed at the Abbey' in 1740;

'Prices of nails in Wolverhampton in 1742 & in Kirkcaldy'; 'Prices of wright work paid to James Buchan at Hopt. House, 1743'; work at the Royal Bank in 1744 and a 'Computation of the expense of Burning Tyles & Bricks at Links' of 1745. The latest entries in the book are from the 1760s: there is a recipe 'To stain marble to give it the Antique Patina' dating from 1764 and a long note, dated 'Lond. 18th December 1765', on the prices of bundles of split laths, when John must have been in London in relation to the affairs of William Adam & Co, the brothers' buildings supplies company, set up in the previous year.

The majority of the items entered in the commonplace book have no dates. On one of his jaunts to England, Adam came across a neat plan for a brewery, showing the disposition of the various pieces of equipment required to make beer. He noted the arrangement in his travel journal and later copied the plan he had made into the commonplace book. He did the same when visiting the Royal Mews at Charing Cross, recording the plan of the horses' stalls and then making a small perspective sketch of the arrangements of the saddle racks with 'pins for hanging bridles, collars etc'. Detail and precise information had real value. Thus in considering the 'Dimensions of the Coach Houses & Stables' for his own yard, Johnnie took the opportunity to record the size of the coaches of Lord Hopetoun and the Duke of Montrose, comparing their height, length, width and axle length. The particular responsibilities of the firm as government contractors at Fort George no doubt accounts for the details, provided by a major of dragoons, of the requirements of each mess and the kit for individual soldiers.

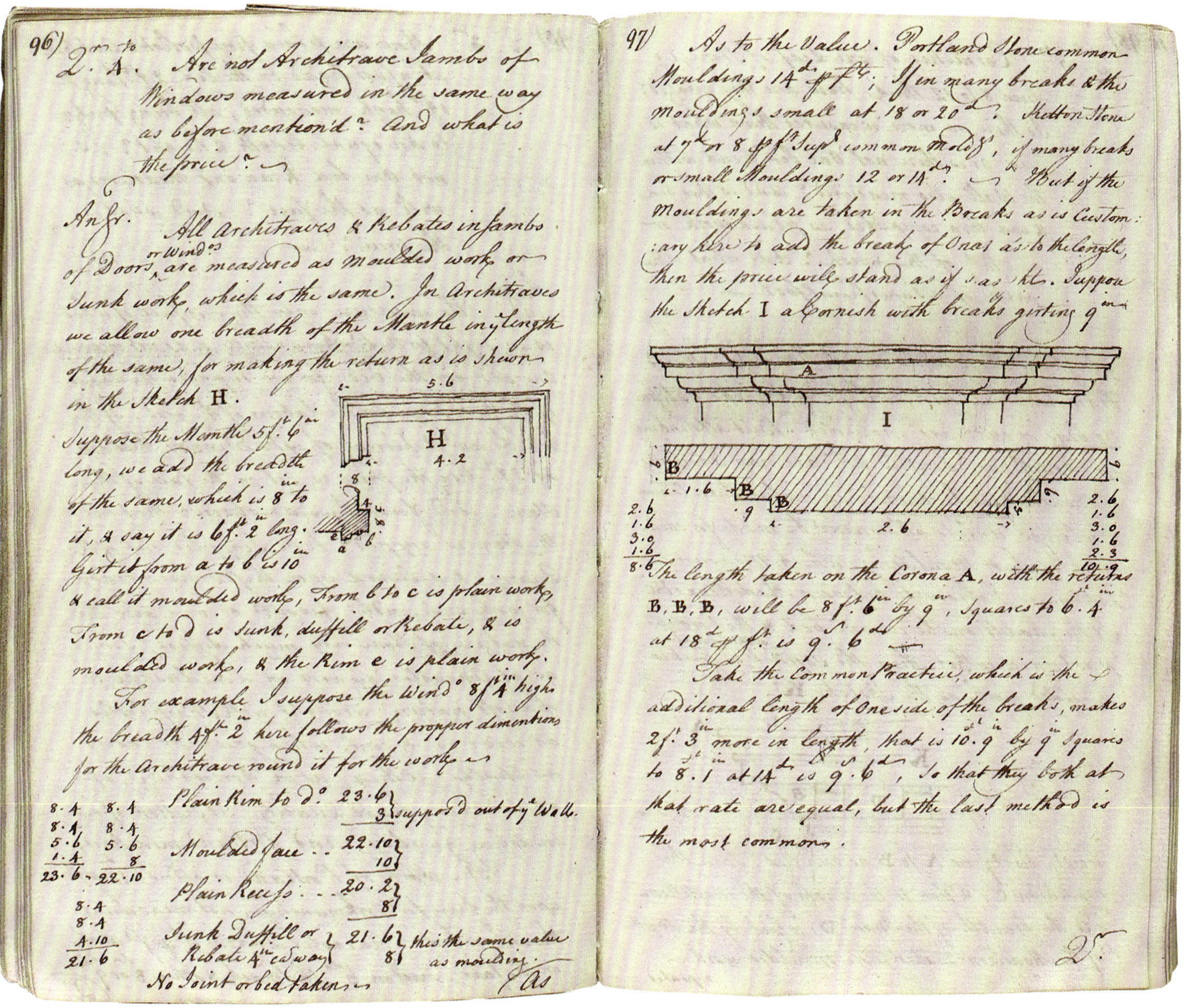
96) 2.nd to A. Are not Architrave Jambs of Windows measured in the same way as before mention'd? And what is the price? —

Ansr. All Architraves & Rebates in Jambs of Doors or Wind.os are measured as moulded work or Sunk work, which is the same. In Architraves we allow one breadth of the Mantle in length of the same, for making the return as is shown in the Sketch H.

Suppose the Month 5ft. 6in long, we add the breadth of the same, which is 8in to it, & say it is 6ft. 2in long. Girt it from a to b is 10in & call it moulded work, From b to c is plain work, From c to d is sunk, duffill or Rebate, & is moulded work, & the Rim e is plain work.

For example I suppose the Wind.o 8ft. 4in high the breadth 4ft. 2in here follows the proppor dimentions for the Architrave round it for the work —

8.4	8.4	Plain Rim to d.o	23.6 / 3	suppos'd out of ye Wall.
8.4	8.4			
5.6	5.6	Moulded face	22.10 / 10	
1.4	8			
23.6	22.10	Plain Recess	20.2 / 8	
8.4				
8.4		Sunk Duffill or Rebate 4in edway	21.6 / 8	this the same value as moulding.
4.10				
21.6				

No Joint or bed taken — As

97) As to the Value. Portland Stone common Mouldings 14d. pr ft; If in many breaks & the mouldings small at 18 or 20d. Ketton Stone at 7d or 8 pr ft. Supl common mold.s, if many breaks or small Mouldings 12 or 14d. — But if the Mouldings are taken in the Breaks as is Custom: :ary here to add the breaks of Ones as to the length, then the price will stand as if s.as kt. Suppose the Sketch I a Cornish with breaks girting 9in

2.6 / 1.6 / 3.0 / 1.6 / 8.6 The length taken on the Corona A, with the returns B, B, B, will be 8ft. 6in by 9in, Squares to 6.4 at 18d pr ft. is 9s. 6d

2.6 / 1.6 / 3.0 / 1.6 / 2.3 / 10.9

Take the Common Practice, which is the additional length of One side of the breaks, makes 2ft. 3in more in length, that is 10ft. 9in by 9in Squares to 8ft. 1in at 14d is 9s. 6d, So that they both at that rate are equal, but the last method is the most common.

25

Fig 1.7
Two pages from John Adam's commonplace book, from a section listing questions put to Robert Morris in London and his answers. Few items show better the care and precision of Johnnie's mind when considering the question of architectural details.

Most of the entries in the commonplace book are clearly copied from other documents and sometimes, as is the case with the plasterer George Dawson at Hopetoun, a number of prices which Adam did not have to hand are left out. Elsewhere there are detailed personal records taken from tradesmen, particularly John Paterson, whose bill for the 'Great Apartment' at Hopetoun had been queried. In this case the joiner maintained that parts of the job had actually cost him more than he had charged and Johnnie subsequently defended the prices to his employer. Two pages are devoted to the prices charged for the 'Estimate of Dumfries House' in 1753. Listing a total of 46 separate measured prices for masonry, carpentry, joinery, lead work, roof work, windows and plasterwork they demonstrate conclusively the extent to which John Adam was master of each process involved in the creation of that remarkable house. The commonplace book even contains a neat measured sketch of a section of V-jointed rustication that Johnnie had seen in Norfolk. The drawing is entitled 'Rusticks in the Basem.t of Houghton Hall, whose proportions had a good effect. The same proportions followed in Ld. Dumfries's house, & look extremely well.'

It seems clear that in discussions with his clients, possibly particularly in the first years that he was head of the firm, Johnnie used the commonplace book as a kind of ready-reference volume to clarify points that related to the business of building. The commonplace book could not of course provide evidence of his competence in architectural design – that had to be taken on trust – unless it were possible that John used another notebook, into which he copied the plans of some buildings he had either designed or on which he had worked, to show to his clients. The nature of his employment throughout Scotland, where he would stay, as his brothers also did, in the houses of his patrons, must have offered ample opportunity in conversation of an evening for plans in a notebook to be passed around as illustrations of Johnnie's ideas and experience. Some role of this kind may explain the existence of a small notebook, in John Adam's hand, with a collection of neatly drawn plans and elevations of buildings, which is preserved at Penicuik House. Not surprisingly the book contains a plan of Inveraray Castle and an elevation of its innovative Gothic front; a plan and elevation of the Garron Bridge, which Johnnie had designed and was proud of; and a 'Building over a Spring at Inveraray', with a section showing a Gothic vault. All this architecture was new and topical in the Scotland of 1750. There is the Adam family tomb in Edinburgh, drawn in elevation and section, with the simple inscription 'Monument', and there are floor plans of Roger Morris's Combe Bank in Kent, which was effectively another Argyll property. The rest of the book is taken up with sets of designs for small houses, several of a type of villa that was becoming fashionable in Scotland around the mid-century.[23]

It is instructive to compare the clean and tidy drawing of this little book with the freehand style of a third manuscript volume, now in the collection of the Royal Institute of British Architects (RIBA), which Johnnie used to jot down the details of several buildings he had visited when staying in London between January and March in 1750. Here the 22 drawings in the book were produced for the architect's own use. At Whitton, in Middlesex, he made a spirited pencil drawing of the elevation of the Duke of Argyll's triangular tower, which dominated the end of a canal. Though the building has long since vanished, Adam's rapid sketch admirably conveys the solidity and authority of the design. There is a similar confidence in the brisk drawings in sepia ink that he made of joinery details at Combe Bank and in his sketch of 'A Seat in Kensington Gardens', a half elevation, busy with egg-and-dart mouldings, which he evidently enjoyed. These lively, practical drawings of English buildings offer a clear proof of Johnnie's understanding and focused engagement with the business of architecture as he settled into his role as head of the Adam family firm.[24]

John Adam's approach to architectural design

We may gain some impression of John Adam at work by considering the villas and little houses recorded in his notebook,[25] to which should be added three other houses published as extra plates in *Vitruvius Scoticus*.[26] Beyond what he learned from his father, John's formation as an architect is grounded in the design preferences put forward by Colen Campbell's *Vitruvius Britannicus*, the three volumes of which were in the library at Blair Adam. Campbell's choice of Palladio as the model for contemporary architecture found instinctive approval in Johnnie's own taste, and whatever he believed to contribute to good design can only have been confirmed by his contact with the accomplished neo-Palladian Roger Morris, the architect of Inveraray Castle and of Combe Bank, the plan of which Johnnie had copied into his notebook. If his younger brothers were later to cock a snook at 'the Renowned Palladio',[27] it is possible that their critical remarks took their origin as much in reaction to their elder brother's respect for Palladian principles as in any real deficiencies in the late Renaissance architect: Palladio was never pretty; in their neoclassical interiors Bob and Jamie often were.

In contrast to his younger brothers, Johnnie gave precedence to proportion and to the rational disposition of the parts of a design. His architecture favours a clear, straightforward solution with a tendency to understatement. It is, nonetheless, an architecture of scrupulous, considered selection, informed by the critical notion that 'less is more'. The type of rational simplicity which Johnnie favoured had found consistent expression in the writings of Robert Morris, the outstanding British architectural theorist of the first half of the 18th century. A kinsman of Roger Morris, he worked as a building surveyor in London and was certainly someone whom Johnnie and his brothers knew.[28] In the commonplace book no less than 14 pages of closely written notes, including carefully drawn details of masonry and joinery, carry the heading 'Questions put to Mr. Morris about the Methods of Measuring & the prices of different kinds of work in London with his answers'. Here is a record of Johnnie at work, arranging a long meeting with the leading architectural theorist of his day. It is inconceivable that, besides the practical answers to Johnnie's list of questions, considerations of style and taste did not feature as well.

The first house in John's notebook is Bardrochwood in Kirkcudbrightshire (Fig 1.8). This villa was built for a Mr Maxwell and is described in 1823 as 'a little convenient house now in ruin'.[29] What is immediately striking about the building is how the architect has contained all its diverse functions within a simple rectangular shell. There are no wings, no extensions or

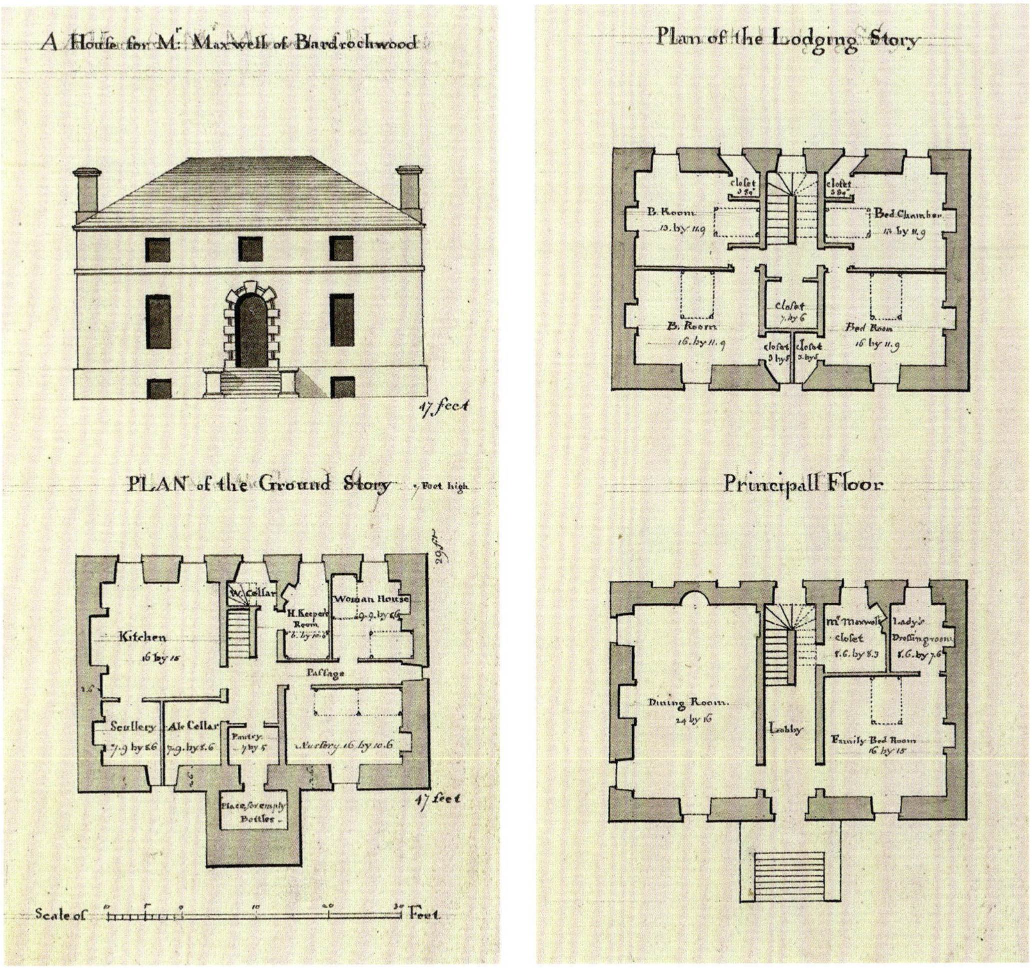

Fig 1.8
Characteristic villa plans from John Adam's architectural notebook at Penicuik House: Bardrochwood, Kirkcudbrightshire.

projections at the sides or the back. Johnnie sets himself the problem of devising a convenient plan within a pure rectangle. The elevation, where the architectural features are limited to a Gibbs surround to the door and a string course between the principal and first storeys, shows how important the simple cubic massing of the house was to its appearance. The dining room, on the left of the hall, is a well-proportioned room in the ratio of 2:3, with a niche on the short back wall opposite one of the front windows. On the bedroom floor the two back rooms have the beds cleverly accommodated within recesses, which also provide lit closets, while the two front bedrooms have a similar pair of closets lit by the window above the front door. A bedroom fitted with a recess for the bed was common enough in an 18th-century laird's house in Scotland. Indeed William Adam had often provided similar rooms, but the type of ingenious planning found at Bardrochwood seems to have been something that John Adam particularly enjoyed.[30]

The second plan in the notebook, for an unknown villa, employs the same criteria (Fig 1.9). The house is contained within a neat rectangle. It is of two storeys above a semi-basement, and any conscious architecture is limited to a pediment over the front door and architraves to the windows. This house is deeper than Bardrochwood, which leaves room for a more elegant stair on one side of the hall and an arched recess opposite the main door, giving access to linked reception rooms on the left. All the elevations are arranged symmetrically with three windows on each side of the house. The architectural treatment of the elevation is comparable to a scheme which Johnnie prepared for John McCulloch of Barholm in Galloway, of which a handsome plate was to be included in *Vitruvius Scoticus*. This house, which was not built, had at its centre a three-bay, two-storey block, connected by straight wings to miniature, single-bay flanking pavilions, with the windows scaled down proportionally in relation to the main house. There is a neatness and rightness about

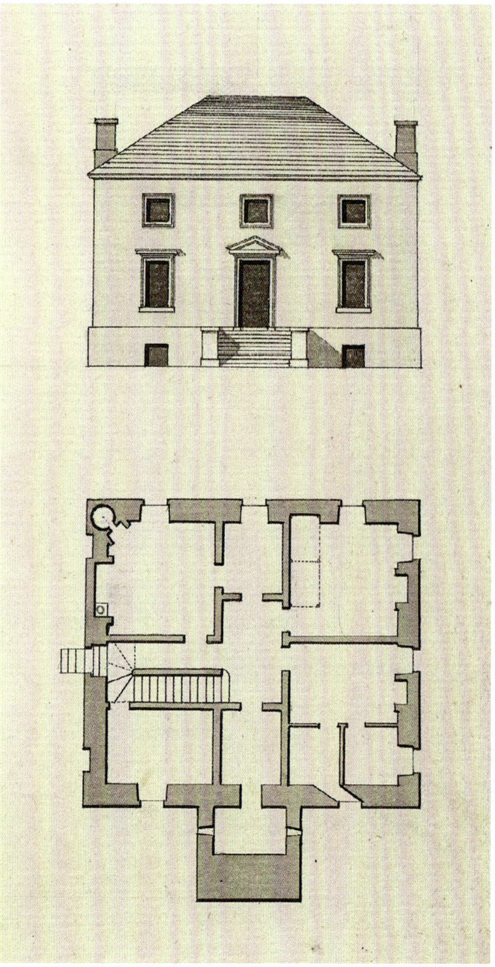

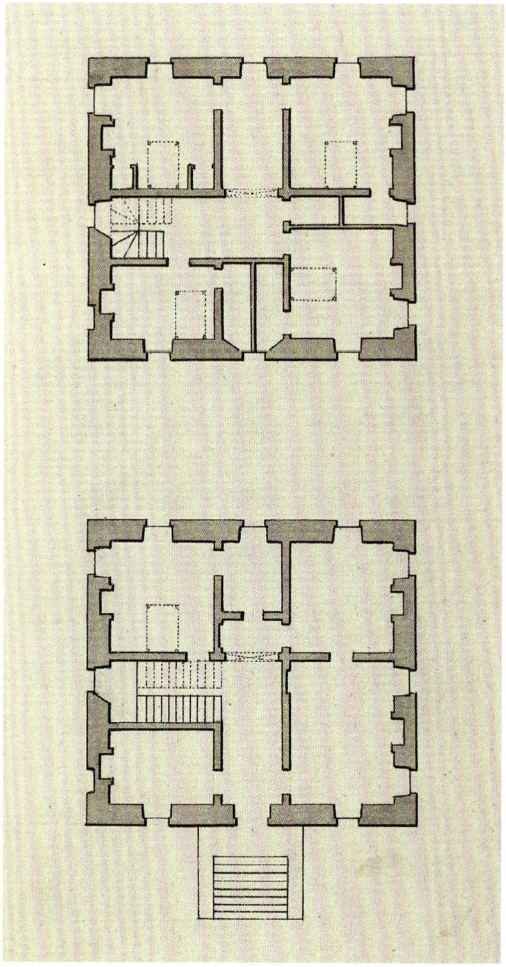

Fig 1.9
Characteristic villa plans from John Adam's architectural notebook at Penicuik House: an unidentified house.

this miniature country house, which gives the building considerable authority even though the architectural means are slight.

The other two villas in the plates of *Vitruvius Scoticus* were each designed for legal clients living in Edinburgh: Alexander Fletcher, Lord Milton, whose advice Johnnie had sought on whether he should go to Italy, and Andrew Pringle, Lord Alemore, a bachelor who built a villa for himself at Hawkhill, near Leith links. Both houses have been demolished. Lord Milton's house, under construction in 1753, stood in the Canongate at the bottom of Edinburgh High Street with its back rooms looking out over Holyrood Park.[31] Its plan shares similarities with the second of the little villas, for there is an elegant stair off the hall to one side and an office at the front. Since the site was restricted, the side elevations could hardly be seen and were irregular, but the front was given considerable architectural importance with busy blocked architraves round the ground-floor windows, two string courses, an eaves cornice and upper-floor windows similar to those on the front of Hopetoun. Evidently the setting in the centre of the Scottish capital prompted a greater degree of architectural display than Johnnie would have felt appropriate in a rural setting.[32]

Hawkhill, the second Edinburgh villa, built relatively late in 1757, shortly after its owner had become solicitor general for Scotland, is an ambitious but also a challenging design (Fig 1.10).[33] Here Johnnie had to cope with the requirements of a house that would be both metropolitan and yet adapted to an open rural setting – *urbs rure* – to invert the familiar tag. The plan of this villa marked a radical departure from 18th-century norms, for Hawkhill was a double-pile house with a substantial spinal wall (which carried no less than 12 chimneys), separating the principal reception rooms on one side from the domestic arrangements on the other. In an intriguing sequence of rooms, Johnnie provided an octagonal entrance hall that led directly into the public rooms, and separated Lord Alemore's business room on the left from his bedroom on the right. A spiral staircase placed behind the splayed wall on the left of the front door led up to additional bedrooms and down to the basement, with a lavatory conveniently placed immediately at the bottom, under the platform of the front steps. The octagonal hall, whose architectural treatment was minimal, must have struck John Adam's contemporaries as completely modern. The room was lit by a single, semicircular window above the front door and, eschewing emphasis, set eight identical doors – only two of which were dummies – within extremely slender half-torus frames, with a shallow frieze and simple shelf-like cornice above. The room cornice was little more than a line of crisply executed Ionic dentils. By contrast, the main reception rooms were handsome; the dining room, 20ft (6.10m) square, opened into a rectangular drawing room, 20ft by 30ft (6.10m by 9.14m), and, unlike the low and understated hall, had 15ft- (4.57m-) high ceilings with substantial room cornices. The drawing room had a full Corinthian cornice and frieze with finely modelled modillions. The proportions, 20 by 30 by 15, had Palladian authority, like the main room at Bardrochwood.

Where the design of Hawkhill seems uncertain is in the switch in scale between the rooms at the back, each 20ft (6.10m) wide, and those at the front, where the unit of width was 16ft (4.88m), so that the dimensions of the front 'pile' of the building were reduced by 20 per cent in comparison with the larger rooms behind. Even more curiously, the front was built as two storeys over a basement, while on the rear elevation there was simply a long single-storey façade of five windows above the basement (*see* Fig 1.10). The decision to provide two storeys of accommodation on the entrance front made for a rather pinched façade, where the windows of the bedroom floor were uncomfortably close to the top of the floor below, while, around the sides, the villa changed, unexpectedly and in a disconcerting way, to read as a single-storey building set over a basement.[34] The choice of uncoursed rubble masonry with random quoins to the corners underscored the rustic nature of the house while setting off the architect's distinctive choice of window surrounds, graded (as so often in Johnnie's designs) from flat freestone surrounds at the sides of the house to regular architraves on the entrance front, and taller windows at the back, with a full architrave surmounted by a flat entablature. Hawkhill was certainly an intriguing house and should possibly be judged as a *jeu d'esprit* in which John Adam aimed to captivate Lord Alemore's visitors by notions of cultivated surprise. Even the front door contributed to this theme, since its Venetian tripartite opening, framed by Ionic pilasters to match the order of the hall inside, tricked the visitor into expecting a room lit by sidelights, whereas one window lit the spiral staircase to the left and the other a cupboard off Lord Alemore's bedroom on the right.

A letter from James Adam to his mother of February 1761 throws an interesting light on Johnnie's professional interest in 'the villa' and suggests that Hawkhill might once have been planned as part of a series of such buildings outside Edinburgh – 'The estate Peggy mentions as purchased near Craigmillar is a sort of forsight of John's when Publick Works are finished to have ground to build villas on, which he imagines will become the taste as soon as the citizens become rich. This was a project talked of before I left home.'[35] Three years later, Johnnie's affairs were

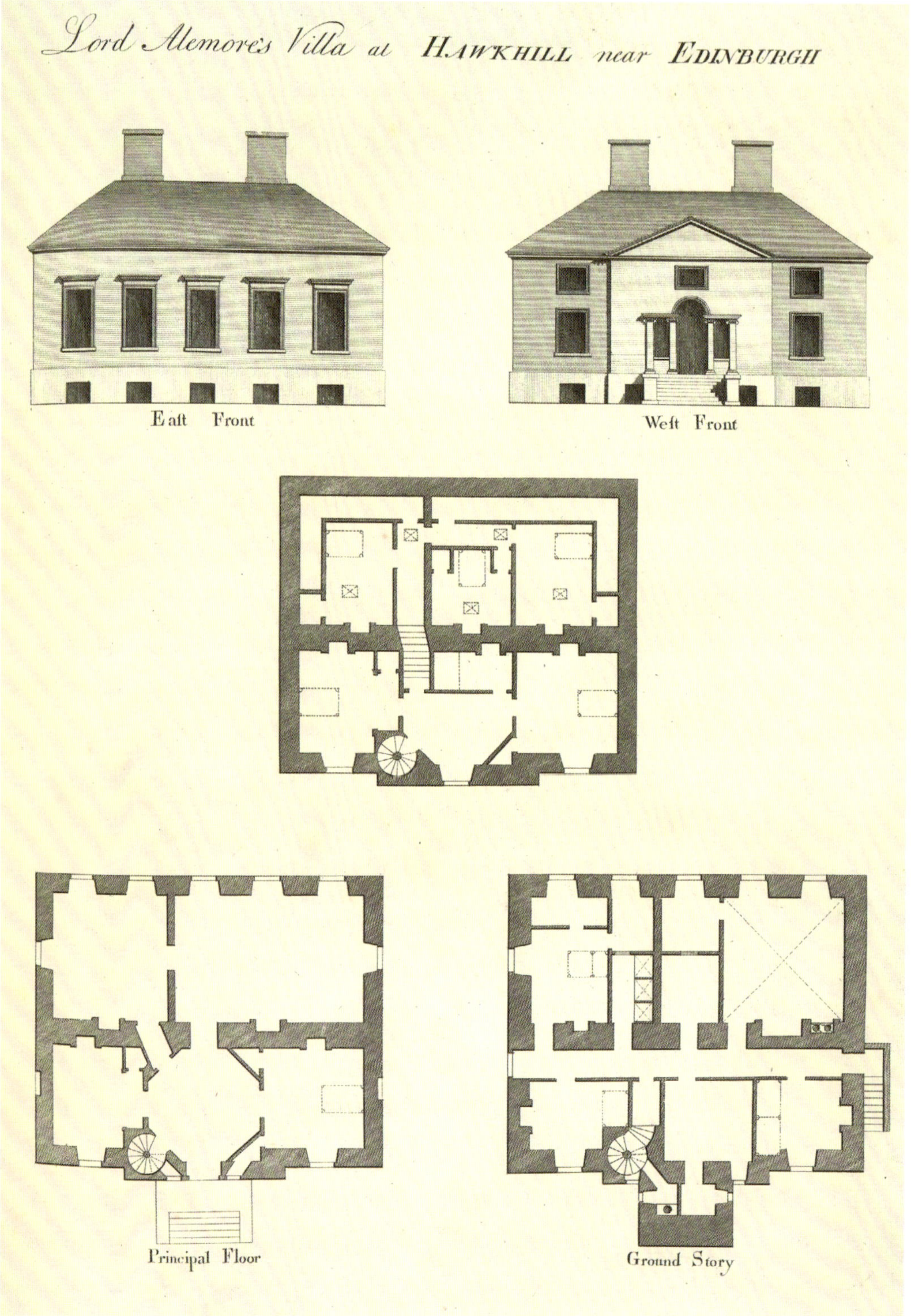

Fig 1.10
Elevations and floor plans of Lord Alemore's villa at Hawkhill, near Edinburgh. From Vitruvius Scoticus, *plate 123.*

to be thrown into confusion by the failure of his bankers, Thomas and Adam Fairholm (or Fairholme); by a curious irony, it would be Robert who, much later, built Sunnyside, an intriguing villa near Liberton and not far from Craigmillar.[36]

The Edinburgh practice

On 18 April 1755 Robert wrote from Naples:

> Let me, D[ea]r Jamie, suppose the best, the most favourable & at the same time the most unexpected thing that can happen, which is, that during the course of my travels I shall acquire that taste I am in search of, that art of designing and drawing which is absolutely requisite to support with dignity the name of a good architect; there are many domestic difficulties which make me, after all I can think, as uncertain how to resolve, how to conquer, as when I began to consider them.[37]

In Edinburgh, Johnnie was never to be perplexed by such expansive future visions as his brothers entertained, nor did he ponder the relevance of 'the true antique' to his Scottish clients. To Bob and Jamie it might appear that Johnnie was part of their problem,[38] yet the visionary ambitions of the younger men should not detract from the extraordinary competence of their elder brother as a builder nor diminish his status as an architect in his own right. Bob might consider Scotland 'but a narrow place', yet it was there than Johnnie had established a sound reputation and where he was widely accepted as a professional man. No architect can work above the expectations of his clients and, if the majority of patrons in Scotland sought practical workaday solutions to their needs, that is what Johnnie would supply. Moreover, the protracted and damaging lawsuit brought by Lord Braco in 1743 against his father for the excessive cost of Duff House – unquestionably William Adam's most ambitious architectural exercise – can only have prompted the eldest son to avoid elaborate display.[39] We might note that the Duke of Montrose never built the ambitious additions which had been planned when he first approached John Adam, and the modest, well-built court of stables at Buchanan might stand as typical of much of Johnnie's work.

There is therefore a dichotomy in the production in the Adam office in Edinburgh between a quantity of plain, almost utilitarian buildings and a more articulated form of architecture which culminated on occasion in a vein of rational neo-Palladianism in which Johnnie excelled. His position as master mason to the Board of Ordnance in Scotland, important as a regular source of income throughout the 1770s and 1780s, meant that Johnnie was routinely responsible for maintenance and building work at government forts and barracks, where plain, reliable structures were the order of the day. Inevitably something of this straightforward idiom was to find its way into the work commissioned by other public bodies and private individuals. Notable examples are: the dour recasting, between 1753 and 1756, of Castle Grant in Morayshire, a bulky Highland house to which Adam added an utterly plain four-storey, seven-bay façade in ashlar masonry; the austere north elevation of the Edinburgh Royal Exchange, now the City Chambers, constructed between 1753 and 1761; and the plain, three-storey granary erected in 1765 for Lord Seafield, above the harbour at Portsoy.

A halfway house between John Adam's plain astylar architecture and the more fully developed Classical houses is provided by his design for Banff Castle, Morayshire, commissioned in 1749 and built over the following two years (Fig 1.11).[40] Johnnie had personal experience of modern Gothic architecture at Inveraray Castle and would later design a vast U-shaped Gothic pile with substantial circular corner towers for the Duke of Douglas at Douglas Castle in Lanarkshire.[41] However, he made no attempt at a medieval style at Banff even though his client, Lord Deskford, heir to the considerable estates of the earls of Seafield and Findlater, had employed him to design additions to an older house built into the north-east corner of the medieval curtilage of the former castle. Here, the existing building was to be kept and Johnnie's commission was to modernise the accommodation. The old house has since disappeared, but the 18th-century addition remains with its interiors intact. As part of the improvements, the entire south wall which had enclosed the castle was removed and replaced by a simplified Palladian formula of central block linked to the front of a stable courtyard, on the west, and to a belvedere room overlooking the sea, on the east. Like the small villas recorded in John's notebook, Banff Castle depends on proportion for the sense of gentility that the house conveys. The architectural accents are minimal: a high roof and regular chimney stacks, greater spacing on either side of the central windows of the five-bay front and a single string course above the ground-floor windows. While the simple design offers modest scope for subtlety, a dressed stone edging to the main block of the house, the chimney stacks and the pavilions reflects the architect's appreciation of good detail in the construction of the building. Unusually, the line of flanking walls linking the house to the wings is given a convex profile.

As in the houses planned and published by Robert Morris, the rooms of Banff Castle are well-proportioned, pleasant spaces, which are always rectangular. John was not drawn to the kind of geometrical planning based on centralised circular spaces with intersecting axes, domes, columnar screens, apses and niches which fascinated his younger brothers, particularly Robert. Complexity was costly and, for the Adam office in Scotland, there had to be good reasons for expenditure. It was primarily the quality of the craftsmen employed which justified such expenditure. Johnnie had particular ideas about the lightness of an interior and laid stress on the profiles of mouldings, in joinery or plasterwork, which were used in his houses. 'Moulds' or full-sized drawings of these details were regularly sent to the tradesmen working on site, with the result that an interior by John Adam, as in the rooms at Dumfries House or at Banff Castle, has a cleanness and finesse quite different from the heavier forms that characterised the 1730s and 1740s. When Robert and James, in the preface to the first volume of their *Works in Architecture*, published in 1773, spoke of the demise of the massive entablature, ponderous ceiling and tabernacle frame, which they had replaced by 'a beautiful variety of light mouldings', the spirit of that process had been initiated in the interiors they had worked on with John. In the front hall at Hopetoun and in the drawing room at Banff Castle, a refined doorcase with an unorthodox lightly bracketed cornice set on a scrolled frieze is introduced,

Fig 1.11
Banff Castle, designed by John Adam c 1749.

while in the bedrooms at Banff the window reveals are formed without architraves and have no more than an elegant cavetto moulding in their place. Given the resources of his father's architectural library and his contacts with Robert Morris, it is clear that Johnnie was completely literate as an architect, content at times to be unorthodox – effectively as Palladian as Palladio – and to adjust the form of any architectural member to make the profile appropriate to its setting. It is this freedom that accounts for the elision of the Doric entablature in the entrance hall at Dumfries House which, though complete with a triglyph frieze and architrave across the columns at the back of the room, dies to no more than a mutule cornice along the side walls and over the top of the front door (*see* Fig 1.6).[42] In the same way, on the front of Hawkhill, where the space was quite restricted, Johnnie felt free to reduce the entablature to no more than a slim architrave, which rose from capitals of a Scamozzian Ionic order to arch lightly across the fanlight of the front door.

For John Adam an important consideration in the design of any building was the disposition and arrangement of the windows. He liked rooms to be balanced and evenly illuminated, a preference that is evident not only in the plans that he made but also in his correspondence in relation to modifications undertaken for James Grant at Moy House, Morayshire, in February 1763.[43] In the dining room and drawing room of the house as he found them, there were three windows along one side and two more at the end of each room. Johnnie thought that was too many and recommended that two of the side windows should be closed. Like many people who fancy that a lot of light is more important than the coherent arrangement of a room, his client questioned whether these changes would be an improvement. In a letter written in April 1763, Johnnie replied that since the dining parlour on the ground floor was a room with a low ceiling, he was prepared to concede that the three windows might serve some purpose there, but this was not the case on the floor above:

> In the drawing room overhead, where the windows are large and not so much overshadowed, I am sure one window on the side opposite to the fire, will be enough with the two windows in the end. I cannot say I would approve of keeping the side window open that is next to the end, as that would be irregular and throw the light too much to that end. But if your objection still has weight with you that the single window would appear *isolée* and make the room look heavie, then the middle one may be shut up and the two extreme ones left open. But as these things depend entirely upon fancy, it is reasonable that you should please yourself in it.

Yet he did not give up, writing later:

> I am sorry to differ from you still in opinion as to the window on the side of the drawing room. I do not think it will look lonely, especially as there are two windows on the end. The spaces on either side of that window will be roomy and convenient. And as there will be great plenty of light in the room they will not look heavie. A Venetian window on the broad side of the room might do very well, but there is not height sufficient to make it of any tolerable size. Upon the whole as you leave it to me, I have taken upon me to give orders to having but one window on the side of the drawing room, which I am convinced you will approve of when finished.

Perhaps supervising the alterations at Moy House – some 260km by road from Edinburgh – was the sort of minor job which John Adam undertook willingly but which Robert regarded as trifling.

Three medium-sized houses by John Adam, which like Banff Castle were each conceived as a central block with flanking wings, are Moy Hall at Dalarossie in Inverness-shire, designed for Aeneas McIntosh, chief of the clan McIntosh, and seemingly built in the early 1750s; Ballochmyle House, in Ayrshire, designed for Allan Whitefoord, lawyer to the Earl of Dumfries, which was built in about 1760; and Moffat House, Dumfriesshire, designed for Lord Hopetoun and built between 1761 and 1763.[44] The plan of each house is closely similar, indeed almost formulaic: there is a central front hall, opening directly into a staircase behind, with the principal rooms set on either side of the hall. Each elevation consists of a five-bay front, with two storeys over a semi-basement at Moy and Ballochmyle and three storeys at Moffat. Apart from a Doric doorcase at Moffat House, which is the finest and most finished design, each house is astylar and, like Banff Castle, conveys a sense of architecture that is orderly, solely through its symmetry and proportion. At Moy and Moffat the pavilions are long wings, kept close to the

Fig 1.12
Gunsgreen House, Eyemouth harbour.

Fig 1.13
Gunsgreen House, Eyemouth harbour: elevation by John Adam from his architectural notebook.

sides of the house to form a deep court, while at Ballochmyle they were laid out at right angles to the front and are linked by quadrant corridors topped with ball finials after a Palladian model.

If these houses represent John Adam's architecture in practical and sober mode, there are others, most notably Gunsgreen House above the harbour at Eyemouth and Paxton House near Berwick-upon-Tweed, where Classical forms are used with confidence and with a marked personal taste. Historians have tended to criticise John Adam for his dependence on pattern-book sources in his dealings with Lord Hopetoun; he lists no fewer than 11 precedents from *Vitruvius Britannicus*, Desgodetz, Gibbs and Ware for the type of straight stair which he proposed to build in place of his father's intended portico and curving flights of steps, while he selects a chimney piece published by Isaac Ware as one that would do well for the great bedchamber.[45] Yet such criticism misses the point. Johnnie was very well read and knew the sources of British Palladianism thoroughly, yet to copy uncritically any one piece of work was not in his nature. In the spring of 1759 he took a 'jaunt into England' to look at modern architecture, accompanied by John McGowan – a Classical antiquarian and architectural amateur who was a friend of the family. The journal of his trip contains a number of plans with comments and criticism on what he saw.[46] Early on the two men visited Belford in Northumberland, 'Squire Dickson's house', a recently completed mansion by James Paine, which he considered 'too heavie', with a front that was 'not in good taste'. Nor did he approve of the lack of closets on the bedroom floor, an aspect of intricate design where Johnnie himself excelled. His note of the treatment of the side elevations – 'the ends are much better' – provides a minute critique of the gradation of the window frames, just such a detail as he himself had recently had to consider at Hawkhill: 'The centre window is dressed like a tabernacle, the side ones in the break have only cornishes [*sic*] supported with trusses & those without the break are quite plain, which we thought too quick a falling off & that it would have been better if it had been more gradual.' Even when he got to Holkham Hall, where he made detailed notes, later transcribed into the commonplace book, on the 'Price of brick making at Lord Leicesters in Norfolk', he thought the house itself was not a success. Though the exterior had 'a pleasant effect from the colour of the brick & neatness of the execution', he did not much approve of the design, particularly on the court front: 'There are too many Venetian windows, breakings & pediments which mince a great thing into too many small pieces and so give it an effect to[o] much trifling.' These are not the opinions of a man who, lacking in critical understanding, would be content to copy blindly the schemes of other architects.

At Gunsgreen (Fig 1.12), John Adam's client was a wealthy local merchant and professional smuggler, John Nisbet, whose house was begun in 1752 and completed two years later.[47] Though the building is not identified by name, the four floor plans and front elevation are recorded in John's architectural notebook (Fig 1.13). While the house has the same cubic quality as the other small villas and is designed to ensure that the side elevations are symmetrical, the front, set prominently above the entrance to the harbour, has considerable architectural authority. Here the reticence that marks so many of John Adam's designs is set aside in favour of a confident display of contemporary Classical forms. Gunsgreen stands on a semi-basement; the entrance-floor level is articulated as a band of V-jointed rustication, above which the five windows of the principal storey are framed with architraves in dressed stone, with blind balustrades under each window to link them to the floor below.

If we look for the origin of this confident design, it would seem that John Adam may have used as a starting point the central block of Chesterfield House in Mayfair, London, designed and recently completed by Isaac Ware, a contemporary whom he particularly admired (Fig 1.14).[48] Under construction from 1748, Chesterfield House would have been entirely new and a building of note when Johnnie and Bob made a family visit to the capital in December 1749. Certainly there is a strong similarity between the tall five-bay, three-storey block by Ware and the front of Gunsgreen. Each house has a high hipped roof (though Ware, in his published plate, cheats and misses out the chimneys from his elevation); the top-floor windows float free on the wall surface and are framed by architraves; on the *piano nobile* the windows are much taller and have full entablatures, and the middle window in each front has console brackets and a pediment. In each design there is a blind balustrade below the windows at this level.

Fig 1.14
Isaac Ware's design for Chesterfield House, South Audley Street, Mayfair. From Ware's Complete Body of Architecture, *1756.*

Adam uses the ubiquitous V-jointed rustication of English neo-Palladianism at entrance-floor level while Ware does not, and there are, of course, other differences of detail. Chesterfield House has a Corinthian eaves cornice, while Gunsgreen has none; Ware places pediments over all the main-floor windows, while Adam has flat entablatures. Adam's door is in a plain round-headed arch, reserved in character, while Ware has a fussy Corinthian aedicule. The purpose of the comparison, however, is not to detect copying but rather to point out how a fashionable design in the metropolis may rapidly become part of the intellectual capital of an architect working in Scotland.

The last commission to consider in reviewing John Adam's architecture is Paxton House (Fig 1.15), 6.5km upstream from Berwick-upon-Tweed. Paxton was built for a wealthy landowner, Patrick Home of Billie, and was begun in the spring of 1758.[49] Johnnie made two designs for the building which are preserved at the house. The façade of the smaller, first scheme is rather like a much reduced version of Dumfries House: it has an understated eaves cornice like that of Lord Milton's house in Edinburgh and seems to sit somewhere between a country house and the smaller villa designs. The plan of this scheme, while it appears to adopt the tripartite division that is typical of Palladio's 16th-century villas – much copied in Britain – is adjusted in an interesting way. At the back of the house are three rooms, placed in an expected sequence, but at the front John divides the central space to create a hall, two windows wide, with a stair hall on the left that occupies a single bay of the entrance front. The planning of the suite of rooms on the west side of the house, arranged around a secondary spiral stair, has much of the ingenious intricacy familiar from the small villa plans.

In the event, Patrick Home decided that a rather grander house might better suit his needs. The original modest plan was made deeper; a portico *in antis* – that is, a portico checked back into the depth of the building – was added to the centre of the entrance front, and the house was given flanking pavilions for a kitchen on the right and stables on the left. Both wings are linked to the house by quadrant corridors at ground-floor level. The central block at Paxton occupies an important position in the history of Classical architecture in Scotland since it provides the earliest example of a free-standing giant-order portico to be built north of the border. This is a remarkable first for the eldest Adam brother to have achieved, and the whole complex, under construction from 1758 to 1761, is thoughtfully designed and beautifully built. It is worth noting how the flanking pavilions are four windows wide – not three or five as might be expected – for the even number establishes in the mind of the visitor that the pavilions are not to be entered from the front and are not in competition with the main block of the mansion house itself. As completed, Paxton was to have a wide symmetrical front with stables and kitchen yards on either side of the pavilions, paired gates with ball finials and a range of pedimented offices at either end to terminate the composition. There is in this long façade a scrupulous gradation of the architectural elements, building from the comparatively plain outer offices to the greater mass of the pavilions and finally the front of the house itself.

For the form of the main mansion, John Adam had a number of precedents to draw on. James Gibbs had published a design for Lord Islay at Whitton near Hampton Court in 1728 which uses essentially the same formula of a seven-bay, three-storey front with a portico.[50] Johnnie would certainly have known Gibbs's design, but it seems possible that he may also have been influenced by a recently published proposal by Isaac Ware of a house for James Murray of Broughton. Ware's house had appeared in the *Complete Body of Architecture* in an issue which appears to have come out in early spring 1757, just when Paxton was under consideration. Certainly there are clear similarities: the shape and massing of the two houses are identical, and Ware proposes a giant-order portico in the centre, with the cornice and frieze continued across the front of the façade just as was done at Paxton, though of course at Paxton the order is changed from Ionic to Doric.

The tendency to use pattern-book sources as a starting point also appears on the garden front. Here we might compare Paxton with the earlier Scottish house of Shawfield Park, Glasgow, designed by Colen Campbell in 1711.[51] Of course Johnnie's façade is not a replica of Campbell's design, but the idiom is the same and the details, the hipped roof and lead flat platform, the number of window bays, the pedimented projecting centre and, particularly, the heavy eaves cornice of deep mutules, are all closely comparable. Yet Paxton is different from both Ware and Campbell. This is the substantial point: Johnnie receives inspiration from a variety of sources, yet he is no mere copyist, and, just as Bob was to do

Fig 1.15
Paxton House, west front, in 1996.

later, he renders back his inspiration, certainly from Palladian models, but as something that is new with his own stamp upon it. The fronts of the two wings recall the simple architecture of the smaller villas, with windows set wide apart and framed by finely moulded architraves. As the design turns into the space in front of the house, two things happen: the number of windows changes from four to five, because each of the pavilions is entered from a central doorway opening into the forecourt, and the frames of the windows are simplified to become flat bands, while within the linking quadrant walls the windows have no frames at all. These are subtleties of which few visitors will be aware, yet Johnnie, when visiting Belford Hall, just one year after Paxton was begun, was quick to pick up errors of judgement in the gradation of the stonework of the windows there. The central doorway leading from the court into each of the two pavilions is given consequence by a heavy and boldly projecting pediment (Fig 1.16). The complexity and the positioning of different elements – voussoirs, console brackets, salient entablatures and cornices – all go back to 16th-century Italy, but their intricate conjunction is Johnnie's own. Moreover, in contrast to this fine display of polished stone, the pavilions at Paxton are built of blocks of squared dressed masonry in which the surface has been finished not as ashlar but as broached stone with dressed margins. The broaching is diagonal and creates the effect of a delicate corrugation across the walls of the quadrant corridors and flanking wings. This is precisely the finish which the masons had been instructed to use at Dumfries House. It is reproduced at Paxton almost as John Adam's signature and was to reappear at Ballochmyle House in Ayrshire some years later.

Ashlar is used at Paxton, as at Dumfries House, only on the central façade where the fineness of

Fig 1.16
Paxton House, detail of pediment and broaching to pavilion walls.

the surface complements the careful drafting of the architraves and the Doric frieze that runs along the front of the house. This is John Adam's finest piece of architecture (*see* Fig 1.15), with a magnificent Doric portico set confidently above a heroic flight of steps, appropriate in scale almost to a temple in antiquity. And at Paxton the columns are arranged with considerable subtlety: while the effect is that of a portico *in antis*, John Adam has brought the line of the columns forward from the face of the wall, so that four free-standing columns make up the temple motif at the centre of the house. He does not project the portico, since that would have created too great a forward movement and disrupted the balance of the front. This may be the first giant-order portico built in Scotland but it is also one of the most sophisticated within the British Isles. Normal practice would have been to place a pair of pilasters to frame the recess of the loggia at the top of the steps, but John Adam has found his own solution to the problem and in doing so creates an entrance space of unique quality. The free-standing columns at either end seem to embrace the volume of the forecourt and yet, at the same time, to define the point of entrance to the building. This is open-air architecture that is at once within and yet without the house.

Bank and business failures

According to the 'Account of a Scotch Family' written by his son, John Adam was 'brought up to business in which his faculties were most conspicuous' and had 'pursued his course in architecture and other concomitant occupations with extraordinary success until 1763'.[52] In that year the failure of Thomas and Adam Fairholm's bank, which had offices in London and Edinburgh, brought chaos to Johnnie's affairs. Adam Fairholm had speculated in English stocks: he had gone bust, burned the company's books and disappeared.[53] A bundle of 15 documents at Blair Adam – 'State of John Adam's affairs in consequence of His Embarrassment by Adam Fairholm' – offers a partial insight into the turmoil caused by these times.[54] In April 1763 John drew up a statement of his 'Accounts in Trade', which, when every penny had been counted, stood at the remarkable figure of £37,552 3s 4¾d. A year later 'John Adam's Stock – March 1764' provides a list with values of no fewer than 16 separate properties that he had either inherited or had acquired in business as an architect and developer.[55] In this period 31 creditors – including a number of tradesmen whom John habitually employed – had claims on the architect, and it was at this time that he took a crucial decision. He met his creditors, explained his difficulties and put his affairs into administration. Allan Whitefoord, the client for whom he had built Ballochmyle, then took on the role of 'Trustee for John Adam and his creditors'. There was a good understanding between the two men and they worked well together. Though much property had to be sold, as a consequence of their joint management the Adam businesses in Scotland were able to continue trading, John retained his position as master mason to the Board of Ordnance, his practice as an architect continued and, over time, every creditor was paid. Whitefoord's draft report of December 1765 is a testament to the probity of Johnnie's conduct: 'A general state of John Adam's affairs is here laid before you from which the trustee has the satisfaction to inform you that matters are still growing better. The funds are increasing & the debts diminishing, in so much that once next year [is completed] he has the greatest reason to hope that every thing will be cleared off.'[56]

This prudent management in a financial crisis not of John Adam's making, which had threatened to become a total disaster, stands in stark contrast to the reckless later behaviour of the younger brothers. At the time of the bank's failure the family was united: Johnnie and Adam Fairholm had been schoolboys together at Dalkeith Academy and there were ties of close friendship between the two men. They had speculated jointly – in the purchase of Royal Bank of Scotland stock and in government funds – while Willie, the youngest Adam brother, had trained in the late 1750s in Fairholm's bank. As an act of solidarity, Robert and James invited John to join them in their London practice, then in a very flourishing condition, and both men immediately advanced some cash to help in the crisis.[57] Not wanting to leave Scotland, Johnnie decided against becoming an architect in London, though he agreed to their proposal that he should act as an equal partner in William Adam & Co, the brothers' London business of builders' providers and merchants. Over time, when the financial clouds had cleared in Scotland, he was to contribute no less than £30,000 towards the working capital of the company, on what was worthless security. It was William Adam & Co that developed the Adelphi and also Fitzroy Square on Lord Southampton's estate in Marylebone, both of which, however innovative and enterprising as architecture, proved to be financially unsound.[58]

It was John's great tragedy that the younger brothers whom he had encouraged and aided financially would not listen to him or take his advice. From Robert's return to London in January 1758 – even more, after James had joined him in his business – an assumption of cultural superiority, a preference, after the Adelphi failure, for keeping Johnnie in the dark and, at times, downright dishonesty poisoned relations between them. For years Johnnie had been forced to live in straitened circumstances and with the constant threat that, because of the debts run up by his brothers' company, Blair Adam would have to be sold.[59] This is where conflicting narratives come into play, confusing the Adam family story. There is the unpublished manuscript account of Robert's life written by John Clerk of Eldin, in collaboration with Margaret Adam, who was housekeeper to her brother throughout the London years. Their primary intention was to celebrate the life of 'one of the greatest architectural geniuses ever', whom they considered had been shamefully neglected in his later years – and they are not impartial. A later account, dictated in old age by Commissioner William Adam to his youngest sister, Louisa, in 1838, tells the Adam story more from Johnnie's point of view, and reflects the experience of a son forced constantly to mediate between his overambitious uncles and a father too long alienated by their recklessness and lack of business acumen. 'It vexes me', John wrote to his son on 7 January 1788:

> that you feel so much for some expressions that I made use of in some of my last letters about myself. No doubt I must be sensitive of the irksome situation I am brought into by their means, but my chief reason for making my expressions so strong was in case you should have occasion to use them with the Gentlemen, that they might see that I had the feelings and the failures of Nature as well as themselves, and as well as Mr. Rob[er]t, who has cried out so much about the annihilation of his Faculties.[60]

Less than two years later, in the summer of 1789, John and Robert bumped into each other unexpectedly, when turning a corner in Edinburgh. They were 'both surprised but behaved with civility to each other'. Their talk did not last long and was ended by Johnnie, who, in response to a conversational overture from Bob, bowed and walked off. They never met again.[61]

Acknowledgements

In preparing this account of John Adam as an architect I would like to record my indebtedness to Keith Adam of Blair Adam and to Sir Robert Clerk of Penicuik, Bt, not only for access to their extensive family papers but also for their assistance in establishing the true nature of certain documents in their possession and for drawing my attention to them. I should also pay tribute to the fundamental work undertaken on John Adam by James Simpson and contained in the introduction to his critical edition of William Adam's *Vitruvius Scoticus* (Edinburgh, Paul Harris Publishing, 1980) and to the excellent essay by Ian R M Mowat, 'John Adam: The forgotten brother', in *The Scottish Genealogist* (vol 48, 2001). Mowat's research was inspired by the aim of 'at last giving John his due', to which I hope this paper may make some contribution.

Notes

1 Fleming 1962, 65.
2 Ibid.
3 NRS, GD220/Box 5/935/16.
4 Adam 1812, plates 136–7.
5 The estimate for Buchanan church, 'agreeable to a plan thereof drawn by Mr. John Adam', was for £399 4s 4d, or £415 16s 4d if the roof were slated. Adam's estimate for 'the new court of offices' has not survived (NRS, GD220/Box 6/836/25).
6 John Adam attended the University of Edinburgh for the academic session of 1736 but gave up his studies to work with his father at Mavisbank House, Midlothian, in 1737 – his first recorded appearance as an architect (Mowat 2001, 64).
7 These sums are taken from the manuscript 'Account of a Scotch Family from 1688 to 1838', written by John Adam's son Commissioner William Adam (1751–1839) and preserved at Blair Adam.
8 'Journal of our Expedition to London from Edinburgh', 1749 (BAM, 1454/4/225).
9 The partnership which John Adam entered into with his brothers is nowhere clearly described. According to Colvin, when John succeeded 'as laird of Blair Adam and as Master Mason to the Board of Ordnance', he

'immediately took his brother Robert into partnership' (Colvin 2008, 42). Fleming speaks of 'the partnership profits' and how Robert's share of the Ordnance contracts was so great that it could pay for his grand tour (Fleming 1962, 106). Yet the arrangement must have been fluid and subject to change, and the 'partnership' seems to have been more informal than modern interpretations of the term might imply. Not every bill is made out 'To John, Robert & James Adam Architects': the bill submitted to Lord Deskford for building Banff Castle, Banff, in 1752 is inscribed 'To John Adam & Comp'y', while the payment recorded in Patrick Home's memorandum book in relation to the design of Paxton House, Berwickshire, is made simply to John Adam.

10 *See* Fleming 1962, 91: 'No dated drawings by John Adam have survived from these years and it seems likely that by 1753 he had withdrawn from the drawing office and concerned himself mainly with the administration and general management of the Adam firm, relying on Robert, and later on James, to provide the artistic direction'. Fleming also attributes the Adam Mausoleum in Greyfriars Kirkyard to Robert, while the elevation of John's Gunsgreen House is described as 'A Villa drawing by Robert Adam, *c* 1753' (ibid, plate 27). The repeated tendency to ascribe John Adam's work to his younger brothers is extensively treated in Mowat 2001.

11 NRS, GD18/4770.

12 NRS, GD18/4981.

13 NRS, GD18/4849. John Adam married Jean, the daughter of John Ramsay, a merchant in Kirkcaldy, on 8 July 1750. The couple had two sons, William and John, who died as a schoolboy at Eton, and three daughters, Jean, Mary and Louisa.

14 *See* Rowan 1984.

15 John Fleming, who is always anxious to give the credit for any innovation to Robert Adam, clearly identified the change in style, though without access to the Hopetoun building accounts, he dated the pavilions to much later than they are. He writes: 'It seems probable that Robert rather than John was responsible for their design. William Adam's Pavilions, or the remnants of them, were concealed within the new quadrangular structures. And no more striking evidence of the advance made by Robert since his father's death, can be found than in a comparison between William Adam's heavily rusticated, scrolled and ill-proportioned towers with their octagonal lanterns and his son's carefully modulated, restrained and much more refined compositions.' For this observant criticism, read *John* for Robert! *See* Fleming 1962, 93.

16 *See* Fleming 1962, 66, plates 17 and 19.

17 Ibid, plates 19 and 20.

18 Green 2014.

19 The first trace of the brothers' involvement at the house is in a note from Lord Dumfries's lawyer, Andrew Hunter, written from Edinburgh on 19 November 1748: 'Mr. Adam who I see now in town will wait your Lop. [Lordship] whenever you desire' (quoted in Green 2014, 6). Coming less than five months after William Adam's death, such a note might possibly suggest there was unfinished building business to be resolved.

20 'John Adam's role in the creation of Dumfries House should not be underestimated or eclipsed by the later fame of his younger brothers ... His role was pivotal' (ibid, 37).

21 Ibid, 19.

22 The pocketbook is kept at Blair Adam and the architectural notebook – one of a group of three listed among the Clerk of Penicuik papers (NRS, GD18/6191) – is kept by the Clerk family at Penicuik House.

23 John Adam's architectural notebook is preserved with two similarly sized sketchbooks, which were certainly used by James Adam. As a consequence, it too has been catalogued – in my view erroneously – as belonging to James. Hence Colvin (2008, 41): 'Three of James Adam's sketchbooks are among the Clerk papers at Penicuik House.' On examination, the architectural notebook here attributed to John is quite different from the other two. It sets out its material in a considered, tidy way, with images arranged on the page carefully (elevations are often placed directly above plans), and the drawing style employs a particular graphic convention which uses a thicker pen line for the lower horizontal lines and a thicker line for all the right-hand verticals, giving a crisp, three-dimensional effect to the drawings. This seems to have been a manner adopted by the draughtsmen in the Adam office in Scotland. James's sketchbooks, on the other hand, contain drawings that are quite clearly sketches: ideas for buildings or parts of buildings which are drawn freehand with a pen in a dashing and rather scratchy way. Moreover they each carry the inscription 'Js Adam Invt' on several pages, which are mostly dated '1753'. There is no signature or date on any of the pages in the notebook here attributed to John. Nor does there seem much point in Jamie making careful copies of buildings at Inveraray with which he had not been involved. Moreover the plans that John Adam drew in 1759 in his 'Journal of a Jaunt into England along with Mr. John McGowan', though freehand, are closely similar in character to those of the architectural notebook (BAM, 4/1590).

24 The sketchbook, now with the RIBA Drawings Collection at the V&A, is signed 'John Adam' on the first page but bears no date. The RIBA Drawings Collection catalogue, vol 1 (A), assumes a date of 1748 on the basis of John Fleming's statement that John took the book with him when he went to London that year to represent his father at a meeting with the Board of Ordnance (Fleming 1962, 82). Fleming ignores the 'Journal of our Expedition to London from Edinburgh', which begins on 13 December 1749, when John remained in England until early March 1750 and travelled in Kent. This longer period seems a more likely time both for the visits and the drawings recorded in the RIBA volume to have been made.

25 Two sets of plans in the notebook are not identified, though the larger house proves to be Gunsgreen House in Eyemouth, Berwickshire. There are three other houses in the book: 'A House for Mr. Maxwell of Bardrochwood' in Kirkcudbrightshire; 'a House for Eneas McIntosh of that Ilk', which was Moy Hall in Inverness-shire; and a 'House for Mr. Davidson', who may have been John Davidson of Stewartfield

and Haltree, Deputy Keeper of the Signet. Davidson's house must have been built, since plasterwork done for it by Philip Robertson features in the commonplace book in November 1757.

26 The plates which John Adam intended to add to his father's 'Book of Scotch Houses', though not published in his lifetime, have numbers which frequently duplicate other plate numbers in the book: the extra plates were of Dumfries House (plates 19–20); a House for Lord Milton (plate 45); Ballochmyle House (plate 63); Inveraray Castle (plates 71–3); Barholm (plate 94); Hawkhill (plate 123); and Douglas Castle (plates 135, 126).

27 Robert spoke condescendingly of Palladio as the 'so much adored master', and claimed Sanmicheli was the better architect. James, when in Vicenza, 'walked out to see the different buildings of Palladio with which the city abounds', but found himself to be 'no admirer' and thought the palaces were 'ill adjusted both in their plans and elevations'. *See* Fleming 1962, 241, 273; *also* Adam 1831, 167–8; and page 14 of the Introduction to the present volume.

28 For Robert Morris, *see* Harris and Savage 1990, 317–24.

29 Symson, A 1823 *A Large Description of Galloway*. Edinburgh: W & C Tait, 142.

30 A letter of 22 February 1763 from John Adam to James Grant concerning improvements at Moy House in Morayshire offers an ingenious solution to the problem of different levels between two parts of the house. Adam proposes making two steps inside a dressing room to accommodate the change in level, 'which steps may be covered up by a breaking into the room like a press' where 'at the other end of the room another press might be made to answer it, which would render the room regular, and in case of a bed being in it, 'twould stand in an alcove. I have sketched it upon the plan of the room' (NRS, GD248/49/3/11).

31 The period of the construction of Lord Milton's house may be inferred from the presence of two designs inscribed 'Drawing room Chimney for Lord Milton' in James Adam's second Penicuik sketchbook, dating from 1753.

32 For an account of the house, *see* Bell, D 2008 *Edinburgh Old Town: The Forgotten Nature of an Urban Form*. Edinburgh: Tholis, 160–1.

33 Hawkhill was gutted by fire in 1971 and subsequently demolished by Edinburgh Corporation. For a detailed account of the building, *see* Simpson, J 1972 'Lord Alemore's villa at Hawkhill.' *Bulletin of the Scottish Georgian Society*, **1**, 2–9.

34 John Adam's completion, from 1754, of the western third of Arniston House, Midlothian, where rooms with high ceilings on two floors replaced the three storeys intended in William Adam's design, created a comparable effect. Here 'rounding the corner from south to west gives a distinct visual shock, with a four-storey house appearing to have become three-storey'. *See* Cosh, M 1984 'The Adam family and Arniston.' *Architectural History* **27**, 214–25.

35 NRS, GD18/4888.

36 Villas designed by Robert Adam from the mid-1780s up to his death show a similarity of interests to Hawkhill. In terms of invention and the disposition of rooms, Hawkhill might be regarded as a prototype for Robert's exploration of the Classical villa type. Sunnyside, built on land which had once belonged to John Adam, was designed for Sir Patrick Inglis, 5th Bt, between 1785 and 1790 (*see* Fig 0.19, right). For the range of Robert's ideas on this form of architecture, *see* Rowan 1985, plates 14, 15, 23, 28.

37 NRS, GD18/4770.

38 In a letter of 6 December 1761 from Naples to his sister Nelly, James talks of 'a consultation and commission from Johnnie about his Great Room and bow windows which is likely to turn out a very Scotch piece of architecture ... Two bow windows are indeed hard to adjust but he mentions the room as a chef d'oeuvre. It is indeed a good size and two years ago I should probably have been an admirer of it likewise, so greatly does one's taste change in a short while.' James hoped therefore that Johnnie might 'fall in with a few of my more Classical ideas' which he would send to his brother (NRS, GD18/4918).

39 John Adam, 22 years old at the time of Lord Braco's lawsuit, must have been aware of the damaging effect on his father's constitution in a way that the younger brothers would not have been.

40 Work at Banff Castle is extensively recorded in the Seafield papers, firstly in John Adam & Co's account as contractors for the building, settled in August 1752 (NRS, GD248/984/4/4/20), and secondly by Lord Deskford's 'Books of Disbursements', which list the minute expenses at the house between March 1749 and August 1752. These record a total cost, excluding work in the park, of £1,475 1s 1d (NRS, GD248/1169–70).

41 Adam 1812, plates 135–6. Though the idiom of Douglas Castle, designed in 1757, is rooted in the style of Inveraray, comparison with James Adam's design of 1767 for a vast castellated house for the Earl of Lonsdale at Lowther in Westmoreland suggests that the scale of the Lanarkshire house – which was to have had six substantial round towers, two Gothic Venetian windows and curious mezzanine floors lit by circular windows not found at Inveraray – may reflect the contribution from James. Only one wing of the castle was built and that was demolished after 1939.

42 Generally where Robert and James adopt a full entablature within a room it is not broken or interrupted but used conventionally as a tripartite motif throughout the interior. In this regard John Adam's hall at Dumfries House is less conventional.

43 Moy House is a ruin of which only the lower walls remain. Adam's letters to Grant are extensive. *See* NRS, GD248/49/3/11, 16.

44 Four floor plans and the front elevation of a 'House for Eneas McIntosh of that Ilk, Esquire' are in John Adam's Penicuik notebook. The fire insurance policy of the Edinburgh Friendly Insurance Society for 1810 describes the building as 'the mansion house of Moyhall ... having a wing at each end thereof adjoining, all stone and slated' (NRS, GD176/1275). Window tax records for 1771–5 are charged for 'one house and forty windows at Moyhall' (NRS, GD176/1267/7–8). In 1873 the building was greatly enlarged to designs of Alexander Ross and given the

appearance of a sprawling Scottish Baronial house. It was demolished in 1958. Ballochmyle was illustrated in *Vitruvius Scoticus* (Adam 1812, plate 63). Like Moy Hall, it was destined virtually to disappear within a Jacobean remodelling by Hew Maitland Wardrop of 1887. Only the plain rear elevation and some internal plasterwork survive at the house, which has been converted into 20 apartments.

45 Fleming 1962, 91. *See also* Macaulay 1987, 136: 'One can accept that John Adam was not an inspired architect; what is distressing is his lack of creative design or an interest in new ideas.'

46 The buildings of which Adam and John McGowan made plans are: Belford Hall, Northumberland; York Assembly Rooms; Devonshire House, Piccadilly; Sir Gregory Page's House (Wricklemarsh) at Blackheath; Burlington House, Piccadilly; and a plan of a Barn at Houghton Hall (BAM, 4/1590).

47 The first recorded notice of Gunsgreen is an entry in the window tax returns for Ayton Parish, on 13 September 1753 – 'Mr. John Nisbet, a new house not yet inhabited' (NRS, Window Tax Records, Berwickshire, E. 326).

48 Ware, I 1756 *Complete Body of Architecture Adorned with Plans and Elevations, from Original Designs.* London: T Osborne & J Shifton, plate 61.

49 *See* Rowan 1967.

50 Gibbs, J 1728 *A Book of Architecture*. London, plates 56–7, 59.

51 Campbell, C 1717 *Vitruvius Britannicus*, vol II, plate 51.

52 See note 7, above.

53 A succinct account of the Adam Fairholm crisis is given in Simpson 1980, 4–6.

54 BAM, 145/4/196.

55 The list of properties illustrates the extent to which John Adam was aware of and participated in development opportunities in Edinburgh and further afield. They are Blair Adam, North Merchiston villa, Highriggs, Ross Park, Sunnyside, 'the house I now live in', Lord Kames's house, 'Middle New house in my area', two adjoining houses, Lady Seton's Brewerie, a house bought from Sir John Inglis, houses in Niddry's Wynd, Newland Close at Leith, houses in the Links of Kirkcaldy and a barley mill. The total value of all his real estate came to £21,410. The estate of Blair Adam was valued at £9,000 and the barley mill, which was at Balbernie Bridge, was worth £80.

56 BAM, 1454/4/196/15.

57 BAM, 1454/4/196/10; 1454/4/1/6–7; 63/box 4, General Correspondence, 1786–8 (I am grateful to Colin Thom for this information).

58 BAM, 1454/4/16. For the Adelphi and Fitzroy Square, *see* Rowan 2007, especially 14–21 ('The Adelphi Buildings') and 32–5 ('The last speculation').

59 *See* particularly Rowan 2007, 36–40 ('Denouement').

60 BAM, 1454/4/221/6.

61 BAM, 1454/4/221/1.

2

'Antique mad': the Adams as dealers and their stock of antiquities

Jonathan Yarker

This chapter builds on work I first began some years ago while working on *Digging and Dealing in Eighteenth-Century Rome*, a survey of archaeological activity in the Papal States undertaken principally by the resident community of British antiquaries and merchants.[1] While researching the lives of the major antique sculpture dealers operating in Italy it became clear that Robert and James Adam were active within the Roman art world and built up a serious stock of antiquities during their respective periods of residence in Rome. It is this 'stock' and the brothers' activities as commercial agents which will be explored further in this essay.

The Adams' collection of sculpture and paintings has been the focus of periodic attention from architectural historians. The 1773, 1818 and 1821 auction catalogues of their property were all published by A T Bolton; John Fleming wrote about James's acquisition of Cardinal Albani's drawings for *The Connoisseur* and a brilliant article entitled 'Messrs Robert & James Adam: Art dealers (I)' – sadly part II never appeared.[2] But Fleming's article nonetheless provided an important assessment of the brothers as active merchants rather than simply collectors. It is an assessment that has been largely overlooked by more recent scholars, who tend to frame the Adams' Italian purchases as a 'collection' made to furnish their London house and office.[3] This is in part because of the limited attempts to identify specific objects which belonged to the brothers. Working on the basis of the four surviving auction catalogues, the general consensus has been that their collecting was not a success in either financial or artistic terms.[4] To quote Alan Tait: 'The brothers' collecting was too ruthlessly practical and unadventurous to be successful, and too hurried to be lucrative.'[5]

The purpose of this essay is twofold: to underline that the Adams were not collectors but dealers who were committed and intelligent agents, and, by tracing some of their purchases, to show that what they bought was of considerable quality and highly desirable on the London art market. This essay will also frame their activities within the immediate commercial context of Rome and London in the 1750s and 1760s, making a case for both brothers as innovative operators within the 18th-century antiquities market.

Robert in Rome

Travel to Italy in the mid-18th century had twin motivations for young artists: education and commerce. The two spheres of activity have been routinely separated in grand tour scholarship, but there is a growing realisation that young artists undertook the lengthy and expensive journey to Rome with an expectation that it would be financially remunerative as well as educationally rewarding.[6] This is certainly true of the Adams. Although Robert on his arrival in Rome in 1755 was viewed more as a 'gentleman' than an architect, his determination to make his time in Italy pay both professionally and monetarily was evident from the outset.[7] Shortly after settling in the city, he wrote back to Scotland that his own commercial success as an architect depended on 'succeeding' William Chambers: 'But still more to cut out one Brettingham who has been two or three years in Italy, has gone to England lately under great protection and has £15,000 or £20,000 from his father. He had commissions, when at Rome, for near £10,000 for statues &ca.'[8]

Matthew Brettingham, the son of a successful East Anglian builder turned architect, had left Italy the previous summer. His trip had been partly sponsored by Thomas Coke, 1st Earl of Leicester, then in the process of finishing the interiors of Holkham Hall. It is clear that Robert's anxiety was only partly based on Brettingham's professional capabilities – it was the 'near £10,000 for statues &ca' which evidently piqued his jealousy. His envy was not without foundation. Brettingham's surviving Roman account book demonstrates that, though he was there ostensibly as an architect, the majority of his time was spent in buying sculpture and old master paintings, and in making moulds and casts of famous antiquities, all for resale back in Britain. With financing from home in place, Brettingham was able to supply an impressive group of antiquities not just to Coke at Holkham, but to Charles Wyndham, 2nd Earl of Egremont, for Petworth. At the same time he capitalised on his position in Rome to serve visiting travellers, such as William Legge, 2nd Earl of Dartmouth, and Sir William Stanhope, as well as acquire a stock of objects for resale at home.[9] Brettingham's activities provided a powerful model for Robert. He showed not only the diversification of the role of the young architect in Italy, but a clear method for financial success.

Once Robert had arrived in Rome, a stream of correspondence flowed back to his family in Scotland describing his activities and life in the city. These letters, preserved in Edinburgh, are a familiar and well-explored resource.[10] Highly evocative, they reveal eminently quotable glimpses of Adam's life as part of the British community in Rome. Among the practical descriptions of his professional training with Charles-Louis Clérisseau and Laurent Pécheux and gossip about fellow travellers, we catch glimpses of Robert's commercial activities. Shortly after his arrival he wrote home: 'I have already pick'd up some antique vases, an antique altar & some lions heads which tho, John Jamie & John McGowan might be ravish'd with, would be thought dirty scabbed stones by the females of our society.'[11] A few months later he mentions losing 'a whole day' at the sale of the painter Pier-Leone Ghezzi's collection.[12] Adam clearly spent a great deal of time in the public auctions and shops in Rome. By November 1755 he was able to write to his sister Margaret: 'My orrow bodles [spare pennies] I spend on Pictures & Drawings & I believe will be master of a very pretty little collection well chosen & cheap, In so much that I am fully convinced in selling one half of my pictures at London I shall have t'other for nothing.'[13]

But in general Adam's correspondence is frustratingly light on detail about his commercial activities. We can infer that he followed precisely the pattern of other visiting British artists: using his position in Rome to cultivate commercial contacts and buy and sell works of art. The people he met place him at the heart of a network of agents, dealers, antiquaries and collectors, providing countless opportunities for him to acquire paintings and sculpture for resale in London. Early mentions of Gavin Hamilton, whom Robert knew from Scotland, and the Abbé Peter Grant introduce two of the key British agents resident in the city. Hamilton was still ostensibly practising as a painter when Robert was in Rome and, like another Scottish friend, Allan Ramsay, was living away from the commercial centre of the city. But Hamilton would later become a central figure in the grand tour trade in old master paintings and antiquities, acting as a conduit between the Roman market and British travellers and collectors.[14] Grant, a Catholic and Jacobite, fulfilled the practical role of *entremetteur* and agent, introducing British visitors to key figures within the city and organising practicalities such as accommodation. It was either through Grant, or more likely Clérisseau, that Robert met Cardinal Alessandro Albani.[15]

Albani deserves far greater attention than he has received by scholars of the British community in Rome.[16] The nephew of Clement XI, he was at the heart of Vatican politics for 50 years, as well as one of the most influential collectors of, and dealers in, antiquities. It was Albani's ability to synthesise statecraft and art dealing which made him of paramount importance to visitors such as Adam. During the 1750s Albani understood the perilous political and economic position of the Papal States. Officially the papacy still supported the exiled Stuarts who were resident in Rome, which made an official British diplomatic presence impossible and travel to Rome potentially compromising for high-profile British figures, such as members of the royal family. Albani realised that British protection and the money spent by British tourists in Rome was key for the state's stability. He made overtures to the Hanoverians and as a result became the unofficial protector of British interests in the city. The principal vehicle he found for serving both visiting tourists and influential figures in Britain was through artistic commissions.[17] This makes Albani's extensive correspondence with the British minister in Florence, Horace Mann, remarkable among diplomatic papers, because of its combination of political intelligence with virtu news.

Robert not unsurprisingly found Albani an enthusiastic supporter on his arrival in the city, writing home at the end of 1756:

> Cardinal Albani and I are turned very thick as he has discovered my hidden talents for the hidden treasures of antiquity. He has given me allowance to mould several things from his originals, shakes me by the hand like an honest goddam and claps my shoulder. In short *son* éminence and me are as grit as dogs' heads.[18]

Albani and Robert undoubtedly shared a number of passions. Albani was in the midst of constructing his great neoclassical villa on the via Salaria (Fig 2.1), designed by Carlo Marchionni, and was also at the height of his activities as a collector of and dealer in antiquities.[19] While in Rome, Albani clearly helped Robert to 'mould several things from his originals', as the 1773 Christie's sale includes several casts of Albani's antiquities. He probably also helped Adam have casts made of sculpture in the recently formed Capitoline museum and elsewhere in the city.[20] But the attention Robert received was entirely typical and a consistent example of Albani's cultural politics. What makes their relationship exceptional is the way Robert used it to his financial and political advantage once he returned to Britain.

After Albani, Giovanni Battista Piranesi was Robert's most important contact.[21] Though most famous as an engraver, designer and architect, it is important to note that Piranesi was also an enormously active dealer. His premises on strada Felice were filled with fancifully restored antiquities as well as his own modern pastiches, prepared expressly for the tourist market.[22] Piranesi's influential approach to the antique was predicated on the excavation, restoration and ultimately trade in sculpture. His own collection, particularly of antique fragments, was central to his work as a designer. His stock would form the basis of his major publication the *Vasi, Candelabri, Cippi* (1778), which consisted of engravings of antiquities he had restored and sold, with plates dedicated to a roster of British and European clients (many of his 'reconstructed' antiquities were acquired by British tourists, such as the Newdigate candelabra and the Boyd vase).[23] Much has been written

Fig 2.1
View of the Great Vaulted Portico of the Villa Albani, Rome, *by Gian Paolo Panini,* c *1760. Pen, grey ink and wash on paper.*

about Adam's direct engagement with Piranesi and the implied impact of the latter's graphic style on Adam's own work, but I would argue that Piranesi's stock of sculpture was of equal importance. The sharp and canny Robert would have grasped quickly from the Venetian the value of acquiring antiquities, both as a potential source for design and as valuable commodities.

We can assume that 'antique mad' Robert was familiar with the studios of all the major sculpture restorers and dealers in Rome. There is growing evidence to suggest that after around 1750 these studios became grand tour destinations in themselves.[24] By the time Robert arrived, the most famous and productive was that of Bartolomeo Cavaceppi (Fig 2.2). Based in a large palazzo on via del Babuino, Cavaceppi was both sculptor and dealer. He had built up a large client base, particularly among British visitors; Brettingham bought a considerable number of

Fig 2.2
Bartolomeo Cavaceppi's studio in Rome. From Cavaceppi's Raccolta d'antiche statue, *vol I, 1768.*

sculptures through Cavaceppi for both Coke and Wyndham.[25] Robert was therefore almost certainly drawn to his workshop and, in turn, recommended Cavaceppi to James; the resulting relationship had a hugely important bearing on the brothers' careers as dealers.

From his surviving correspondence, it is also clear that Robert established working relationships with agents who could manage the practicalities of buying, exporting and shipping works of art from the Papal States back to Britain. Robert's correspondence with Innes & Clerk, the London merchant house that represented the Adam office in Edinburgh and helped provide finance for his trip, gives an indication of the range of these contacts. They included a corresponding financier in Rome, Francesco Barazzi, who ran a successful banking house much used by British travellers.[26] It was either through Barazzi or Innes & Clerk that Adam was introduced to Ragueneau & Co, merchants based in Livorno, who organised the shipping of Robert's purchases back to Britain.[27] Transporting his acquisitions to London was fraught, with the Seven Years War at its height and the French navy on the lookout for merchant shipping.[28]

As we have seen, Robert identified a key professional advantage in Brettingham's grand tour over his own – namely liquidity. Brettingham was able to call upon his father's clients, particularly Thomas Coke, as well as his own cash reserves to facilitate the advantageous acquisition of antiquities. Robert seems not to have had similar opportunities. In line with most British artists who established themselves in the city for any period of time, he was able to pick up commissions from wealthy travellers to broker the purchase of works of art.[29] Robert was employed by Francis Egerton, 3rd Duke of Bridgewater, to negotiate the acquisition of several contemporary paintings, including works by Anton Raphael Mengs, Gavin Hamilton and Joseph Vernet, and he also undertook commissions for old Scottish clients.[30] On his return journey through Italy, Robert prolonged his stay in Florence in the hope of 'fulfilling my Lord Hopetouns last commission of 7 pictures for Arnistone'.[31] But these seem to have been isolated examples. Adam's major commercial achievement was to make contact with the personnel who could serve the purposes of his business once he had established himself back in London.

By January 1756 Robert described his lodgings in the Casa Guarnieri as being so full 'of antique cornices, friezes, figures, bas-reliefs, vases, altars, so that I now have a room as full as it can stick from roof to the floors'.[32] By the time he had decided to return to Britain, this stock of antiquities had assumed a significance beyond their resale value. Robert had resolved to develop a highly personal Classical language of ornament, using the ancient sources and models gathered in Rome, to help establish a successful architectural practice in London. The antique contents of the crates Ragueneau & Co shipped home for him undoubtedly served as inspiration and source material for his 'fund' of decorative motifs, but they also provided potential furnishing for the antique interiors Robert was proposing to create. It was partly to fulfil these complex purposes that James was dispatched to Italy in 1760.

James 'on the ground'

Before James had even left for the Continent, Robert was plotting possible purchases he could make in Italy. Newly established in Mayfair, Robert was beginning to forge important contacts among patrons and collectors. In a letter dated November 1758, he described to James a conversation with the great numismatist and collector Charles Frederick, about sculpture.[33] Robert had seen a colossal statue of Marcus Agrippa in the Grimani collection in Venice, which he then attempted to acquire, it being, as he observed to James, 'worthy of a crowned head'. Robert wrote to Clérisseau about trying to secure the statue, explaining to James his strategy should he succeed: 'I would make a vast Puff by advertising the Arrival of it, By fixing a day of the week in which it was to be seen by all the Nobility & Gentry in My Court Yard from Such an Hour to Such an Hour, which would have a fine effect.'[34] Despite Robert's, and later James's, attempts to purchase it, the sculpture remained in Venice and is now in the Museo Archeologico.[35] But this letter establishes Robert's ambitions as a dealer and his need for a reliable agent on the ground.

Jamie's commercial activities are far better documented than those of Robert, as his letters home are filled with details of possible projects and acquisitions. On reaching Florence in December 1761, he wrote that he had 'bought a good many fragments of bassoreliefs & little statues' from the collection of the antiquary Baron Philipp von Stosch.[36] In the same letter he reports his success in acquiring marbles from Stosch's collection, and mentions the possibility of purchasing '12 or 14 columns of porphery that are to be sold ... They are about 14 or 15 ft [4.3 or 4.6m] high & would ornate a charming temple, they are in good order & might be had cheap.'[37] Such a purchase would have presented insuperable practical difficulties; speculatively exporting large architectural elements, such as antique columns, was prohibitively expensive and rarely attempted.[38] By the time he settled in Rome, James's enthusiasm was tempered with an acute sense of the commercial realities of the complex Italian market.

We can assume that James, furnished with letters of introduction from Robert, entered effortlessly into Rome's commercial and antiquarian circles. Shortly after his arrival, he records having met 'the Scarpelino of the Campo Vaccino of whom Bob had his granite tables &c'; this was Antonio Vinelli, a notable sculptor and stonemason.[39] His letters begin to record the virtu news, the typical round of gossip relating to dealers, their stock and wealthy British travellers.

A fragmentary journal kept by James in Italy, published in two parts in the 19th century, records his dual activities as tourist and consumer. It was in an acquisitive mode that James noted the things he admired in Florence:

> The Pedestal in the Gallery at Florence I propose having Harwood to copy in marble; and perhaps a bust of Homer, and one of Sappho. The scagliola is curious, and could be made to answer for different purposes; for instance, for columns resembling different marbles, for tables resembling mosaic work, and for the most elegant floors for baths ... and likewise for imitating different marbles in cabinet work.[40]

James's observations underline the need to make his grand tour a commercial success. In the summer of 1761 he recorded a visit to the villa of Cardinal Domenico Passionei at Monte Porzio Catone, near Frascati. Passionei had recently died and the reason for James's visit was to see the 'antiquities he had collected and incrusted in the walls'. A few weeks later James returned to the villa 'to look at some of the antiquities of Cardinal Passionei, and made [an] offer of 205 crowns for the Urn of the Villa Adriana and some other fragments'.[41] Several antiquities with a Passionei provenance appeared in the 1773 Adam sale.[42]

With the London practice established and Robert attracting an expanding group of patrician clients south of the border, having someone 'on the ground' in Italy was extremely valuable. As James himself noted in a letter to his sister Margaret from Rome in January 1762: 'I wrote Bob formerly of my strong desire of having some commission for the purchasing of Antiquitys, if any palace or villa is built or gardens certainly these things wou'd be of immence use & they never will find a better opportunity than that of my being here.'[43] Syon, Kedleston, Bowood and Bute (later Lansdowne) House were all under way by this date, and each project directly benefited from James's time in Rome.

James immediately began assembling a group of sculptures Robert could offer to clients back in London. By February 1763 he had evidently bought a considerable collection. In a letter to Elizabeth Percy, later Duchess of Northumberland, the travelling tutor Jonathan Lipyeatt noted: '[I] sounded Mr. (James) Adam about his statues and find that some are bought for commission and some on his own account. The boxer which I mentioned to his Lordship, is for Lord Shelbourne.'[44] As well as confirming that James had amassed a significant stock of sculpture, this letter introduces William Petty, 2nd Earl of Shelburne, later 1st Marquess of Lansdowne. In 1762 Shelburne had commissioned Robert Adam to remodel Bowood, his Wiltshire house, and at the same time he evidently commissioned the brothers to supply antiquities. The statue referred to was a Roman copy (1st century BC or 1st century AD) of a lost bronze sculpture of an athlete (or *Apoxyomenos*) by Lysippos, which had been fancifully restored as a boxer (Fig 2.3).[45] This acquisition demonstrates the powerful modus operandi which enabled Robert to recommend James's sculptural purchases to patrons of his architecture.

Robert's few surviving letters to James in Rome are filled with commissions and give a sense of how he was leveraging Jamie's presence there. In February 1762 Robert wrote from London:

> I have given in a paper to Lord Bute in which I represented that as you was at Rome you could make purchases for his Lds of all sorts of antiquities as also pictures, tables &c ... Lord Northumberland beg'd I would write you to see if there were any ancient mosaic paving to be found at Rome that would answer for making a hearth as he thinks one of that kind cleverly composed would be pretty for the Drawing room.[46]

Robert had, in effect, achieved through James the kind of commercial dynamic he had envied Matthew Brettingham for forging during his grand tour.

James responded to Robert's requests, filling his letters with schemes for purchasing whole collections of antiquities en bloc for clients back in London. The following letter is worth quoting at length as it gives a sense of both how focused James was on fulfilling Robert's commissions and how well acquainted he was with the Roman art world. Written to his sister Betty in June 1762, the letter informs the family of the possibility of acquiring a major collection of antiquities:

> [A] sale that is just now propos'd, of a collection of antiquity's belonging to Cle Furietti & which being above private purses, I think might not improperly be mentioned to Ld B – Surely he will not attribute my mentioning these things to any motive but that of being, if possible, of service by my intelligence. The collection consists of two beautiful centaurs of Bassalts or black marble dug up at the Villa Adriana, of most elegant workmanship a piece of mosaick finely executed representing pidgeons on the lips of a Bowl or Cup & mention'd by Pliny the Elder ... Besides there are three other pieces of mosaic representing landskips &c. & 10 mosaick tables, 4 of which are very handsome. The sum here demanded by the family for the whole is 30,000 cns but it is thought ... that they will not sell them under 20 or 21.000 cns or £5250.[47]

James added: 'In giving the earliest intelligence at home I think I do my duty, & this sale is, I am sure, not yet publickly known here. If there is any inclination for making a purchase, let me know.' Giuseppe Alessandro Furietti formed a notable collection of sculpture in the first half of the 18th century. Granted permission to excavate on the land around Hadrian's Villa at Tivoli, Furietti had discovered several hugely important antiquities, including the famous statues of centaurs and mosaic of four doves drinking mentioned by James. Like Albani, Cardinal Furietti was highly active on the antiquities market: Brettingham, for example, had acquired numerous antiquities for Holkham from him.[48] But James's letter makes clear that he had early intelligence of Furietti's plans to dispose of his entire collection and was keen to alert John Stuart, 3rd Earl of Bute – either for Bute himself, or for his protégé, George III, who was then in the midst of acquiring artworks through agents in Italy.

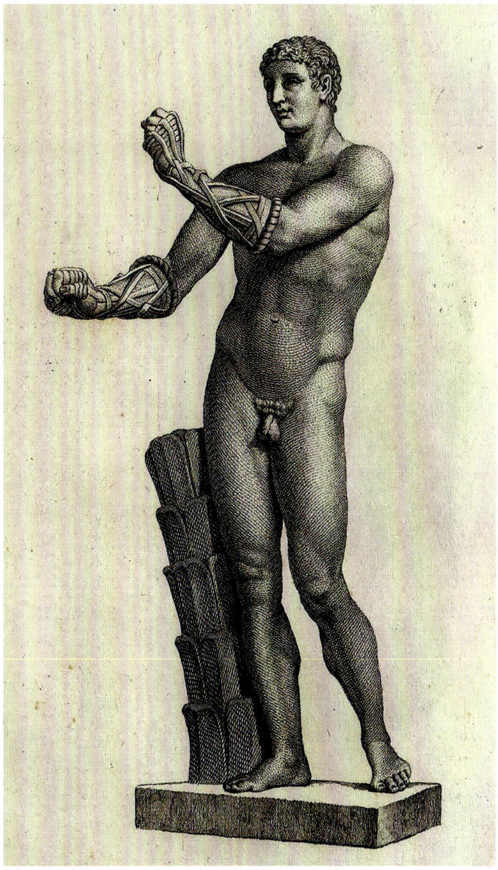

Fig 2.3
The Lansdowne Athlete, *Roman, 1st century BC or 1st century AD (far left); and Cestiario, 'esistente in Inghilterra' – the same statue, restored in the 18th century by Cavaceppi as a boxer – from Cavaceppi's* Raccolta, *vol I, 1768, plate 21 (left).*

James's early 'intelligence' of the dispersal of Furietti's collection demonstrates how embedded he was within the Roman art world. A memorandum written at the back of one of his Roman notebooks, entitled 'Good Antiquitys yet to be procur'd at Rome', itemised established collections which might offer potential purchases.[49] The list underlines how in tune James was with the shifting fortunes of contemporary Roman collections and how alive he was to the potential for acquiring sculpture for resale in London.

The first name on the list is 'Collection Mattei'. Situated on the Caelian Hill, the gardens of the Villa Mattei contained a notable collection of antiquities. This is the earliest mention by a British dealer of a possible dispersal from Villa Mattei, and James seems to have been unusually early in his awareness of the collection's vulnerability. It was the eventual sale en bloc of the Mattei marbles in 1770 by Don Giuseppe Mattei to the British dealer Thomas Jenkins which finally prompted the foundation of the Museo Pio Clementino.[50] This was a situation repeated in the other names on the list. Albani's collection is mentioned, as it was in perpetual flux and Albani acted as much as a dealer and agent as a collector.[51] James lists a number of other patrician collections which saw significant dispersals during the latter half of the 18th century, including those from the Rondinini, Giustiniani and Lancellotti palaces.[52] Another vulnerable collection was the Massimo marbles at Villa Montalto Negroni. Formed in the 17th century by Cardinal Camillo Massimo, it was one of the great assemblages of antiquities in the city.[53] Again its dispersal in the 1770s was a major factor in prompting the foundation of the Museo Pio Clementino.

Evidence survives of James using his list to make an acquisition. He bought five cinerary urns from the Villa Montalto Negroni – the only Adam purchase for which it has been possible to identify an export licence.[54] The Reverenda Camera Apostolica (the papal treasury), keen to protect its artistic patrimony, had instituted a series of restrictive export laws and all antiquities leaving Rome required an export licence. Preserved in the Archivio di Stato in Rome, the licence was issued in September 1763 in the name of Francesco Barazzi.[55] This confirms that James had built on the network of contacts established by Robert, and it also shows that the brothers had an agent in Rome who was capable of organising the practicalities of shipping and export, raising the possibility that more of the licences issued to Barazzi were applied for on behalf of the Adam brothers.[56] It is evident that acquisition drove the list. James noted the 'Musaicks Furietti', suggesting that, despite Bute's reluctance to purchase the marbles en bloc, James was still intent on buying objects from the collection. In the end James did acquire a number of antique mosaic table-tops from the Furietti collection, which are specifically mentioned in the Christie's sale catalogue of 1773.[57]

James was a sagacious and highly active merchant, and well placed to maximise commercial opportunities. His most spectacular enterprise was not in the field of antiquities, but drawings. In 1762 James managed to secure the remarkable collection of Cardinal Albani, which included the 'Paper Museum' of Cassiano dal Pozzo, along with a hugely important body of 17th-century and earlier Italian drawings initially assembled by Pope Clement XI. This was one of the most impressive and audacious artistic purchases made by a British dealer during the 18th century and deserves far greater attention.[58]

The Adams and Cavaceppi

Along with Lords Shelburne and Bute, James was also seeking sculpture and works of art for Hugh Smithson, 1st Duke of Northumberland. Smithson, a Yorkshire landowner, had in 1740 married Elizabeth Seymour, daughter of the 7th Duke of Somerset and co-heiress to the historic Percy lands, collections and fortune. By the 1760s the couple had embarked on a protracted period of lavish construction and collecting. Robert was busy remodelling the interiors of Syon House, the Northumberlands' great villa in Middlesex, and James's correspondence is peppered with commissions for that project.[59] James was in Italy at the same time as the Northumberlands' son Hugh Percy, Lord Warkworth, whom he befriended and advised.[60] Together they attempted to find Italian columns for Syon, and it was for this commission that James was in contact with Antonio Vinelli, mentioned on p 68. Among the commissions for Northumberland that James coordinated were a relief of the *Three Graces* after an antiquity in Palazzo Mattei by the French sculptor Luc-François Breton, works by the sculptors Filippo della Valle and Francis Harwood, the search for mosaic table-tops, and various casts and antiquities. The Adams' extensive works for the

Northumberlands are to be discussed in detail by Adriano Aymonino in his forthcoming book on the couple as collectors, and so will not be addressed here, but one of the commissions is significant for its wider contribution to our understanding of the brothers as sculpture dealers.[61] In a letter to Lord Northumberland's agent in August 1761, James listed the sculptures intended for the niches in Hugh Smithson's dining room at Syon, among them the 'Cerere Borghese and Muse in the Capitol by Cavaceppi'. This is the only direct reference to Cavaceppi in the brothers' surviving correspondence, but I want to suggest that there is considerable evidence that the Adams worked closely with him – and it is a relationship I want to consider in some detail.[62]

As we have already seen, Cavaceppi's vast workshop on via del Babuino was one of the centres of the antiquarian trade in the city.[63] His name has become synonymous with the elaborate restorations that characterise 'grand tour' antiquities, a status earned, in part, from the close relationship he fostered with British dealers and excavators.[64] Cavaceppi appreciated the potential of the British market from the early 1750s; we have already seen that Brettingham purchased a large number of sculptures for Holkham and Petworth from him. Plates from Cavaceppi's three-volume *Raccolta d'antiche statue*, published between 1768 and 1772, commemorate the antiquities he restored and the collections he helped form, demonstrating the scale of his operation.[65] Among the sculptures celebrated in the first volume is the statue of *Ceres*, the 'Cerere Borghese' mentioned by James in his letter to Northumberland and one of two figures Cavaceppi completed for the dining room at Syon. Also among the plates in this first volume is the *Boxer* mentioned by Jonathan Lipyeatt as having been sold by James to Lord Shelburne, confirming that the statue was restored by Cavaceppi, presumably under the supervision of James, before being dispatched to London (*see* Fig 2.3).

A number of other plates in the first volume of the *Raccolta* are identifiable as antique sculptures which Cavaceppi restored for the Adams and which they retained in London until the 1773 auction. These include a sculpture of *Silenus* (Fig 2.4), discussed on p 74, a statue of a *Muse* and a statue of a *Consul*.[66] Cavaceppi usually proudly advertised the location of the sculptures

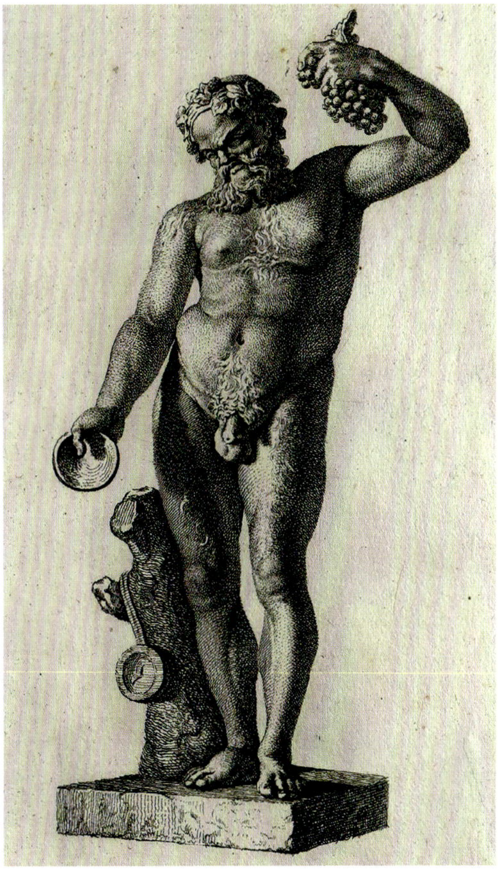

Fig 2.4
Silenus, *Roman, 2nd century AD, at Nostell Priory (far left), and* Sileno, *'esistente in Inghilterra', from Cavaceppi's* Raccolta, *vol I, 1768, plate 16 (left).*

he worked on, whereas each of these sculptures is simply captioned 'In Inghilterra'. The anonymity of the Adam marbles is therefore notable. It raises the possibility that other plates from the *Raccolta* labelled as simply being 'In Inghilterra' reproduce marbles handled by the Adams. Further evidence of the brothers' access to Cavaceppi's workshop can be found among the drawings they acquired while in Rome.

One of the activities central to both brothers' periods in Italy was the commissioning of drawings, a large body of which survive in Sir John Soane's Museum. The finished sheets became a major constituent of the 'material' both James and Robert were gathering to fuel their new language of ornament. Among the more characteristic studies by the artists who formed their circle in Rome – Laurent Pécheux, Charles-Louis Clérisseau and Antonio Zucchi – there are a number of sheets depicting antiquities, principally architectural fragments, altars and cinerary urns. These sheets are more 'archaeological' in approach and are largely attributed to the French artist Nicolas François Daniel Lhuillier. Lhuillier's drawings show sculptures from famous Roman collections, some inscribed with the location of the sculpture being copied, revealing he had access to a roster of patrician collections. Lhuillier drew antiquities at the Villa Albani, Palazzo Mattei and Palazzo Barberini – all collections mentioned by James in his memorandum 'Good Antiquitys yet to be procur'd at Rome'.[67] Lhuillier also drew antiquities belonging to Cavaceppi. For example, a highly finished red-chalk drawing of a section of a Roman architectural frieze depicting a panther, a putto and a torchère, which was published by Cavaceppi in the third volume of his Raccolta, is listed simply as 'ornati in Inghilterra' (Fig 2.5).[68]

Fig 2.5
Nicolas François Daniel Lhuillier (attrib), Relief with Putto, Candelabra and Panther (below)*, and 'Ornati in Inghilterra' from Cavaceppi's* Raccolta*, vol III, 1772, plate 39 (bottom).*

Examining Lhuillier's drawing from the perspective of someone interested in the 18th-century Roman antiquities trade, it is striking how much it looks like the kind of sales drawings produced by Roman dealers to tempt clients purchasing sculpture by correspondence. Much of the grand tour trade in antiquities was conducted through letters and drawings, and the practice of recording antiquities for potential purchase was key to the business of most dealers.[69] The fact that Lhuillier made four meticulous drawings of the fragment which Cavaceppi subsequently published as being 'in Inghilterra' raises the possibility that the sculpture in fact belonged to the Adams. It is therefore tempting to speculate that Lhuillier's drawings may have been made pre-purchase to be sent back to Robert for his approval.

Working through the third volume of Cavaceppi's *Raccolta*, a number of plates depicting architectural elements which are listed by Cavaceppi as being in England are also represented by drawings by Lhuillier in the Adam volumes. These include a relief of putti holding a garland surmounted by a helmet and a relief with two eagles holding a garland, surmounted by a mask (Fig 2.6).[70] More work needs to be done, but Cavaceppi's *Raccolta* offers potentially fascinating evidence for the brothers' substantial stock of decorative and architectural antiquities, as well as illustrating one of their key commercial relationships with a leading Roman sculptor and dealer.

Lhuillier's drawings in the Adam volumes offer a further link that deserves exploration. Several of the highly finished red-chalk drawings attributed to Lhuillier appear to be counter-proofs of drawings which have recently been identified in an album preserved in the

Staatliche Kunsthalle in Karlsruhe.[71] The albums contain a large number of studies associated with Piranesi and his studio, suggesting that Lhuillier may well have worked for both James and Piranesi and that there may have been some form of coordination in the commissioning of drawings. The realisation that there is an overlap between the Adams' drawings and collection, Cavaceppi's publication and Piranesi's archive also points to a powerful axis between three of the pioneers of European neoclassicism.

The stock in London and the 1773 sale

Some of the brothers' stock seems to have remained in Rome after James's departure in May 1763. Left in the hands of other Italian dealers – possibly Cavaceppi – pieces were gradually sold over the following decade. When Gavin Hamilton was commissioned to supply sculptures for Lord Shelburne's garden, he sent a bas-relief of the *Rape of Proserpine* and a 'sepuchral stone', which he noted had belonged to the Adams.[72] The great collector Charles Townley was in Rome in 1773 and bought a number of small antiquities from Thomas Jenkins, which he listed as having originally belonged to the brothers. Now in the British Museum, Townley's purchases include a bust of Minerva and two portraits of young boys.[73] This suggests that, in line with other dealers of the period, Robert and James assembled a large *magazzino* of antiquities and marble fragments which could be either restored or traded with sculptors, and that this stock was still being dispersed a decade after their departure from Rome.

Meanwhile the sculptures that did return to London were used, in part, to decorate the Adam residence and offices in Lower Grosvenor Street. Preserved in Sir John Soane's Museum there is a famous and frequently reproduced sketch, labelled: 'the manner of placing the antient marbles under the Room in my area'.[74] This study underscores the dual purpose of the stock: to act simultaneously as an advertisement for the Adam style and the brothers' ability to supply the ancient sculptures necessary to achieve the look.

Robert's purchase of paintings had been specifically speculative, and shortly after James returned from Italy the brothers organised an auction of their stock. The three-day sale was held at Prestage's in Savile Row in February 1765 and included 218 paintings. This was a well-established strategy: artists who had spent a prolonged period in Italy frequently returned with a collection specifically selected to sell well on the London market. The group consisted largely of landscapes attributed to popular figures – for example Claude, l'Orizzonte (Jan Frans van Bloemen), Salvator Rosa; also modish contemporary painters, such as Zuccarelli, Canaletto and Sebastiano Ricci; and also a few classic 'old masters'. Only one partially priced catalogue survives, but it suggests that the enterprise was a moderate success.[75] At the end of the third day was a *St Catherine* by Guido Reni described as 'very capital'. This was a picture Robert had described in hyperbolic terms to his sister Betty: 'I value it at a great price, at least 200 guineas, though I had it for much less money, not having cost me £20.'[76] The painting sold for 55 guineas; though Robert may not have achieved the profit he was hoping for, he more than doubled his money.

It was the Adams' financial difficulties, precipitated by the Adelphi crisis, that forced the brothers to disperse much of their stock of antiquities at auction in London and by lottery.[77]

Fig 2.6
Nicolas François Daniel Lhuillier (attrib), Relief with Putti (top)*, and another 'Ornati in Inghilterra' from Cavaceppi's* Raccolta*, vol III, 1772, plate 34 (above).*

At least one fully priced copy of the Christie's sale catalogue survives, giving a full list of the purchasers, thus offering a remarkable insight into the brothers' commercial activities, the success of their strategy and the calibre of their stock.

The sale began with more paintings. A number are identifiable as unsold from the earlier Prestage sale, and many of the lots were works by the brothers' mentors, Clérisseau and Pécheux. Others showed evidence of the brothers' grand tour activities: the final lot of the second day, for example, was a series of drawings by the great 17th-century antiquarian and draughtsman Pietro Santi Bartoli of Trajan's Column in Rome, a spectacular album which James had acquired as part of the Albani drawings.[78] Having a record of the purchasers' names has made it possible to trace a number of the paintings in the sale. For example, four pictures acquired for £6 16s 6d by Sir Justinian Isham survive at Lamport Hall in Northamptonshire: these include a pair of paintings by the Dutch artist Jan van Bijlert depicting an old man and old woman with a glass and bottle. Isham also acquired a pair of canvases by Sebastiano Ricci for £24 3s, also still preserved at Lamport. Some purchasers had direct or indirect links to the brothers: for example, Lord Bute's wife, Mary, is recorded as purchasing a drawing by the Venetian painter Giovanni Battista Piazzetta, and their daughter Lady Mary Lowther acquired a number of works too. Other purchasers were also notable architectural patrons of the Adams: Sir Watkin Williams-Wynn, for example, acquired a pair of seascapes by Vernet, which Robert had acquired at the Ghezzi sale.[79] Adam was in the midst of working on Williams-Wynn's London house, 20 St James's Square. This seems to establish a formula repeated throughout the sale: many of the most expensive lots were acquired by existing clients of the brothers' architectural practice.

Fig 2.7
Francis Harwood, A Pedestal, *Audley End, Essex.*

The penultimate lot of the fifth day of the sale was described as: 'lot. 52 Sylenus holding a bunch of grapes in one hand and a cup in the other, with his Cymbols hung on the trunk of a tree, of Grecian workmanship and exquisitely fine'.[80] The *Silenus*, as we have seen, had been acquired by the brothers from Cavaceppi, who included the sculpture in the first volume of the *Raccolta*, listing it simply as being 'in Inghilterra'.[81] It was acquired at the sale by Sir Rowland Winn for the enormous sum of £210. Robert was in the midst of remodelling Winn's Yorkshire house, Nostell Priory, and it is tempting to speculate that the brothers suggested the purchase to Winn as appropriate furnishing for one of their interiors (*see* Fig 2.4). An early plan for the upper hall at Nostell, for instance, shows four full-length Classical sculptures set into niches, for which Winn may have purchased the *Silenus*.[82] The upper hall plan was amended after 1773, when projected decoration commissioned from Antonio Zucchi was cancelled. The *Silenus* has long been paired with a full-length sculpture of *Artemis*, which Seymour Howard has identified as having been restored by Cavaceppi and which was probably acquired directly from the brothers, as it is not listed in the sale.[83]

During the 1770s the Adams were also engaged at Audley End in Essex, remodelling a suite of rooms for Sir John Gifford Gifford.[84] On the fourth day of the sale, Gifford purchased lot 24, a 'square Pedestal, a copy from the famous Bronze pedestal in the Medicis Gallery at Florence', for the substantial sum of £99 15s. This was a replica of the pedestal of the so-called *Idol*, a Roman copy of a Greek sculpture in the Uffizi, and was an item mentioned by James in his travel diary as one he had commissioned from the sculptor Francis Harwood (Fig 2.7).[85] The purchase, which is recorded in Sir John Gifford Gifford's accounts, was again almost certainly suggested by the Adams.[86] Harwood, as we have already seen, was one of the sculptors the brothers had employed to complete the statue of Diana for the dining room at Syon. It is perhaps therefore notable that Sir John Gifford Gifford went on to commission a series of urns for the Adam interiors at Audley End from Harwood.

Another purchaser at the sale was David Murray, Viscount Stormont, who acquired a terminal herm of *Jupiter Ammon* for £14 14s.[87] Adam was at the time remodelling Kenwood for Stormont's uncle the Earl of Mansfield. Remarkably the bust remains at Kenwood, having been acquired by the Iveagh bequest in the 1940s (Fig 2.8). Again, the purchase of a decorative bust of Jupiter entirely accords with the brothers' designs for the library at Kenwood, where small antiquities are included in the office drawings of the room.[88]

But by far the most prodigious purchaser, in terms of expenditure, was the Duke of Northumberland. The Adams were in the midst of their lengthy work for Northumberland: as well as Syon, the brothers had also been employed to work at Alnwick Castle and Northumberland House in London.[89] Once again, it would seem that Northumberland's purchases were made on the advice of the Adams. On the fourth day of the sale, Northumberland acquired a 'bas-relief representing a combat, perfectly entire, and forming a panel of a proper size for the Tablet of a Chimney, very fine', for £26 5s, which was sold in 2014 from Alnwick Castle. He also picked up full-size casts of the Borghese and Medici vases, described as 'thought to be the only casts of these vases to full size in England', for £19 19s each, and a number of other small antiquities.[90]

It was the Adams' work at Syon that prompted Northumberland's most expensive purchases. The first in the grand sequence of rooms Robert had designed there – the marble hall – had been

Fig 2.8
Terminal bust of Jupiter Ammon, *Roman, 1st century BC, at Kenwood House.*

specifically conceived for the display of sculpture (Fig 2.9). James, in his correspondence, had despaired at being able to acquire full-scale figures for the room. Writing from Rome in August 1761 he noted:

> The consular figures of 7 ft. 6 high, are still more difficult to find, those of Pyrrhus, J. Caesar etc. are immense and never could answer unless they were modelled to the size wanted, which would come high, [Filippo della] Valle says that the model, form and cast of these figures would come to 400 crowns or £100 each figure.[91]

But in the end the Adams were able to supply four large antique standing figures for the hall at Syon. There has been some confusion about their purchase and installation.[92] With the evidence of the priced copy of the Christie's catalogue, it is now possible to establish with certainty that the duke acquired three of the figures at the Adam sale. The first, 'The Empress Livia in the character of Juno in flowing robes, of Grecian workmanship, remarkably fine, from the Pope's collection in the Vatican', appeared at the end of the fourth day and cost Northumberland £246 15s.[93] He bought two more sculptures the following day: 'The Consol Scipio with his Consular robes and a Volume or roll in his hand, in the action of speaking in the Senate House, of Grecian workmanship and exquisitely fine', for

£210, and 'The Emperor Augustus in his Consular Robes in the act of speaking in the Senate, remarkably fine, of Grecian workmanship', also from the pope's collection, for £283, making a total of £740. These three sculptures joined a fourth, full-length sculpture of a *Priestess*, which possibly James and Robert had already supplied to Northumberland. As Seymour Howard has already noted, the sculptures were almost certainly restored by Cavaceppi.[94] The *Priestess* and the *Scipio* were constructed from unrelated antique fragments and then fancifully reconfigured. Despite Cavaceppi's imagination, the sculptures were an undeniably impressive purchase. The so-called statue of *Livia* had been restored as a *Ceres* and had been recorded in the

Fig 2.9
The marble hall at Syon Park.

gardens of Palazzo Cesi on the Janiculum in Rome since the early 16th century.[95] The exportation of a sculpture of this scale, quality and provenance from the Papal States at this date was a very considerable achievement.

There were of course professional purchasers at the sale. The sculptors John Bacon and Joseph Nollekens also made substantial purchases.[96] Then there were the agents and dealers who were undoubtedly working on commission. Robert Ansell, an exporter of pictures, dealer and auctioneer who entered a partnership with James Christie, made a number of purchases, as did Robert Udny, a merchant and collector. Joseph Rose, the plasterer, who had a close working relationship with the Adams, made a number of significant acquisitions. Another big buyer listed as 'Duval' can probably be identified as John Devall (1728–94): he and his father were leading stonemasons and sculptors responsible for many of the decorative chimney pieces used by Robert in his interiors.[97] Devall paid 125 guineas for James's mosaic tables acquired from Cardinal Furietti.[98]

Were the Adam brothers successful?

Before concluding I think it is important to make some assessment of the Adams as dealers. Financially it is relatively difficult to quantify their success. As far as I am aware we only have the confirmed cost price of one antiquity. In 1762 James wrote that he had acquired 'a Venus loosly attir'd & adjusting her hair after bathing' for 300 Roman crowns, or about £75.[99] In a longer letter to his younger brother, William, written on the eve of his departure from Italy, James valued his collection of 'Drawings, Antiquity's medals, gesso's &c' at £1500.[100] While we cannot be sure how much of this material ended up in the 1773 sale, adding up the prices achieved for the antiquities alone we get a total of £3,531 12s 6d. This suggests that the brothers, at the very least, recouped what they had spent in Italy and probably made a respectable profit, even when forced to sell their stock en bloc at public auction. It was certainly one of the most profitable auctions of antique sculpture held in London in the period.

As to their success as agents, I think the most revealing document is the short list of Roman antique collections made by James Adam at the back of one of his Roman notebooks.[101] It was dismissed by Alan Tait as 'disappointingly … cautious' for concentrating on established Roman collections rather than listing sculpture from the excavations run by James's 'antiquarian associates' Gavin Hamilton and James Byres.[102] Apart from the obvious anachronism – Byres was not an active excavator and Hamilton only began his speculative archaeology in around 1770, long after James had left Italy – Tait's remark fails to recognise how prescient and informed the list actually was.[103] Extracting ancient marbles from Rome was incredibly difficult, as the papacy had instituted a complex set of laws to restrict their exportation. The Adams managed to remove large quantities of ancient material from established patrician collections – including those of the Mattei, Montalto-Negroni, Furietti and Passionei – frequently decades before other British dealers. They in turn sold to some of the most important and demanding British collectors of the period. The furnishing of the marble hall at Syon alone is testament to their abilities as merchants.

Fig 2.10
Lansdowne Amazon,
Roman, 1st century AD.

As to the quality of their stock, many of the pieces I have identified were impressive of their type. For example, the so-called *Lansdowne Amazon*, now in the Metropolitan Museum of Art in New York, which was sold by the Adams to Lord Shelburne, is considered one of the best surviving Roman marble copies of the lost Greek bronze original (Fig 2.10).[104] It is true that much of their stock was heavily restored by Cavaceppi, but this accorded with the taste of the period which privileged complete figures over fragments. But the brothers themselves were conscious of shifting trends in connoisseurship. The 1773 catalogue shows that they were familiar with the taxonomic definitions developed in the previous decade by Johann Joachim Winckelmann, whom both brothers would have encountered in Rome in his position as librarian to Cardinal Albani.[105]

I want to go further and suggest that they were at the forefront of taste, not just in Britain but Rome as well. The concentration on architectural elements, decorative reliefs, cinerary urns and altars all directly fed their work as designers. The collection provided a sourcebook of ornament, as well as a reminder of the diversity and profusion of ancient design. It was also one of the earliest collections of these types of antiquities formed in Rome, contemporaneous with Piranesi's *museo* in Palazzo Tomati and long before that formed by Cavaceppi. It was also the largest collection of these types of antiquities to be imported into Britain during the 18th century and would go on to have an enormous impact on later collectors, such as Sir John Soane, who acquired numerous pieces at their 1818 sale.[106]

The Adams were intelligent and innovative merchants whose place in the 18th-century antiquities market deserves reassessment. While more evidence is needed for Robert's active role in the cut and thrust of dealing when he was briefly resident in the city, it is clear that the antiquarian business models he encountered did have a profound effect. It was in Rome that Robert Adam realised that he could act as a neoclassical designer, decorator and, most crucially, supplier of antiquities to a cosmopolitan domestic clientele. James's time in Italy also needs to be reconsidered. For all the unrealised enthusiasm apparent in his letters, there is a very real sense that he was not only at the heart of the antiquities market in Rome, but that he was a shrewd and acute participant. James's acquisition of Cardinal Albani's drawings is evidence of this fact. There is more work to be done to reconstruct the Adam brothers' commercial networks and recover more of their stock. This will undoubtedly in turn offer valuable insight into their activities as architects and designers.

Acknowledgements

I would like to thank the following for their wisdom and insights in navigating the world of the Adams: Eileen Harris, whose peerless scholarship is matched only by her remarkable generosity; Alistair Rowan, Mia Jackson and Jerzy Kierkuć-Bieliński, who all supplied invaluable information; Colin Thom, who has been unfailingly encouraging and patient; but most of all Adriano Aymonino, a great friend and scholar, to whom I dedicate this essay.

Notes

1 For my earlier account of the Adam brothers in Rome, *see* Bignamini *et al* 2010, vol I, 224–6.

2 Bolton 1922, vol II, 324–39; Fleming 1958b, 1959.

3 *See* particularly Tait 1993, 71–87.

4 The dispersal of the Adams' collection is complex. There was an initial, speculative sale of paintings in 1765. Following the Adelphi crash the brothers were forced to sell much of their personal collection and Roman stock at Christie's in 1773. More stock was dispersed by public lottery. *See* Rowan 2007, catalogue no 41. Some remaining Adam property was sold at Christie's on 20–2 May 1818 after the death of their unmarried sisters, and a further sale of Robert Adam's possessions took place there on 9 July 1821.

5 Tait 1993, 81–2.

6 *See* for example Yarker, J and Hornsby, C 2012 'Buying art in Rome in the 1770s', *in* Wilcox, S and Sánchez-Jáuregui, D (eds) 2012 *The English Prize: The Capture of the Westmorland, an Episode of the Grand Tour* (exhibition catalogue). Oxford: Ashmolean Museum, 63–87.

7 This was an observation made in Richard Hayward's list of artists in Rome. He noted a painter called 'Mr Capel' and 'Mr Adams Architect from Scotland these two last Rather as gentlemen Rather than studiants'. *See* Stainton, L 1983 'Hayward's list: British visitors to Rome, 1753–1775'. *Walpole Society* **49** (1983), 10.

8 Quoted in Fleming 1962, 161.

9 *See* Kenworthy-Browne, J 1983 'Matthew Brettingham's Rome account book, 1747–1754'. *Walpole Society* **49** (1983), 37–132.

10 Clerk of Penicuik MSS (NRS, GD18). This correspondence forms the basis of Fleming 1962.
11 NRS, GD18/4768.
12 NRS, GD18/4774. Robert Adam to John Adam, 31 May 1755: 'I was all day yesterday at a sale of Cavalier Gezzi's Drawings, Prints, Pictures & other things But though I lost a whole day amongst them I made no purchases, this was the man famous for his caricatures of which you have seen those Andrew Hay had, he dyed lately above 80 year old.' The mention of Andrew Hay (d 1754) confirms that Robert knew one of the most successful traveller-dealers of the previous generation. George Vertue noted Hay had become 'a great dealer in pictures, marbles busts – books &c. bronzes – had been 6 times in Italy on that account, he walk there once or twice afoot'. *See* Vertue, G (1933–4) *Vertue Note Books: Volume III*, Volume of the Walpole Society 22. Oxford: Printed for the Walpole Society by J Johnson at the University Press, 125.
13 NRS, GD18/4792.
14 Robert describes dining with Hamilton at Villa Lante on the Janiculum, where he was then living (NRS, GD18/4833). The most comprehensive account of Hamilton's activities is Cassidy 2011, but *see also* Cesareo, A 2002 'Gavin Hamilton (1723–1798): "A gentleman of probity, knowledge and real taste"'. *Saggi e Memorie di Storia dell'Arte* **26** (2002), 211–322. For Grant, *see* 'Grant, Peter (1708–1784)', *Oxford Dictionary of National Biography*; Lewis 1961; Ingamells 1997, 420–22; Skinner, B C 1966 *Scots in Italy in the 18th century*. Edinburgh: Scottish National Portrait Gallery, 17–18.
15 Fleming suggests that it was Clérisseau who introduced Robert to Albani (Fleming 1962, 141).
16 The most important and enduring account of Albani and the British remains Lewis 1961, on which this paragraph is largely based.
17 *See* for example Albani's work for one of the Adams' most important patrons, Hugh Smithson, 1st Duke of Northumberland: Wood, J 1999 'Raphael copies and exemplary picture galleries in mid eighteenth-century London'. *Zeitschrift für Kunstgeschichte* **62** (1999), 394–417.
18 Quoted in Fleming 1962, 165.
19 For the construction of Villa Albani and Albani's cultural activities, *see* Debenedetti, E (ed) 1993 *Alessandro Albani patrono delle arti: Architettura, pittura e collezionismo nella Roma del '700*. Studi sul Settecento Romano **9**. Rome: Bonsignori; Bol, P C (ed) 1989–98 *Forschungen zur Villa Albani: Katalog der antiken Bildwerke*, 5 vols. Berlin: Mann.
20 For Albani and the Capitoline *see* Arata, F P 2008 'La diffusione e l'affermazione dei modelli artistici dell'antichità: Il ruolo del Museo Capitolino nella Roma del Settecento', *in* d'Agliano, A and Melegati, L (eds) *Ricordi dell'Antico: Sculture porcellane e arredi all'epoca del Grand Tour* (exhibition catalogue). Rome: Musei Capitolini, 60–71; *also* Mangiafesta, M 2013 'Antichità dei Colli Albani ai Musei Capitolini, dallo scavo al collezionismo'. *Bollettino dei musei comunali di Roma* **26**, 5–20.
21 On the relationship between Adam and Piranesi *see* for example Tait 1984 and Wilton-Ely 1995 and 2006.
22 The dealer Thomas Jenkins noted in 1770: 'The Cavalier's [Piranesi's] house is really the most Curious thing that Ever was Seen, I really Wonder it does not Tumble down, the Land lord with reason has Entered a Protest in Case of Accidents.' *See* Bignamini *et al* 2010, vol II.
23 For Piranesi and his British clients, *see* for example McCarthy, M 1972 'Sir Roger Newdigate and Piranesi'. *Burlington Magazine* **114** (1972), 466–72, and Scott 2004, 104–7.
24 At the end of his stay in Rome in 1788, Goethe records visiting sites he had neglected, 'particularly Cavaceppi's house' on via del Babuino, where he mentions being 'especially delighted with two casts of the heads of the colossal statues on Monte Cavallo', adding that 'at Cavaceppi's they can be seen close up in their whole grandeur and beauty'. *See* Goethe, J W 1994 *Italian Journey* (trans R R Heitner). Princeton, New Jersey: Princeton University Press, 425.
25 Kenworthy-Browne, 'Matthew Brettingham's Rome account book'.
26 Robert wrote angrily to Innes & Clerk: 'There is no living here without money, Barazzi is just now in Germany & perhaps may be in England before he returns to Rome if you should see him in London use him as he deserves that is to say like a Dog.' Bolton 1922, vol II, 318. For Barazzi, *see* Hornsby, C 2009 'Serving the lovers of virtù – Barazzi, Batoni and the British dealers', *in* Barroero, L (ed) *Intorno a Batoni* (proceedings of an international conference held in Rome, 3–4 March 2009). Lucca: Edizioni Fondazione Ragghianti Studi sull'Arte, 49–60.
27 Bolton 1922, vol II, 318–20. Further correspondence between the Adams and Innes & Clerk is in London Metropolitan Archives (hereafter LMA), CLC/B/227/MS03070.
28 'Concerning the manner of my sending my purchases to London, of which I have also got advice from Leghorn and am told that it is surely best to send them by a British bottom that sails with convoy... I shall begin this winter to send the boxes to Leghorn, but order they may keep them there till a very safe and advantageous way casts up to send them by parcels so that if some are taken others may be saved in another vessel that may be more lucky' (LMA, CLC/B/227/MS03070, fol 22).
29 This was the natural role for most travelling artists. *See* for example the activities of the painter James Russell, who was in Rome from 1747 until 1753, in Kelly, J 2012 'Letters from a young painter abroad: James Russell in Rome, 1747–53'. *Walpole Society* **74** (2012), 61–164.
30 Fleming 1959.
31 Robert wrote to Jenny Adam from Florence concerning the purchase: 'I have done in part, & hope to his satisfaction, as to the pictures over doors they are difficult to be found for which reason he can't do better than make old norrie (if yet alive) paint 3 apurpose for the pannels, & pray to the great Earl of Heaven and Earth to prevent the French seizing the 4 I am to send him from this town' (NRS, GD18/4838).
32 NRS, GD18/4797. The Casa Guarnieri was on the via di Malta, on a site that fronted onto the present via Sistina and via Francesco Crispi.

33 NRS, GD18/4853.
34 Fleming 1959, 171.
35 Hamilton in partnership with the engraver and dealer Giovanni Volpato also attempted to purchase antiquities from the Grimani collection, and reported his failure to the collector Charles Townley in 1781. *See* Hamilton to Townley, Rome, 10 January 1781 (BM, Townley archive, TY7/645), and Fleming 1959, 171.
36 For Philipp von Stosch, *see* Lewis 1961, 63–91.
37 NRS, GD18/4878.
38 A notable exception is the eight antique columns installed by Henry Holland in the Sculpture Gallery at Woburn. For their history, *see* Yarker, J and Hornsby, C 2010 'A speculative grand tour excavation: Aubrey Beauclerk, Thomas Brand and Thomas Jenkins at Centocelle'. *British Art Journal* **11** (3), December 2010, 21–9.
39 NRS, GD18/4928.
40 Adam 1831, 173.
41 Ibid, 177.
42 *See* Adam Christie's Sale (hereafter ACS) 1773, 4th day, 1 March, lots 20, 21 and 35, candelabra and a griffin from Cardinal Passionei's collection.
43 NRS, GD18/4921.
44 Fleming 1959.
45 Kansteiner, S 2003 'Ein Apoxyomenos des 5. Jahrhunderts: Überlegungen zu einer von Cavaceppi ergänzten Statue in Los Angeles', *in* Grossman, J B, Podany, J and True, M (eds) *History of Restoration of Ancient Stone Sculptures*. Los Angeles: J Paul Getty Museum, 45–60.
46 NRS, GD18/4926.
47 NRS, GD18/4939.
48 Brettingham recorded that he acquired a pair of mosaic tables from Furietti for Holkham. *See* Kenworthy-Browne, 'Matthew Brettingham's Rome account book', 82; also Angelicoussis, E 2001 *The Holkham Collection of Classical Sculptures*. Mainz am Rhein: Philipp von Zabern, 160–1.
49 NRS, GD18/4954: 'Collection Mattei/ Collection Albani/ Rondanini/ Giustiniani/ Apollo Verospi - Hamilton/ Sleep Faune Barbar - d°/ Obelisk - Campus Martius/ Massimi/ Villa Negroni/ Basrelivos & Pompy palazzo Spada/ Musaicks Furietti & Gabrieli/ Palazzo Lancelotta Vestale/ Friezes Palazzo della Valle/ Collumns portico St Apostoli/ A bachus villa Casole near the Temple of Jaunas. Antique chairs in the Quire St John Lateran/ St P in Vicoli St Agnese.'
50 For the dispersal of the Mattei marbles, *see* Hautecoeur, L 1910 'La vente de la collection Mattei et les origines du Musée Pio-Clémentin'. *Mélanges d'Archéologie et d'Histoire* **30** (1910), 57–75.
51 Albani was a serious collector and dealer in Classical sculpture and it was in this capacity that he was most useful to British travellers. In 1749 he aided Brettingham with the export of sculpture for Lord Leicester. Brettingham complained that Albani was 'so great a dealer himself', purchasing from him the statue of a *Faun* from his own collection that had recently been excavated at Tivoli and restored by Cavaceppi. Albani supplied six sculptures for 1,500 scudi to furnish a purpose-built gallery designed by Giovanni Niccolò Servandoni at Brandenburgh House in Hammersmith, which was owned by the wealthy Whig politician Bubb Doddington. Albani sold his first collection of antiquities to Augustus the Strong of Saxony, his second to the pope for the Musei Capitolini, and part of the third, assembled on the strictest museographic principles, was seized by the French and eventually formed the nucleus of the Munich Glyptothek.
52 For the dispersal of the Rondinini collection, *see* Candilio, D (ed) 2011 *I marmi antichi del Palazzo Rondinini*. Rome: De Luca Editori d'Arte. For the Giustiniani collection, *see* Gallottini, A 1998 *Le sculture della collezione Giustiniani* (2 vols). Rome: L'Erma di Bretschneider. Brettingham had managed to acquire a statue of 'Agrippina' from the Lancellotti collection for Holkham. *See* Barbanera, M and Freccero, A (eds) 2008 *Collezione di antichità di Palazzo Lancellotti ai Coronari: Archeologia, architettura, restauro*. Rome: L'Erma di Bretschneider, 63.
53 For this collection, *see* Barberini, M G 1991 'Villa Peretti-Montalto-Negroni-Massimo alle Terme di Diocleziano: La collezione di sculture'. *Studi sul Settecento Romano* **7** (*Collezionismo e ideologia. Mecenati, artisti e teorici dal classico al neoclassico*), 15–90; Rausa, F 2005 'L'album Montalto e la collezione di sculture antiche di Villa Peretti Montalto'. *Pegasus* **7**, 97–132.
54 In the ACS 1773 sale: lot 30, 'A Cenerario or Cenerary Urn, representing a soldier carried to the funeral pyle, by his companions, Massime Collection and very Fine', as well as lots 32–4 and 39.
55 Rome, Archivio di Stato di Roma, *Camerale I, Diversorum del Camerlengo*, b 676, fol 96r, 26 Sept 1763: 'un Frammento di una Statua giacente rappresentante Endimione riattata ... cinque urnette Cinerarie, un vasetto, un piccolo piedestallo'. Published in Bignamini *et al* 2010, vol I, 233. It is possible that the other licences published under Barazzi's name refer to sculpture belonging to the Adams. For the crisis in Roman patrimony and its impact on the formation of the Museo Pio Clementino, *see* Collins, J 2004 *Papacy and Politics in Eighteenth-Century Rome: Pius VI and the Arts*. Cambridge: Cambridge University Press, 132–92.
56 I am grateful to Adriano Aymonino for pointing out that Barazzi was also responsible for securing an export licence for the bronze cast of the *Dying Gaul* by Luigi Valadaier for the hall at Syon.
57 The ACS 1773 sale, lots 28 and 49. But it is also instructive to consider the names in James's list in the context of the Adams' interest in design and display. An entry such as 'Friezes Palazzo della valle' points to James's interest in strategies for display in Rome. The courtyard of the Palazzo della Valle, with its combination of free-standing statues in niches and antique friezes, offers a possible model for the same elements as reconstituted by the Adams in any number of interiors.
58 Fleming 1958b. The Adams retained six volumes from the dal Pozzo collection which are now in Sir John Soane's Museum. *See* McBurney, H 1989 'The later history of Cassiano dal Pozzo's Museo Cartaceo'. *Burlington Magazine* **131** (1989), 549–53.

59 'Lord Northumberland beg'd I would write you to see if there were any ancient mosaic paving to be found at Rome that would answer for making a hearth as he thinks one of that kind cleverly composed would be pretty for the Drawing room' (NRS, GD18/4926, copy letter, Robert Adam to James Adam, 8 February 1762).

60 It is notable that Warkworth used Barazzi as a banker and it may well have been on the advice of the Adam brothers. For Warkworth, *see* Ingamells 1997, 978–97.

61 *See* Aymonino forthcoming.

62 'The figures of marble fixed on for the Dining room are, The Diana Venetrix of Florence by the English Statuary there. The Flora Farneze and Apollo Medici by Filippo Valle. The Cerere Borghese and Muse in the Capitol by Cavaceppi.' Archives of the Duke of Northumberland, Alnwick Castle, DNM: G/1/2, James Adam to Robert Adam, 14 August 1761.

63 Cavaceppi wrote a brief description of his studio in vol III of his *Raccolta d'antiche statue* of 1772 (see note 65): 'Discorso che spiega il frontespizio del libro' (Prospetto dello studio di scultura ideato da Bartolomeo Cavaceppi Romano), having already illustrated the industrious interior in vol I (*see* Fig 2.2).

64 Much has been written on Cavaceppi recently but the most authoritative account remains: Howard, S 1982 *Bartolomeo Cavaceppi: Eighteenth-Century Restorer*. New York: Garland.

65 Cavaceppi, B 1768–72 *Raccolta d'antiche statue busti bassirilievi ed altre sculture restaurate da Bartolomeo Cavaceppi scultore romano* (3 vols). Rome.

66 ACS, 4th day, 1 March, lots 44 & 46, and 5th day, 2 March, lot 45, appear in: Cavaceppi, *Raccolta*, vol I, plates 20, 15 and 18.

67 SM, Adam vol 26/46, 26/49, 26/59.

68 SM, Adam vol 26/131, 26/132, reproduced in Cavaceppi, *Raccolta*, vol III, plate 39.

69 The most important group of sales drawings is preserved among the papers of Charles Townley in the Department of Greek and Roman Antiquities in the British Museum. For their use, *see* Coltman, V 2009 *Classical Sculpture and the Culture of Collecting in Britain since 1760*. Oxford: Oxford University Press.

70 Cavaceppi, *Raccolta*, vol III, plates 34 and 47.

71 Kabierske, G 2015 'A cache of newly-identified drawings by Piranesi and his studio at the Staatliche Kunsthalle Karlsruhe'. *Master Drawings* **53** (2), 147–78.

72 For Adam's purchases for Shelburne, *see* Cassidy 2011, 356–8. The 'sepuchral stone' was sold at the Lansdowne sale in 1930 and acquired by the Ny Carlsberg Glyptotek, Copenhagen. For the sculpture the brothers sold Shelburne, *see* Bowood House Archives, box AA2.

73 BM, museum nos 1805,0703.242, 1805,0703.111 and 1805,0703.111.

74 SM, Adam vol 54/3/28. This sketch is reproduced in Tait 1993, 83 (fig 65).

75 'Catalogue of the genuine, large and valuable collection of Italian, Flemish, French and Dutch Pictures, belonging to Robert and James Adam ... which will be sold by Auction, by Mr Prestage, at his Great Room, the End of Savile-Row, next Conduit Street,' February 1765. The auctioneer's own copy is preserved in the archives at Christie's, London. I am very grateful to Lynda McLeod, archivist at Christie's, for allowing me to view the sale catalogue.

76 Fleming 1959.

77 The Adam Christie's sale (25–27 Feb and 1–2 March 1773) was followed by the Adelphi Lottery sale, which included the residue of the collection of antiquities. Rowan 2007, catalogue no 41.

78 ACS, 2nd day, 26 February, lot 73, 'From the Collection of the Pope Albani. An Elevation and Section of Trajan's Column at Rome, with Drawings of all the Bas-Reliefs, representing the various military Enterprises of that Great Emperor. These magnificent drawings are the original ones done by Santo Bartoli, with the utmost precision from the Column itself, at the time that Scaffolds were erected to make casts of the whole Sculpture by orders of Louis Quatorze for the use of the French Academy. These drawings were afterwards reduced to a smaller scale, and engraved by Santo Bartoli and form a Volume of the works of that ingenious Antiquary.'

79 ACS, 2nd day, 26 February, lots 38–9, 43–4; 3rd day, 27 February, lots 21, 26, 13.

80 ACS, 5th day, 2 March, lot 52.

81 Cavaceppi, *Raccolta*, vol I, Plate 16.

82 Harris 2001, 196–211. Winn also acquired five full-length casts of an *Antinous*, the *Juno*, *Marcus Aurelius* and *Caius Marius* from the Capitoline and *St Susanna of Fiamingo* from Santa Maria di Loreto (ACS, 4th day, 1 March, lots 1, 6–9).

83 Howard, *Bartolomeo Cavaceppi*.

84 For Adam at Audley End, *see* Williams, J 1966 *Audley End: The Restoration of 1762–1797*. Chelmsford: Essex County Council.

85 Haskell and Penny 1981, 240–1; Adam 1831, 173.

86 Chelmsford, Essex Records Office, D/DBy A31/11: James Ansell, 1773, 'Bot. at Messrs. Adams Sale, 4 Day 24 A square, Pedestal, 99.15.0'. I am very grateful to Mia Jackson for drawing my attention to this reference.

87 ACS, 5th day, 2 March, lot 34.

88 Harris 2001, 180–95.

89 Ibid, 64–103.

90 ACS, 4th day, 1 March, lots 14–15. For the relief, *see* Sotheby's, London, 9 July 2014, lot 16.

91 Northumberland, Alnwick Castle, DNM G/1/2, James Adam to Robert Adam, 14 Aug 1761. I am grateful to Adriano Aymonino for supplying this reference.

92 *See* for example Guilding 2014, 103–6.

93 ACS, 4th day, 1 March, lot 51.

94 Howard, S 1970 'Bartolomeo Cavaceppi and the origins of neo-classic sculpture'. *Art Quarterly* **23** (Summer 1970), 131 n 10.

95 This statue is actually a *Venus*, a 1st-century Roman replica of a Greek original of the 5th century BC. This was sold at Sotheby's, London, on 9 July 2014, lot 17, fetching the record price of over £9.3m. It has been replaced in the hall at Syon by a poor-quality cast. *See* Aymonino forthcoming.

96 ACS, 4th day, 1 March, lots 26, 38 and 5; 5th day, 2nd March, lots 13, 20, 35, 42.

97 I am very grateful to Colin Thom for identifying John Devall. The Devalls made chimney pieces for Kenwood House, Harewood House, Nostell Priory and 20 St James's Square (Sir William Watkin Wynn's London house) – all designed by the Adams.

98 The tables were listed as lot 28 on the 5th day: 'Two Tables of ancient Mosaick, dug up in Adrian's villa near Tivoli, from Cardinal Furietti's Collection and very fine'.

99 NRS, GD18/4929. For an approximate table of currency conversions, *see* Cassidy 2011, vol I, 123.

100 NRS, GD18/4937.

101 NRS, GD18/4954. For James's list *see* note 49 above.

102 Tait 1993, 81.

103 The great age of British-led excavations does not begin until the late 1760s. *See* Bignamini *et al* 2010, vol I, 1–16.

104 It has long been assumed that the *Lansdowne Amazo*n was one excavated at Tor Colombaro by Gavin Hamilton and sold to Shelburne in 1772. But Brendan Cassidy is surely correct to observe that the sculpture was in fact supplied by the Adams but later returned by Shelburne to Hamilton in 1787. *See* Cassidy 2011, vol I, 218–19 n 5.

105 The most expensive lots in the 1773 sale are described as being 'of Grecian workmanship', suggesting that the Adams were conscious of the importance attached by collectors to Greek sculpture.

106 Soane acquired four cinerary urns, a sarcophagus and the torso of a seated muse from the 1818 sale, along with numerous architectural casts, books and drawings. SM, museum nos A44, M841, M777, M837, M438, M471.

3

Context and attribution: Antonio Zucchi's portrait of James Adam, 1763

Jerzy J Kierkuć-Bieliński

The role of Antonio Pietro Zucchi (1726–95) as a draughtsman and decorative painter in the office of Robert and James Adam is relatively well documented, and so he has been included in many studies and publications in relation to the Adams and their architecture.[1] One of the earliest was the first volume of the Adam brothers' own *The Works in Architecture of Robert and James Adam*, of 1773, where, in the section detailing Lord Mansfield's suburban villa at Kenwood, we learn that: 'The paintings [at Kenwood] are elegantly performed, by Mr. Antonio Zucchi, a Venetian painter of great eminence.'[2] Following the lead set by *The Works in Architecture*, subsequent writers have treated his *oeuvre* mainly in terms of its place in architectural history. Zucchi's paintings, physically and programmatically, were intended to be read within the broader context of Robert and James's work as architects and designers, and it is in this manner that they have generally been approached. Yet, Robert Adam's description of Zucchi suggests that he set some store by the artist's status – or at the very least felt that potential clients to whom the publication was in part addressed would do so. Either way – and perhaps because he has been regarded essentially as a decorative painter for the Adams – Zucchi's career as a fine artist has been obscured for art historians. With comparatively few known works outside the Adam *oeuvre*, he is not so easy to engage with as an artist as he is for architectural historians to understand as a decorator.[3] Perhaps this reflects a certain divide between art history and architectural history – though they are, of course, sister disciplines. However, the recent reattribution of a significant portrait of James Adam (Fig 3.1) to the Venetian's hand should offer some guidance as to how we might approach Zucchi in future if we wish to assess his qualities and career as a gifted painter in his own right. It should be noted that, to date, this is the only portrait attributed to the artist. However, a number of sketches also now attributed to Zucchi, such as *An English Family Group, Four Connoisseurs Seated at a Table* (*see* Fig 3.8) or *The Vendor of Silk Stockings*, tantalisingly point to his possible engagement in that genre at some length.[4]

Antonio Zucchi: *pittore dà Venezia*

Zucchi's career, as indeed implied by Robert Adam, did not begin when he came to England in 1766 at the invitation of the Scottish architect. In fact, a little-known biography of the artist was included by Alessandro Longhi (1733–1813) in his 1762 publication *Compendio delle vite de' pittori veneziani istorici* (Fig 3.2).[5] This placed Zucchi beside 24 other esteemed contemporary Venetian painters, including well-known figures such as Giovanni Battista Pittoni, Sebastiano Ricci, Giovanni Baptista Piazzetta, Giovanni Battista Tiepolo and Guiseppe Nogari, as well as lesser-known painters like Gaspare Diziani. Alongside these, Longhi included the two masters under whom Zucchi had trained – Jacopo Amigoni (1682–1752) and Francesco Fontebasso (1707–69). Longhi's biography of Zucchi comprised little more than a paragraph but it does mention the fact that he came from an artistic family, and it also indicates the type of paintings that initially drew attention to him:

> Antonio Zucchi nacque in Venezia del 1726 da Franco. Celebre Incisor in rame, che conoscendo esser bene che in una famiglia vi siano varie applicazioni, da Economo prudente le suggeri la Pittura, e fu tosto dal Figlio abbrac-

ciato il suo volere; e postosi a disegnare, si scoprì d'un abilità tutta spirito; e considerando esser necessario al Pittore lo studio d'Architettura, e Prospettiva, vi s'applicò con tutti' il fervore; e sotto la direzione di Fontebasso si mise a dipingere; ma poi ottenuta l'amicizia del nostro rinomato Amigoni, e l'ingresso nella di lui Scuola, con ogni diligenza cercò imitare quel suo vago colorire, come s' osserva in S. Giobbe di Venezia in una Tavola d'Altare con S. Bonaventura, S. Pietro d'Alcantara, ed altre Ste Francescane, dipinta con tant' arte, che s'acquistò nome di valente Pittore. Viaggia al presente in compagnia d'un celebre Archittetto Francese, e d'un dilettante Inglese, par osservar, nelle Città più famose d'Italia, come Roma, Napoli, ed altre, ciò che vi si trova di più raro si antico, come moderno, tanto d'Architettura, Scoltura, e Pittura, quanto altri famosi avanzi d'Antichità.[6]

Antonio's family were known as printmakers. His father, Francesco Zucchi (1692–1764), one of several artistic brothers, was primarily an engraver of topographical views, including of Venice. He also seems to have had some connection with the British architectural world, having produced an allegorical print showing Dorothy Boyle, Countess of Burlington, after a design by William Kent.[7] On at least one occasion, Antonio collaborated with his father, Francesco, on a publication of prints, entitled *Varie fabriche antiche e moderne della città di Verona* (1753), contributing four plates showing the portrait busts of the Roman emperors and antiquities, and after drawings by Giovanni Battista Tiepolo (Fig 3.3).[8] However, it is clear that printmaking was not the mainstay of Antonio's career, unlike the other members of his family. Antonio's uncles on his father's side were particularly distinguished in that profession. Andrea Zucchi (1679–1740) shared his younger brother's interest in topographical views; he moved as a scene painter in 1726 to Dresden, where he would die. Lorenzo Zucchi (1704–77) joined his father, Andrea, in Dresden, where he was appointed court engraver in 1738 to Augustus III, King of Poland, Grand Duke of Lithuania and Elector of Saxony, and was subsequently made professor of engraving at the Academy there in 1764. Notably, Francesco's other brother, Carlo Zucchi (1682–1787), would be remembered for teaching perspective (and possibly etching) to the towering figure in the history of 18th-century Italian printmaking, Giovanni Battista Piranesi. Antonio's older brother – Giuseppe Carlo Zucchi (1721–1805) – was also a printmaker; in 1766 he would join his younger brother in London, where, among other works, he produced the engravings of the Great Room or library at Kenwood for the Adam brothers' *The Works in Architecture*.[9] In the

Fig 3.1 (opposite)
Antonio Zucchi, Portrait of James Adam, *oil on canvas, 1763.*

Fig 3.2 (left)
Alessandro Longhi, Antonio Zucchi, *etching, 1762. From* Compendio delle vite de' pittori veneziani istorici, *plate 21.*

Fig 3.3 (below)
Antonio Zucchi Antinous *(left) and* Bacchus *(right), etchings from* Varie fabriche antiche e moderne della città di Verona, *1753, plate 22.*

preface to the section dealing with Kenwood, Antonio is, of course, mentioned as the author of the decorative wall and ceiling paintings for the very same room that his brother Giuseppe would record. Given that so many Zucchi males were simultaneously involved in printmaking in Venice, it is not surprising that later architectural and art historians have had difficulty identifying which member of the family was responsible for specific printmaking commissions.[10]

It is clear from Longhi's account that Antonio Zucchi's work in Venice consisted mainly of religious paintings, such as the 'Stations of the Cross' for the church of San Giobbe, Venice, which he undertook around 1750 (Fig 3.4). It is also clear that he had absorbed the style and characteristic subject matter of his two earliest teachers, Fontebasso and Amigoni.[11] Although we are faced with a slim body of work by Zucchi, particularly for the earlier part of his career, his better-documented commissions for the Adams show how closely, at least at the start of his association with them, his work was still dependent upon Amigoni's late baroque style in terms of colour and composition. Zucchi's use of a light, pastel palette and 'dynamic' forms in his ceiling painting of *Hercules between Glory and the Passions,* executed in 1769 for the Great Room or library at Kenwood, is comparable in handling to Amigoni's *Crowning of Homer*, of 1730–9 (Figs 3.5 and 3.6).[12] Under Amigoni and Fontebasso, Zucchi would also have been taught portrait painting, as both artists worked in that genre. Given that the 1763 *Portrait of James Adam* is the only known such work by Zucchi, it is difficult to evaluate his approach to portraiture. Yet, a comparison with Fontebasso's luxuriant depiction of textiles, furs and details of jewels in his *Portrait of F Tadini* (1761), now in the State Hermitage at St Petersburg, perhaps indicates areas of artistic interest that were passed from master to pupil.

Fig 3.4
Antonio Zucchi, Christ Carrying the Cross, *or* Christ Encounters Veronica; *one of Zucchi's 'Stations of the Cross' series in the Church of San Giobbe, Venice. Oil on canvas, c 1750.*

Amigoni's role as a portraitist may also have suggested to Zucchi a future career path. From 1729 to 1739 Amigoni was resident in England, where, among other things, he found patronage as a portraitist to Queen Caroline (d 1737), Frederick, Prince of Wales (d 1751) and Anne, Princess of Orange (d 1759). Although we do not have precise dates for Zucchi's training, we know that by 1747 Amigoni was back in Venice (when Zucchi would have been 21). Coincidentally, while he was in England, Amigoni also executed a series of decorative paintings for the hall at Moor Park (where Robert Adam would later work in the 1760s), showing the *Story of Io,* dating to c 1732. Here Amigoni made use of a specific technique, painting not in fresco but on canvases, which were mounted on the walls of the hall in architectonic frames.[13] Zucchi would use the same technique when he came to Britain to work as a decorative artist for the Adam brothers, executing his paintings on canvas or paper to fit into the architectural framework of an Adam-designed room. That Zucchi had already undertaken decorative painting for architectural schemes before his work for the Adam office is noted in James Adam's grand tour journal, where on 1 October 1760 he records a visit to the house of a noble Venetian at Massinsaga, where Zucchi had painted the ceiling frescoes, 'and very well'.[14]

Jacopo Amigoni was Antonio Zucchi's first connection with England, the country where he would make his home and pursue his career for some 15 years. Possibly it was Amigoni who first drew Zucchi's attention to Britain as a suitable place for the artist to find patronage and acclaim – though it should be remembered that his uncle Andrea and his cousin Lorenzo had also shown the possibilities travel offered to the career prospects of a Venetian artist if he were willing to relocate abroad.

From other surviving works by Zucchi it is evident that he also came under the influence of other significant Italian and Venetian painters. Notably, the four large canvases with capriccios that Zucchi executed in 1771 for Robert Adam's music room at Harewood (*see* Fig 0.10), are clearly comparable to the work of Giovanni Paolo Panini (1691–1765), such as his 1741 painting *A Capriccio of the Roman Forum* (now at Yale University Art Gallery).[15] Indeed, two capriccios showing *Ruins with Figures* at the Fitzwilliam Museum in Cambridge (nos 119 and 204) have been ascribed to both Panini and Zucchi in the past.[16] Our lack of a full understanding of Zucchi's *oeuvre* also makes it difficult to assess

Fig 3.5 (above)
Antonio Zucchi, Hercules between Glory and the Passions, *oil on paper laid on plaster, 1769, Kenwood House.*

Fig 3.6
Jacopo Amigoni, The Crowning of Homer, *1730–9, oil on canvas, private collection.*

his work in relation to his fellow Venetian painter the great *vedutiste* Antonio Canaletto. Two paintings now attributed to the Circle of Canaletto, again in the Fitzwilliam Museum, Cambridge – *Ruins with Figures* and its companion canvas of the same subject – highlight this anomaly. In the 1902 catalogue of the Fitzwilliam's collections by F R Earp they are attributed to Zucchi.[17]

Zucchi's ability to absorb the styles and approaches of others and adapt them to his own needs is also evident in his drawings. But again, the problems of fully understanding Zucchi as a draughtsman hamper our ability to securely attribute drawings to his hand. In his *Italienische Reise,* Johann Wolfgang von Goethe reminisced about his friendship with the painter Angelica Kauffmann (1741–1807) and Zucchi, the husband she married in 1781. Goethe wrote on the 15 February 1787 that Zucchi:

> arbeitete indes an einer großen architektonischen Zeichnung, die er in Dekorationsart vortrefflich zu machen versteht. Er war mit Clérisseau in Dalmatien, hatte sich überhaupt mit ihm assoziiert, zeichnete die Figuren zu den Gebäuden und Ruinen, die jener herausgab, und lernte dabei so viel Perspektive und Effekt, daß er sich in seinen alten Tagen auf eine würdige Weise auf dem Papier damit vergnügen kann.[18]

This reference by Goethe has subsequently been taken to refer to the trip undertaken in 1757 by Robert Adam in the company of the French draughtsman Charles-Louis Clérisseau to draw and survey the ruins of the Emperor Diocletian's Palace at Spalato (modern-day Split) for subsequent publication.[19] Although this assertion merits re-examination, being based on Goethe's remembered conversations with Zucchi about his youth, what is certain is the close artistic relationship between Zucchi and Clérisseau, which has also led to problems of attribution. That Zucchi provided the staffage for certain capriccios by Clérisseau can be seen in a gouache signed by both men showing *Roman Ruins with a Pyramid* (c 1766, private collection).[20] The palette of pastel blue and pink robes on the figures seated among the ruins, and the oblique line used to describe the eyes of the figures in profile, are similar in the well-known decorative paintings undertaken by Zucchi for the Great Room or library at Kenwood in 1769. Also, Zucchi was learning to

Fig 3.7
Charles-Louis Clérisseau, Temple of the Sibyl, Tivoli, *(1750s or 1760s) formerly attributed to Antonio Zucchi.*

adapt Clérisseau's approach to depicting antiquities, in particular his use of ochre tones to describe decayed travertine contrasted against a blue sky. Such a technique can be seen in another drawing, of a recognisable ruin, *The Temple of the Sibyl, Tivoli* (British Museum; Fig 3.7).[21] Previously attributed to Zucchi, this is now thought to be by the French draughtsman. Even the better-documented period of Zucchi's career, which encompassed his work for the Adam office from 1766 to 1781, still poses questions. A sketchbook in the Henry E Huntington Library and Art Gallery, for example – generally thought to be by Zucchi because it contains several drawings relating to his schemes of decorative painting for the Adams at Saltram House, Osterley Park, David Garrick's house at the Adelphi and Harewood House, among others – has, on the basis of one watercolour, also been attributed to Francesco Bartolozzi (1727–1815).[22]

However, misattribution of Zucchi's work most commonly arises in relation to drawings produced by his Swiss-born wife Angelica Kauffmann, under whose influence Zucchi's style moved away from a late baroque sensibility towards an international neoclassicism. A recent example of this question of attribution can be seen in the pen and wash drawing showing a *Design for a Painting depicting Fame and some Genii Presenting to Great Britain the Portraits of Alfred the Great and Queen Elizabeth,* which was a presentation drawing for the overmantel of the chimney piece in the Adam-designed library at Home House, Portman Square, London. Although it relates to one of the Adam office's most significant projects, little was known about the history of this drawing and, having entered the collection of William Mayor, it was attributed to Kauffmann.[23] Similarly, a *Study for the Ceiling of the Ballroom, Home House showing Virgil Reading the Aeneid* has also been given to both the Venetian painter and his wife.[24]

The *Portrait of James Adam*: a question of attribution

As with the Fitzwilliam 'Canalettos' and the drawings Zucchi produced following the style of Clérisseau or Kauffmann, so too the question of the attribution of the *Portrait of James Adam* has been problematic. It resurfaced at a sale at Christie's on 5 December 1863, where it appeared as lot 67, described as 'Pompeio Battoni [*sic*] *Portrait of Sir William Chambers* – admirably painted (31 gns. to Anthony)'.[25] Four years later, in 1867, the portrait was included in the *National Portraits* exhibition at South Kensington, London, where the sitter was now identified as being Robert Adam.[26] The attribution to Batoni rested in part on its ambitious qualities as a grand tourist portrait, but also (more persuasively to some) on the inclusion of a specific antiquity in the background, known to be a favourite motif of Batoni (*see* p 95). Subsequently, the sitter was finally recognised as James Adam. There are obvious similarities between Zucchi's portrait and the only other *Portrait of James Adam*, painted by Allan Ramsay in *c* 1754 (now in the Laing Art Gallery; *see* Fig 0.3); but the inclusion of a capital of James Adam's own British Order is a clear visual clue to the sitter's identity (discussed at further length on p 98, and *see* Fig 3.15). Nevertheless, the attribution to Batoni persisted, even though in the 1980s an alternative suggestion was made that the work might be by the Scottish painter George Willison (1741–97), who was in Rome at the same time as James Adam in 1760–7.[27] Only now has recent cleaning shown unequivocally that we are dealing with the work of Antonio Zucchi: the compelling evidence is the artist's partial signature and an inscription 'Ant[...] Zu[...] Fecit Roma Anno 1763', running prominently along the top edge of the classical altar upon which James rests his elbow.

Antonio Zucchi and Robert Adam's *Ruins of the Emperor Diocletian's Palace at Spalatro*

In certain respects Zucchi's *Portrait of James Adam* falls between the ambiguous early period of the painter's career and his far better documented later career as a decorative artist in Britain. Yet, the question of how Zucchi first entered the Adam circle is not so straightforward as previously thought. It has always been assumed that he first came into contact with the brothers through a project that Robert Adam began during his grand tour of Italy in the 1750s. Antonio Zucchi has sometimes been identified as one of the engravers of the plates for Adam's *Ruins of the Emperor Diocletian's Palace at Spalatro*, published in 1764.[28] Goethe seems to have confirmed this.[29] Furthermore, the pioneering German classical art historian Johann Joachim Winckelmann (1717–68) claimed that James Adam was later in Rome in the company of 'un bravissimo intagliatore Veneziano'.[30]

An examination of these plates reveals that 21 are signed: 'Zucchi Sculpt'. Furthermore, Antonio is listed as one of the foreign subscribers to the publication, where he appears as: 'Il Sign. Antonio Zucchi pittore dà Venezia'. But it should be noted he is described there as a *pittore*, not as a *celebre incisore in rame*, as his father, Francesco, had been in Longhi's biography of Antonio.[31] In fact, Longhi is silent on Antonio's printmaking.

Turning to the other major publication to emerge from the Adam office, *The Works in Architecture* (1773), Antonio Zucchi's name once again makes an appearance. In this instance it graces the frontispiece of *A Student Conducted to Minerva,* which was engraved by Bartolozzi after a design by Zucchi. But if Zucchi, the 'Venetian painter of great eminence', was also an adept printmaker, then it would seem odd that Adam did not employ him to engrave some of the plates for this important volume – especially his own work. Also, there would then be no need to employ Giuseppe (Joseph) Zucchi for this task. But the truth is that Antonio Zucchi was not really a printmaker at all. In a little-known manuscript study of Venetian engravers, Gianantonio Moschini (1773–1840) covers the Zucchi family's printmaking, including the works of Francesco, Carlo, Lorenzo, Andrea, Giuseppe and Antonio. Basing his account of their careers on the description in Guiseppe Zucchi's *Memorie chronologiche della famiglia Zucchi,* written in Venice in 1786, he provides the answer: it was Giuseppe who had been employed by Robert Adam to engrave plates for the *Ruins*, not Antonio:

> Egli [Giuseppe Carlo] aveva anche apparata la prospettiva e l'architettura da Carlo [Zucchi], suo zio, quando venne chiamato dal fratello Antonio in Londra per intagliarvi le antichità di Spalato, che là si voleano pubblicate da' tre fratelli Adam. E fu questa la principale opera intagliata, che ci restò di Giuseppe.[32]

Antonio is barely mentioned as a printmaker in Moschini's account. It now seems certain that it was Giuseppe who was also responsible for the plates in *The Works in Architecture,* and that he came to London with his brother to carry out this work because of an existing relationship with the Adam office through his engagement on the *Ruins* folio.[33] Conclusive though this new evidence is about the Adam engravings, it does not, however, explain how and when James Adam met the man who not only painted his portrait but who, once in England, would contribute so much to the success of his and his brother Robert's 'true style of the antique'.

Collecting paintings: Antonio Zucchi and James Adam in Venice

James Adam followed his brother Robert's example in undertaking a grand tour of Italy, arriving in Venice on 25 June 1760 in the company of the talented Adam draughtsman George Richardson (*c* 1738–1813), whom the pair had somehow persuaded their brother John in Edinburgh to part with, possibly in exchange for one of their less satisfactory assistants in London.[34] Jamie's tour is partially documented in his surviving letters to his brother Bob and other family members (in the Clerk of Penicuik papers in the National Records of Scotland), and also in a journal he wrote covering mainly the period 1760–2.[35] On their arrival, the visitors were greeted by one of Robert's close contacts from his earlier tour: Charles-Louis Clérisseau (whom Goethe claimed had been accompanied by Antonio Zucchi in Dalmatia). In a letter to Robert, Clérisseau wrote of James: 'I knew him immediately from your description but he is infinitely younger looking than I imagined ... one of the most obliging creatures I ever saw.' James found suitable accommodation, taking a 'spacious' apartment from James Grant, the nephew of Lord Deskford, and from a Mr Wynn, which James described as being 'chez la Mingotte' (a Signora Mingotte was the landlady). His description of the lodgings also indicates how he had begun to acclimatise to his stay in the Serene Republic. The apartment was:

> large and *bien meublé* at £6 a month, with seven great rooms ... Clérisseau and I dine and sup in a portico about 60 feet [18m] long which is cool and agreeable, and this settled manner of life and custom begins by degrees to reconcile me to Venice. I pay visits, see churches and palaces and study the language hard. This last I feel the want of greatly but have no remedy but patience.[36]

Zucchi's sketch of *Four Connoisseurs Seated at a Table* (Fig 3.8) calls to mind this image of growing ease and intellectual curiosity delineated by Jamie in his letter to Bob.

It was not just 'supping' that occupied James Adam and Clérisseau while in Venice. Ostensibly, James was there to learn, and he began to draw the antiquities that were embedded in the walls of the Doge's Palace. He also undertook expeditions to the *Terra Firma*, for example measuring the arena in Verona and correcting the earlier mistakes made by the French architect Desgodetz when he had surveyed the ruins there. He also met several significant collectors of antiquities while in Verona, such as Count Pozzo and Monsignor Torelli.[37]

But James was also in Italy to collect. In a letter of 2 July 1760 to his brother Robert, he writes:

> Pictures I can supply with the utmost facility they abound at a very cheap rate. All the delectantes, I find, were gaping for my arrival as a purchaser, but as everything here must be done by stratagem, I have told them all that at present I am resolved to make no purchases, that I have hitherto sufficient knowledge to do it (which by the by is very true), but that against my return from Rome I shall probably lay out some money in that way. In the meantime Clérisseau and I have our eyes upon what we wish to have and he employs a third person to purchase as for himself. And thus I *évite* the high prices they would charge to me as an Englishman.[38]

James in his letter to Robert implies that the 'third person', initially unnamed, who is employed to purchase paintings on his and Clérisseau's behalf, is an Italian (if not a local Venetian). As a painter connected with a prominent family of Venetian printmakers, and an artist who had already undertaken work that drew upon the wealth of Venice's artistic heritage, Antonio Zucchi would have been well placed to act as a local go-between for the Scotsman and Frenchman. That Zucchi had impressed James is made clear in a letter to his younger brother, Willie, dated 8 January 1763, where James describes the Venetian as 'a worthy honest lad, a most singular character in this degenerate country', and gives a glimpse of the esteem in which James held the 34-year-old artist (hardly a lad!).[39] And it is in the role of an 'honest' and adept adviser on the Venetian art market that Zucchi arguably first made contact with James Adam in Venice in 1760 – not as a printmaker. Zucchi was central in advising James when he

Fig 3.8
Antonio Zucchi, Four Connoisseurs Seated at a Table, *pen and brown ink, brown wash, with white heightening, late 18th century.*

purchased £900 worth of old masters in Padua in August 1760. In fact, as James stated in a letter (dated 13 August 1760), it was Zucchi who alerted him to the fact that a collection of 'upwards of 300 pictures' was available there. It was also Zucchi (along with Clérisseau) who advised James that 'several of the sample [of paintings] we have already seen are by very great masters though a little spoiled', and with James 'took note of the numbers of those that were good and very discretely putting them at about one half the price of the whole collection', which enabled James to make a reduced offer. Thus, through Zucchi's advice, James acquired £900 worth of paintings for only £450, including works by Veronese, Bassano, Ricci, Poussin and Salvator Rosa. (Unfortunately, one work by Correggio was described as 'sick' and it is not clear whether Adam included it in his purchase or not.[40])

James Adam in Rome: Zucchi and the *personaggi della commedia*

As James made his progress through Italy on his grand tour, his entourage grew. After he left Venice, Rome ultimately beckoned. He arrived there on 24 February 1761 and in a letter to Robert described the 'Personaggi della Comedia', as he termed his retainers. With his Italian somewhat improved he described himself as a 'Cavaliere Inglese alter volte archittetto Scozzese', with Clérisseau as the 'segretario al Cavaliere'. Zucchi, of course, formed part of the *personaggi*, which by this stage included five draughtsmen and four servants: George Richardson, Domenico Cunego (1727–1803), Antonio Zucchi, Giuseppe Sacco (fl. 1760s), Agostino Scara (fl. 1760s), a footman, cook, valet and coachman. A little later a violinist called Hay from Bath also joined the *personaggi della commedia*. James clearly admired Zucchi's skills as a draughtsman, describing in a letter to Robert three 'very fine drawings' by Zucchi 'of antique ceilings in what they call Livia's Baths: they are painted so pretty and so rare that I thought I could not have done them with too much care'.[41]

It was Zucchi whom James now turned to with regard to a project he began in 1762–3 during his stay in Rome: his designs for the rebuilding of the Houses of Parliament in London, which had long been considered unfit for purpose. That James wished to tackle such a challenge – how to rehouse the British Parliament satisfactorily at the moment of its 'Augustan' zenith – shows the scale of his ambitions. The problem had frustrated far greater architects, such as Inigo Jones and William Kent. The building that James envisioned would have been lavishly adorned with decorative sculpture on the theme of British history, couched in the style of the 17th century or in a neoclassical idiom. Zucchi had trained in both the late baroque and neoclassical styles, and 66 drawings in his hand for this grand scheme survive in Sir John Soane's Museum. Although other *personaggi della commedia*, such as Agostino Scara, were also involved in the project, it is clear that James favoured Zucchi to realise his vision, as the bulk of the drawings are executed by him in a distinctive palette of brown wash with white heightening. The rich iconographic sculptural programme associated with the Parliament project would also find its way into one more work that Zucchi undertook during this period: the *Portrait of James Adam*.

Antonio Zucchi's portrait of James Adam, 1763

Few letters by James survive from his tour after 1762, and those that do become increasingly less informative about his activities. It is unfortunate that the crucial letter discussing the *Portrait* does not seem to have survived (it is assumed that Jamie must have written to Bob on the matter, given the nature of the painting). Similarly, the last significant entry of Jamie's tour journal ends on 20 January 1762.[42] In his account of Robert and James's activities in Edinburgh and Rome, John Fleming states that James was depicted in the *Portrait* 'in the role of dilettante and prime macaroni'.[43] Yet, Fleming's somewhat laconic description of the painting belies the fact that this is a bravura example of a grand tourist portrait and is indeed a testament to the ambitions of both painter and patron, and to the latter's enhanced status as a leading dealer in Rome. It is furthermore an erudite work of antiquarian scholarship.

The three-quarter-length *Portrait* shows James Adam in a relaxed (dare one say, almost louche) manner, resting his left elbow on an architectural capital (*see* Fig 3.1). In his right hand, which rests on the back of his hip, he holds a pair of architectural dividers; his left hand grasps a loosely rolled sheet of paper. He wears a loose banyan, lined with sable fur, of a very vivid blue silk or velvet; beneath this one can see a rich waistcoat in silver silk with gold embroidery, its top buttons undone to reveal a length of lace cravat about his neck. A florid cuff of lace can be seen beneath the left-hand sleeve of the banyan. Just behind James and to the extreme left is a curule chair with a rich, plump, red velvet tasselled cushion. Zucchi clearly had much delight in describing the textures and rich hues of fur, silks, lace and gold thread (there is an almost calligraphic quality, for example, to the intricate description of the lace of James's shirt cuff). The passages describing the startling blue silk of the banyan also indicate a considerable investment in the raw materials used by the artist in the painting: this vivid shade of blue could only be achieved through the use of ultramarine rather than the less expensive (and less vivid or stable) Prussian blue.[44] However, it is also clear that James Adam took as much delight in wearing the clothes depicted in the *Portrait* as his painter did in describing them. Allan Ramsay's earlier portrait also revealed this to an extent, in Jamie's elaborate cuffs and brocaded waistcoat, but Zucchi's work illustrates very clearly this important aspect of the sitter's personality: his vanity.

That James loved clothes and was a man of fashion is well documented, and Zucchi's portrait calls to mind James's account of a gala night at the opera in Naples, where he wore his 'best French suit', of 'middle season velvet, embroidered with silver, in prodigious good taste'. James described in a letter to his brother Robert how he had called on some English friends just before the opera, who:

> were so struck with my suit that they surrounded me to ask where I got it, who made it &ca. They lamented that no such thing was to be had here and, what was very droll, its fame was so great that next morning I had a levee to see my suit for those who had not seen it the night before. 'Mr Adams', says Lord Warkworth, 'it is the prettiest suit I ever saw. It will make a vast figure in England'. 'I doubt that, my Lord', says I, 'I Believe it will be very little worn there'. So much for the history of a suit of clothes which, I am sure you will allow when you see it, to be the handsomest you ever saw.[45]

But the portrait is intended to show James Adam as much more than just a 'prime macaroni'. It also points to his ambition to be perceived as a serious architect, on a par with his brother Robert, and also as an erudite antiquary. While in Venice and Padua, James was already laying plans for his own unrealised folio publication on the antiquities of Magna Graecia, for which he was to engage a boat and travel to Greece and the Levant in the company of Richardson, Clérisseau and Zucchi. First-hand knowledge of the monuments of Greece, the Aegean and Near East – which Robert Adam had never reached – was important to his plans for the future development of his London practice and was seen by the two brothers as an essential element of James's tour, but for various reasons the trip never took place. Their younger brother William in the end chartered a Danish ship (the *Flying Angel*) for Jamie to use, but it was seized by the Spanish in July 1761 en route to Italy; Jamie afterwards bemoaned to Bob the loss of 'the fruits of Willy's labors' and the 'usefull effects I had aboard'.[46]

In the background and foreground of the painting Zucchi has arranged a number of recognisable antiquities, painted in a monochrome manner that further emphasises the richness of the textiles James wears but also echoes the monochrome drawings in brown ink with white heightening that Zucchi was producing at this time for James's Houses of Parliament project. That James was amassing a collection of antiquities alongside his paintings is also documented in his letters. The collection included an *Apollo* and a *Germanicus*, the latter 'perfectly entire unless in one hand which is restored'.[47] He also considered acquiring two black basalt centaurs from Cardinal Furietti's collection that were valued at 30,000 scudi. That he had an eye on profit is also evident. Jamie hoped that he could recoup the costs of his grand tour by selling the collection back in London; as he explained to Bob, 'no doubt the taste of the times will run on these things'. Indeed, he seems already to have had specific clients in mind when, for example, he entered into negotiations for two large classical urns 'with bas-reliefs on them, not excellent but passable [which would be] ... excessively saleable in London, particularly should Bob have any ruined temples to adjust such as he proposed for Sir Nathaniel's garden'.[48] The antiquities would act as suitable additions to the building projects that the Adam office was undertaking back in Britain.[49]

Yet, James's antiquarian interests should not be taken purely as a means to turn a profit. In a telling letter written to his sisters on the occasion of his birthday, James laments the lack of progress in his career and also his slipping youth.

He does this by taking on the imagined persona of an antiquary and sees his future self as 'like one of your grey-headed patriarchs, [that] wander with my flocks and tents till tempted by some febrile spot where time has spared the crop of antiquities'.[50] For James, one could say that the antiquary formed an aspect of his self-image; in the letter this takes on a negative connotation, yet in the portrait the opposite is true. Zucchi shows his sitter as youthful, debonair and, surrounded by the fruits of his collection, successful.

The most prominent of the antiquities in the *Portrait* is the Medici Vase to the left of the canvas (*see* Fig 3.1). James would have had an opportunity to see this in Rome in the gardens of the Villa Medici, and as a well-known antiquity it was often used as a 'prop' in grand tourist portraits – in 1793 Zucchi's wife, Angelica Kauffmann, for example, painted Thomas Noel-Hill, 2nd Baron Berwick of Attingham, in Rome, while he was on his grand tour, with the vase in the background.[51] However, in this instance, the vase depicted by Zucchi in his *Portrait of James Adam* is not a generic representation but an identifiable copy. By July 1762 James could boast that he had acquired some 500 paintings (worth £2,000), 'besides drawings, antiquities, modellos, gessos &ca valued at £1500 more'.[52] From a later inventory of objects sold on the fourth day of the 1773 sale of Robert and James's collection there is a description of one of these 'gessos', which appears as lot 14: 'The Vase of the Medicis. Representing the sacrifice of Iphigenia, the workmanship of Phidias.' It is this very same plaster cast, which James acquired in Italy, that Zucchi has depicted in the *Portrait*. Brought to England by James when he returned in 1763, it was the first full-scale cast of this important antiquity to enter the country.[53]

Behind James, further antiquities can be discerned, set within an architectural framework of an aedicule, above which runs a frieze of Vitruvian scrolls, flanked by two vertical panels or pilasters adorned with acanthus scrolls. The form of the acanthus scrolls is another specific reference to Jamie's interests and success as an antiquarian and dealer. The tendrils of the acanthus tuck under and over the main scroll in a pronounced 'knot-like' manner. This distinctive feature is taken from the pilasters of the *Arcus Argentariorum* in Rome, a heavily decorated marble monument of 204 AD, dedicated to Septimus Severus, which Jamie and Zucchi would no doubt have seen. More specifically these pilasters were recorded in a drawing that was part of the most significant acquisition that James made while in Italy – the Albani collection.[54] Brought together by Cardinal Albani's uncle, Pope Clement XI, this mostly comprised the riches of Cassiano dal Pozzo's Museum Chartaceum and the collections of Carlo Maratta, and included drawings by Domenichino, Carracci and Pietro da Cortona, as well as others attributed to Raphael, Mantegna, Michelangelo and Correggio.[55] The drawings of the Museum Chartaceum included records of antique sculptures, monuments and buildings in Rome and its vicinity. The collection was held in the library of Cardinal Alessandro Albani's town house – now the Palazzo del Drago – where J J Winckelmann was installed as librarian; James and other connoisseurs considered it to be 'the finest in Rome and that it is next to impossible ever to make such another'.[56] It was these drawings that James acquired in 1762 for George III, the cardinal having been persuaded by his mistress to sell, reluctantly, in order to provide a dowry for his illegitimate daughter. Negotiations with the cardinal (or rather with his mistress the Contessa Cheroffini) also occupied members of the *personaggi della commedia*, as Jamie related to Bob:

> I then went to my friend the Cheroffini, whose eloquence I was sure was superior to all entails, and procured from her an order to go with two other persons to examine [the drawings]. This Clérisseau, Zucchi and I set about and after turning over a prodigious number of folios we were all three of opinion that the reputation of this collection was really well founded and that its extent was immense, containing nearly betwixt drawings and prints, 200 volumes in folio.[57]

However, not all the drawings from the Albani collection went to the Royal Collection at Windsor Castle. Several were held back by James and entered the Adam brothers' own collection. Among these were four volumes of drawings by Giovanni Battista Montano (1534–1621), one of the great recorders and interpreters of the fabric of ancient Rome, which were no doubt viewed by Zucchi, Jamie and Clérisseau and which included a study of the pilasters from the *Arcus Argentariorum* (Fig 3.9, far left; these drawings are now at Sir John Soane's Museum).[58] Moreover, the very same detail can be found recorded in one of the many other drawings collected by Jamie during his time in Rome, now with the Adam drawings at the Soane Museum – a pen-and-ink study, attributed by Alan Tait to the French sculptor Nicolas François Daniel Lhuillier (1736–93), a pupil of Clérisseau's (Fig 3.9, left). Furthermore, Jamie purchased several plaster casts of similar pilaster reliefs in Italy, which appeared in the 1818 sale of antiquities from the Adam collection held at Christie's.[59]

The aedicule in the *Portrait* shelters a statue of another recognisable antiquity – and again, one which James and British grand tourists would have had the opportunity to acquaint themselves with while in Rome: the *Minerva Gustiniani* (Fig 3.10). In fact, the statue seems to have had a particular resonance with British visitors to the Eternal City. Goethe related an anecdote told to him by the keeper of the collections at the Palazzo Gustiniani, where the sculpture was kept, that apparently one of the Minerva's hands was whiter than the other, having been blanched by the fervent kisses of so many admiring visiting *Inglese*.[60] That the work was both widely known and admired in this way is evidenced by the fact that it appears in the modified form of a bust on numerous occasions in Pompeo Batoni's portraits of British grand tourists, such as Sir Wyndham Knatchbull-Wyndham, Sir Gregory Turner (1768) and Sackville Tufton, 8th Earl of Thanet (1753).[61] In Zucchi's *Portrait of James Adam* it is clear that what is represented is not the original statue of the Minerva whose hand had been kissed white; neither is it based on a plaster cast of the original that James had acquired as part of his collection, as with the Medici Vase. In Zucchi's painting the Minerva Giustiniani is shown reversed. In the antiquity, Minerva holds her spear in her right hand, while in Zucchi's painting her empty left hand is raised

Fig 3.9
Giovanni Battista Montano, Pilasters from the Arcus Argentariorum, *pen and brown ink with brown wash (far left), and Nicolas François Daniel Lhuillier (attrib),* Detail of a Pilaster from the Arcus Argentariorum *(left).*

Fig 3.10
Detail of the figure of Minerva from Zucchi's Portrait of James Adam.

Fig 3.11
François Perrier, Minerva Giustiniani, *etching from* Segmenta Nobilium Signorum e Statuaru, *Rome, 1638, plate 54.*

to grasp the weapon. Similarly, the distinctive aegis, adorned with a gorgon's mask, is reversed from left to right. This suggests that Zucchi was basing his rendition on a print, rather than on a study of the statue in the collections of the Palazzo Giustiniani, which would account for the reversal of the composition. It is hard to define exactly which prints were available to him at the time, but he may have used a reproduction of the Minerva such as that published by François Perrier as plate 54 in his *Segmenta Nobilium Signorum e Statuaru*, of 1638 (Fig 3.11).[62]

Above the Minerva is yet one more identifiable antiquity or, to be more accurate, a motif based upon it. In this instance it is a Roman gem, in red jasper, showing Bacchus and Ariadne mounted on the back of a panther from the grand-ducal collections in Florence. In the *Portrait*, Zucchi has taken the theme and design of the gem but enlarged it greatly to resemble a decorative sculptural bas-relief (Fig 3.12). Zucchi and James Adam may have seen the original gem while studying the collections of antiquities in the various grand-ducal residences or in the Uffizi while they were in Florence and its environs from November 1760 to January 1761.[63] Also, another version of the relief appears on the bronze pedestal belonging to the Medici family, which James had copied in Florence in 1760–1 (*see* p 75 and Fig 2.7). However, it is equally if not more probable that, as with the Minerva Giustiniani, Zucchi based his composition on a publication, and one that proved especially influential in the formation of the international neoclassical style: Baron Philipp von Stosch's *Gemmæ Antiquæ Cælatæ*, published in Amsterdam in 1724.[64] The plates for this study of antique gems, engraved by Bernard Picart, became an important source-book for artists wishing to emulate closely the art of antiquity; plate 22 shows the Bacchus and Ariadne gem upon which Zucchi based his composition (Fig 3.13). James was clearly aware of the baron's activities as an antiquary and as a collector. While in Florence in 1760 he purchased 'a good many fragments of bas-reliefs and little statues that were in Baron Stosch's collection ... very cheap'.[65]

Of this careful grouping of visual references, only one object can be understood as a representation of an actual antiquity – a Roman altar, the top of which is glimpsed in the foreground of the image. This altar, with its corners terminating in rams' heads and with its front decorated by a festoon, is adorned with a very distinctive frieze of flowered ribbon-guilloche of Roman acanthus

Fig 3.12 (far left) Detail of Bacchus and Ariadne *from Zucchi's* Portrait of James Adam.

Fig 3.13 (left) Bernard Picart, Bacchus and Ariadne, *etching and engraving, from Baron Philipp von Stosch's* Gemmæ Antiquæ Cælatæ *(Amsterdam, 1724, plate 22), a publication that proved influential in the formation of the international neoclassical style.*

Fig 3.14 (below) Detail, giltwood armchair, designed by Robert Adam and executed by James Lawson, 1764.

foliage below an egg-and-dart moulding. The altar depicted by Zucchi appears to be the antiquity described as having been acquired in Rome and subsequently installed in the Adam house at 75 Lower Grosvenor Street, where it formed the plinth for a reduced bronze version of the *Farnese Bull*.[66] An altar, probably identifiable as the very same one, appears in the 1818 sale of antiquities from the Adam collection at Christie's, where it appears as 'Lot No. 81. A marble altar, ornamented with festoons, 4ft.2 high and about 3ft.4 square'.[67] The presence of this altar in the brothers' own collection directly influenced one of the most significant suites of neoclassical furniture designed by Robert Adam: that supplied in October 1764 to Sir Lawrence Dundas, Bt (d 1781), for the Banqueting Hall at Moor Park, Hertfordshire. These pieces incorporated rams' heads masking the transition between the legs and the seat rails of the chair, which copy the distinct flowered guilloche motif of the altar (Fig 3.14). This same motif can also be seen in a slightly later preliminary design for a long stool for the Hall at Osterley Park, London, dating to *c* 1767–79.[68]

The final object that draws our attention in the painting is not an antiquity but rather something that exhibits James Adam's ambitions as a modern architect. Installed on the altar top, and acting as a support for James's elbow, is the bronzed capital of a column which, though reminiscent of those that adorned the Temple of Mars Ultor in Rome, is in fact the product of James's own architectural invention (Fig 3.15): it is a capital of the 'British Order' that he devised in Rome in 1762–3 as part of the extensive iconographic programme of sculptural decorations for his proposed new Houses of Parliament, in this instance specifically to embellish the building's main entrance portico. As part of his work for Jamie on the project, Zucchi had made a number of drawings showing the design (Fig 3.16).[69] James also had a *modello* made of the capital, which he hawked around Rome to show to select individuals, including the painter Charles-Joseph Natoire (1700–77), director of the French Academy in Rome at the time. As James explained in a letter to Robert:

> [The Capital] ... is formed in this way. Lions and unicorns alternately support the angles of the abacus, sustaining themselves on the acanthus leaves, their hinder parts, converted into chimeras, form twenty stalks that produce roses and thistles. The Middle capital is adorned with a sceptre bearing the dove (the emblem of peace), over it the crown of Britain, instead of the Corinthian rose, and on the under part of the moulding of the abacus is the collar of the Order of the Garter. This capital is modelled in wax and bronzed and has altogether the air antique, and has surprised everybody that has seen it who, indeed, are very few besides Natoire, my two English friends Crispin and Richardson. This same celebrated capital is for the portico of my project, as I suppose you may guess.[70]

Fig 3.15
Detail of capital from Zucchi's Portrait of James Adam.

In particular James was delighted at Natoire's response to the wax *modello*, telling Robert that he 'would have taken it for the antique'. Indeed, Natoire's supposed response was so complimentary that James regretted his largesse in showing off his idea to him, going on to suggest that Natoire might prove 'capable of borrowing the thought and calling the invention his own'. There has been speculation that the rolled-up sheet of paper that James holds in the *Portrait*, in the hand that rests against this capital, must also relate to the Parliament project, the logical supposition being that this is a drawing of a plan or other design connected with that scheme.[71] Although the recent cleaning of the *Portrait* has revealed Zucchi's signature and the date and place of its creation, there is no evidence of any marks on the roll and examination under ultraviolet light has shown that none has ever been present.

The presence of the capital in the *Portrait* also points to the engagement of both artist and sitter with the work and theories of Piranesi, Zucchi's fellow Venetian. Piranesi had, of course, met Robert Adam in Rome in 1755, and the two men developed a professional and personal regard for one another.[72] That Piranesi was also familiar with James when he in turn undertook his grand tour is evident in the inclusion of James's British Order in Piranesi's 1765 publication *Parere su l'architettura*. In that publication's polemical 'Dialogo Protopiro, e Didascalo', Jamie's design for the rebuilding of the Houses of Parliament is singled out:

> M. Adams se distingue autant par la grandeur de ses idées que par la maniere dont il rends. Cet artiste a exposé il y a quelque temps un dessein qui a réuni les suffrages de tous les Connoisseurs. Ce dessein présente le projet

Fig 3.16
Antonio Zucchi, Design for a Capital of the British Order, *pencil, pen, brown wash and white heightening on buff paper, c 1762.*

> d'un magnifique édifice qui conviendroit non seulement aux assemblées du Parlement de Londres ... Ce vaste projet dont l'exécution seroit digne de la magnificence d'une grande nation, est particulierement remarquable par le caractere de sagesse, & de gravité qui regnent dans toutes ses parties.[73]

The capital was praised by Piranesi as an example of how an antique design could be modified imaginatively for a contemporary application.[74] However, this is not the only way in which Zucchi's *Portrait of James Adam* engaged with Piranesi's theories of the study of antiquity and its potential role in modern architecture. His proposition of the antique as a relevant source for architects and designers would be crystallised in another polemical work – his 1769 publication *Diverse maniere d'adornare i cammini.* In the text that accompanied the plates there – entitled 'An Apologetical Essay in Defence of the Egyptian and Tuscan Architecture' – Piranesi summed up his approach:

> What I pretend by the present designs is to shew what use an able architect may make of the ancient monuments by properly adapting them to our own manners and customs. I propose shewing the use that may be made of medals, cameos, intaglios, statues, bassorelieves, paintings, and such like remains of antiquity, not only by the critics and learned in their studies, but likewise by the artists in their works, uniting in an artful and masterly manner all that is admired and esteemed in them.

This was exactly the approach to design and architecture that had been adopted by Robert and James Adam and was no doubt encouraged in them by Piranesi even before the publication of the *Diverse maniere*. Also, the style and layout of Piranesi's polemical texts and folio publications of designs seems to have influenced the way that the Adams presented their work in print, most notably in *The Works in Architecture*.[75] It is clear that Zucchi was also working in the spirit of Piranesi's theory of a contemporary design that drew from the antique, and that saw its ultimate expression in the *Diverse maniere*. Indeed, he can be seen to anticipate the imagery called to mind by Piranesi's later text through his own use of 'intaglios, statues, bassorelieves ... and such like remains of antiquity' as the symbolic background to Jamie's portrait.

There is one more feature of the *Portrait* that merits attention, and that is the evident 'plastic' quality of the antiquities and architectural fragments depicted in such a plausible manner by Zucchi. This points to another aspect of the artist's talent – his innate understanding of the medium of sculpture. For the Adams relied on Zucchi not only for his schemes of decorative painting that were such an intrinsic element of their style, but also for designs of scenes from classical mythology to be carved by their sculptors for chimney-piece tablets and other decorative reliefs. For example, at 20 St James's Square the two very fine marble bas-reliefs of the drawing-room chimney pieces, *Aurora Going Before the Sun and the Different Hours* and *The Triumph of Venus*, were based on drawings commissioned from Zucchi.[76]

The *Portrait* in context

At first glance, the *Portrait of James Adam* stands apart from the architectural capriccios and mythological paintings that Zucchi undertook as a decorative painter for the Adam office. As has been noted already, it is the only recognised portrait attributed to the artist to date. Yet this does not mean that it does not share a common context with those works. Firstly, the almost monochrome, ochre background bears comparison with the drawings Zucchi was producing in Rome in 1762–3 for James Adam's

Houses of Parliament scheme, as well as with some of his later decorative paintings, such as the ceiling to the entrance hall at Kenwood. Zucchi's approach in synthesising different styles of painting and draughtsmanship accords with Robert Adam's own abilities as a young architect on the grand tour, assimilating the influences of the various draughtsmen he was instructed and encouraged by, such as Clérisseau, Piranesi, Jean-Baptiste Lallemand and Laurent Pécheux. To a lesser degree, James Adam showed a similar aptitude in his use of 17th-century historic and visual references alongside neoclassical styles in his Parliament project. In the *Portrait*, Zucchi brought together various recognisable antiquities to create an architectural and sculptural context within which to express James Adam's ambitions as an erudite and inventive architect, and also as a modish *cavaliere* – but above all as a leading dealer in Rome at that time. In essence Zucchi's approach corresponds to that of Robert and James in creating a 'true style of antique decoration' through a synthesis of motifs taken from antique buildings, monuments and works of art. It also shows Zucchi's ability to engage with the progressive theories of architecture and design then being proposed by Piranesi.

However, the *Portrait* speaks also of Zucchi's personal ambitions. It is clearly a work of intellectual and artistic sophistication. James reported to Robert that Zucchi was 'hankering' after England. The manner in which the painter has aligned himself with the sitter's intellectual aspirations and self-importance might indicate that he had 'pulled out all the stops', emulating the grand manner of Batoni to curry favour with his patron and potential ticket to Britain. There is also the consideration of the financial investment that the raw materials of the painting called for; the use of ultramarine, for example, and of such a deep hue, over large areas of the canvas would have been costly. One would hope that the painting was a success and no doubt it travelled to Britain with the returning James in 1763. It would be tempting to think it was initially hung in the Adam brothers' house and offices at Lower Grosvenor Street among the antiquities and paintings that they had gathered on their respective grand tours.

As we know, Zucchi did eventually satisfy that 'hankering' to come to England. However, he did not follow the painting immediately but only in 1766, after a three-year sojourn back in his native Venice. What caused this delay is not clear. What did the artist undertake in Venice in those three years? It is hoped that the reattribution of Jamie's portrait to Zucchi will help us reattribute further paintings to his *oeuvre* – as undoubtedly there must be more hanging unrecognised both in this country and in his native Italy. Despite the lack of a full art-historical context, the *Portrait* is not mute. It acts as witness both to Jamie's pre-eminence as the conduit for the flow of important artefacts and paintings between Rome and the Adam office in London, and to one of the most fruitful collaborations between the Adam brothers and an artist, which contributed in no small way to their success as purveyors of the 'true style of the antique'.

Acknowledgements

I would like to thank Adam Williams, Adam Williams Ltd, New York, and Theo Johns, Theodore Johns, London, for their generosity in lending Zucchi's *Portrait of James Adam* to the Iveagh Bequest, Kenwood. I am also indebted to the following individuals for their expertise and help in researching this chapter: Ian Thompson; Dr Adriano Aymonino; Dr Jonathan Yarker; Dr Frances Sands, Curator of Drawings and Books, Sir John Soane's Museum; Sue Palmer, Archivist, Sir John Soane's Museum; Stephen Astley, former Curator of Drawings, Sir John Soane's Museum; Colin Thom, Senior Research Associate, Survey of London, UCL Bartlett School of Architecture; and Don Lickley.

Notes

1 For example *see* Sands 2011 and Harris 2001.

2 Adam and Adam 1773–1822, vol I, no II, 8.

3 Zucchi's role as an independent artist has been further obscured by the commonly held view that Robert Adam dictated the subject matter to be included by the painter in his commissions. For a discussion of this, *see* Sands 2011.

4 *An English Family Group* (pen and brown ink, brown wash) and *Four Connoisseurs Seated at a Table* (pen and brown ink, brush and brown wash with white heightening), both late 18th century, Metropolitan Museum of Art, New York, Robert Lehman collection, 1975.1.554 and 1975.1.555, and *The Vendor of Silk Stockings* (red chalk, pen

and brown ink and brown wash), undated (*c* 1766–81), private collection.

5 Longhi, A 1762 *Compendio delle vite de' pittori veneziani istorici più rinomati del presente secolo con suoi ritratti tratti dal naturale delineati ed incisi da Alessandro Longhi veneziano*. Venice (unpaginated).

6 Ibid: 'Antonio Zucchi was born in Venice in 1726. His father was Francesco, a renowned copper engraver, who, as someone who knew how to run a household, realised that it was a good thing that there were various occupations within a family and suggested that he take up painting, and his son immediately embraced his suggestion. On taking up drawing, Antonio discovered that he had a natural talent for it and, deeming the study of architecture and perspective necessary for a painter, he devoted himself to it with all his energy. He began to paint under the guidance of Fontebasso but later, having gained the friendship of our renowned Amigoni and entered his school, he attempted to imitate with the greatest care his use of delightful colours, as can be seen in the altarpiece in San Giobbe in Venice portraying S. Bonaventura, S. Pietro d'Alcantara, and other female Franciscan saints, which he painted with such skill that he obtained the reputation of being a gifted painter. At present he is travelling in the company of a renowned French architect and an English dilettante with the intention of inspecting every refined work of architecture, sculpture and painting, ancient or modern, that they can find, as well as famous ruins of antiquity in the most illustrious Italian cities, such as Rome, Naples and others' (trans Paolo Gattavari).

7 *See* BM, Department of Prints and Drawings, Y,4.161.

8 Ibid, 1952,0206.1.21.

9 Adam and Adam 1773–1822, vol I, no II, plate 5. Zucchi also engraved the following plates in *The Works*: vol I, no III, plate 6, Luton Park House, details, and vol II, no I, plate 6, showing 'Two Chimneypieces in the Countess of Derby's Dressing Room', Grosvenor Square. These are signed 'J. Zucchi sculp. London'.

10 I refer especially to the identity of the 'Zucchi' described as having been employed by Robert Adam to engrave plates for his 1764 publication *Ruins of the Palace of the Emperor Diocletian at Spalatro in Dalmatia*.

11 Rodolfo Palluchini describes Antonio Zucchi as 'a modest, but significant episode in the transformation of seventeenth-century Venetian art to the international language of Neoclassicism'. *See* Palluchini, R 1960 *La pittura veneziana del Settecento*. Venice: Fondazione Giorgio Cini, 232.

12 *See also* Bryant, J 2003 *Kenwood: Paintings in the Iveagh Bequest*. London: English Heritage, 164–71.

13 This technique had a long tradition in Venice, where decorative paintings executed in oil on canvas and mounted on walls and ceilings could tolerate the damp of the Venetian lagoon better than those carried out in fresco.

14 *See* Adam 1831, 167.

15 Yale University Art Gallery, 1964.41.

16 See Delgou, G 1930 *Pittori veneti minori del Settecento*. Venice: Zanetti.

17 Earp, F R 1902 *A Descriptive Catalogue of the Pictures in the Fitzwilliam Museum, Cambridge, Italian School*. Cambridge: Cambridge University Press, 226–7. This attribution was maintained by Delgou, *Pittori veneti minori*.

18 Goethe, J W 1914 *Goethes italienische Reise*. Leipzig: In Insel, 177: 'works on a large architectural drawing, which he knows how to make superbly well in a decorative style. He was with Clérisseau in Dalmatia, he had very much associated with him, drawing the figures [illustrations] for buildings and ruins that he [Clérisseau] published and by doing so learned so much about perspective and effects that in his old age he could enjoy himself with it on paper in a worthy way' (trans Marion and Robin Page).

19 Croft-Murray 1962–70, vol II, 296. That it was Antonio Zucchi who accompanied Robert is also repeated in Rykwert and Rykwert 1984, 99, and Rowan 1993, 44.

20 Exhibited in *Angelika Kauffmann: Retrospektive*, Kunstmuseum, Düsseldorf, 1998–9, catalogue entry 67 (*see* note 24).

21 BM, Department of Prints and Drawings, Ff,3.218.

22 *See* Osborne, C M 1979 'The Zucchi sketchbook'. *Huntington Library Quarterly* **42** (3, Summer 1979), 263–9. Full understanding of this sketchbook is hampered by the fact that, at the time of writing, it has not yet been fully catalogued. *See also* Huntington Library, Art Collection and Botanical Gardens, European Art, 67.27.

23 The drawing was reattributed to Zucchi by Dr Eileen Harris. See Christie's 2002 *British Art on Paper*. London: Christie's, 17 (lot 7).

24 For a discussion of the artistic relationship between Angelica Kauffmann and Antonio Zucchi, see Baumgärtel, B 1998 'Angelika Kauffmann und Antonio Zucchi – Die Dekorationsmalerei in England', in *Angelika Kauffmann: Retrospektive*. Kunstmuseum, Düsseldorf (exhibition catalogue). Ostfildern, Germany: Gerd Hatje, 179–86.

25 At this sale, the *Portrait* was acquired by J Lupton Lister, Deans Court, Doctor's Commons, London. The history of the *Portrait* till then is unclear. It later entered the collection of Baron John Bonde (1918–2009), a descendant of the Adam family, who acquired the painting at Christie's, London, 26 June 1964, lot 106.

26 This was the first public exhibition of the portrait and it would not be seen again in a museum setting until 2015–16, when it was lent to the Iveagh Bequest, Kenwood.

27 The work was described as by Batoni in the following publications: Swarbrick 1915, 141–2, fig 101 (where Swarbrick identifies the sitter as Robert Adam); Steegman, J 1946 'Some English portraits by Pompeo Batoni'. *Burlington Magazine* **137** (March 1946), 60; Smart, A 1954 'An unknown portrait of James Adam'. *Burlington Magazine* **96** (April 1954), 102; Fleming 1955, 117 (Fleming repeats the erroneous identification of the sitter as Robert but also suggests that it may be James); 1956, 83, fig 9; 1962, 261, plate 88; Colvin 2008, 41; Wilton-Ely 2006, 231; Tait 2008, 139. However, A M Clark's 1985 publication *Pompeo Batoni* (Oxford: Phaidon) questions this

attribution. The attribution to Willison is made in Rykwert and Rykwert 1984, 61, based on the fact that the *Portrait of Robert Adam*, c 1770–4, in the National Portrait Gallery (NPG 2953; *see* Fig 0.1) is thought to be by that artist.

28 For a discussion of the *Ruins* publication *see* Brown 1992a.

29 Longhi also states that Zucchi was then engaged in travelling with a French architect and English dilettante in order to study modern and antique sites. However, the date excludes the possibility that this refers to Robert Adam, who was back in England by January 1758, and must refer instead to his brother James, who undertook his grand tour from 1760 to 1763.

30 Fleming 1962, 282.

31 However, Iain Gordon Brown in his catalogue accompanying the exhibition in 1992 to mark the bicentenary of Robert Adam's death stated that Francesco Zucchi had been employed to engrave the plates (he gives no source for this information). *See* Brown 1992a, 31; *also* 1992b, 78, 79, 81.

32 Moschini, G A 1924 *Dell' incisione in Venezia*. Venice: Zanetti, 63 (printed copy of the 1895 manuscript copy of Moschini's original and lost manuscript): 'He [Giuseppe Carlo] had also learnt perspective and architecture from Carlo [Zucchi], his uncle, when he was summoned to London by his brother Antonio to make engravings of the antiquities of Spalato, which the three Adam brothers wanted to publish there. And this was the most important set of engravings that Giuseppe left us' (trans Paolo Gattavari). In this study of Venetian printmakers it is notable that Gianantonio Moschini lists very few engravings attributable to Antonio Zucchi in comparison to the larger output of Francesco, Andrea, Carlo and Giuseppe Zucchi.

33 From Moschini's account it seems that the few prints Zucchi made were mostly after paintings by earlier Venetian old masters.

34 Richardson had joined John Adam's Edinburgh office in 1756, before heading south to join Robert and James. He stated in 1774 that he had worked for the Adam brothers for 18 years. See Rowan 1993, 43; Tait 1993, 87; NRS, GD18/4853.

35 *See* Adam 1831, 167. This journal appeared in the *Library of the Fine Arts or Repertory of Painting, Sculpture, Architecture and Engraving* **II** in October–November 1831 under the original title 'Journal of a Tour in Italy. By Robert Adam, Esq'. The publisher of the volumes was under the mistaken impression that the diary had been written by Robert.

36 Fleming 1962, 268.

37 Ibid, 272–4.

38 Ibid, 270.

39 NRS, GD18/4955.

40 Fleming 1962, 275, 302, 371–2.

41 Ibid, 280, 282, 371.

42 Ibid, 301; Adam 1831, 245.

43 Fleming 1962, 281. Fleming also reiterates the attribution of the work to Batoni.

44 This use of vivid pigment places Zucchi's *Portrait of James Adam* in the long tradition of Venetian *colore* in painting. I am grateful to Theo Johns for confirming this observation.

45 Fleming 1962, 294.

46 Ibid, 266, 271–2, 277, 294–5; NRS, GD18/4940.

47 Fleming 1962, 297.

48 Ibid, 278, 296, 297.

49 For more on James Adam's contribution to the work of the brothers' architectural practice through his dealings in antiquities in Rome, *see* Jonathan Yarker's account of the Adams as art dealers in Chapter 2 of this volume.

50 Fleming 1962, 272.

51 Attingham Park, The National Trust, NT 608052.

52 Fleming 1962, 296.

53 I am indebted to Dr Jonathan Yarker for drawing my attention to the specific identification of the cast.

54 For a discussion of James Adam's acquisition of the Albani collection, *see* Fleming 1958b.

55 These drawings originally formed part of the Maratta collection.

56 Fleming 1958b, 166–7.

57 Ibid, 167.

58 SM, Adam vol 125/104.

59 SM, Adam vol 26/70 – I am very grateful to Colin Thom for drawing my attention to these drawings and for Dr Frances Sands' and Sue Palmer's help. The casts appeared as lot no 40 and were acquired by Sir John Soane for 10s 6d. They are probably to be identified with casts A2–5, A7 and A8 in the museum. I am grateful to Helen Dorey, Deputy Director and Inspectress of Sir John Soane's Museum, for drawing my attention to these pieces.

60 'Im Palast Giustiniani steht eine Minerva, die meine ganze Verehrung hat ... Als wir die Statue besahen und uns lange dabei aufhielten, erzählte uns die Frau des Kustode, es sei dieses ein ehemals heiliges Bild gewesen, und die Inglesi, welche von dieser Religion seien, pflegten es noch zu Verehren, indem sie ihm die eine Hand küsten, die auch wirklich ganz weiß war, da die übrige Statue bräunlich ist.' Goethe, *Italienische Reise*, 13 January 1787.

61 *See* Haskell and Penny 1981, 270: Bowron, E P 1982 *Pompeo Batoni and His British Patrons* (exhibition catalogue). London: Greater London Council, 37–8; *also* Bowron, E P 2016 *Pompeo Batoni: A Complete Catalogue of His Paintings*. London; New Haven: Yale University Press. It should be noted that Bowron does not include the Zucchi in the section dealing with paintings formerly attributed to Batoni.

62 Perrier, F 1638 *Segmenta nobilium signorum e[t] statuaru[m]: Quæ temporis dentem inuidium euasere Urbis æternæ ruinis erepta, typis æneis ab se commissa perpetuæ uenerationis monumentum. Rome.*

63 From his journal we know that Jamie visited the Uffizi and the Palazzo Vecchio as well as the villas at Poggio Caiano, Pratolino and Poggio Imperiale. See Adam 1831, 172–3, 175–6. James also recorded in his journal that he visited other Medici properties in the company of Zucchi during the period of the carnival in Florence. *See* Fleming 1962, 276–7.

64 It is interesting to note that the catalogue for the sale of the Adam collection at Christie's on 22 May 1818 lists: 'Bacchus and Ariadne by Collins, a pair in *alto relief*' (3, lot no 11). Cited in Bolton 1922, vol II, 332.

65 Ibid, 376. He would also urge Robert to acquire the baron's collection of drawings when they appeared on the market.

66 I am grateful to Stephen Astley for this information.
67 A note in pencil records that at the Adam sale this altar was acquired by the sculptor Westmacott for £24 3s. Christie's sale catalogue, Friday, May 22 1818, 6.
68 SM, Adam vol 54/1/66.
69 Ibid, vols 7/69, 7/13 and 7/193.
70 Fleming 1962, 305–6.
71 *See* for example Rykwert and Rykwert 1984, 59.
72 *See* Wilton-Ely 2006.
73 Piranesi, G B 1765 'Dialogo Protopiro, e Didascalo', in *Parere su l'architettura*. Rome: 'Mr Adams is remarkable as much for the grandeur of his ideas as for the manner in which he performs. This artist exhibited a short time ago a drawing which attracted praise from all the connoisseurs. This drawing presents the project for a magnificent edifice which would be suitable not only for the assemblies of the London Parliament ... This vast project, the execution of which would be worthy of the magnificence of a great nation, is particularly remarkable for the quality of wisdom & gravity which reigns throughout all its parts' (trans Annie Pritchard).
74 Ibid.
75 James would play a significant part in devising the prefaces that accompanied each section of *The Works in Architecture*. How closely they echoed in spirit Piranesi's approach to design can be seen in the preface to fascicle no IV of vol I: 'The master, who has not the opportunity to distinguish himself by displaying his abilities in works of real greatness, will naturally betake himself to other resources, and, following the most approved examples of Greece and Rome, endeavour to call forth the admiration of mankind by the beauty and variety of his forms, by the richness and fertility of his invention, and by the elegance and delicacy of his ornamented decoration.' Adam and Adam 1773–1822, vol I, no IV, 'Public buildings', preface.
76 Harris 2001, 261, 263–5.

4

'The true style of antique decoration': Agostino Brunias and the birth of the Adam style at Kedleston Hall and Syon House

Adriano Aymonino

The genesis of Robert Adam's 'true style'

On 17 January 1758, a few months before turning 30, Robert Adam arrived in London after several years spent in Rome on his grand tour. Two and a half years before, during his life-changing stay in the Eternal City, he had already decided not to return to his native Edinburgh and instead open an office in London.[1] Now, setting foot in the English capital, he ambitiously aimed at conquering the architectural scene and establishing himself as the foremost architect in Britain. But what were the secret weapons that he intended to use to market his style as radically new and exciting? What were the characteristics of this new language of decoration that eventually made him one of the most successful of all 18th-century British architects? Adam could present his novelty through carefully designed contrast in colour, shape and dimensions to create 'movement', as he famously described at the beginning of his *Works in Architecture* more than a decade later.[2] However, first and foremost he could offer a radically new experience of the antique, based on a manifold principle of quotation from architectural, sculptural, decorative and ornamental Roman prototypes, all of which were well known to his clientele. This 'architecture of reference', which alluded to a Rome experienced by the vast majority of his patrons, both in the flesh and through the antiquarian publications that filled their libraries, constitutes the essence of Adam's striking success. It was a 'revolution' in the whole system of architecture, which introduced 'the true style of antique decoration' through a 'careful search into the purest sources of antiquity' – as Adam affirmed in a typical self-celebratory manner in the *Works*.[3] In the following decades his innovative conception of the architectural interior crossed the Channel to be imitated and reinterpreted all over Europe, both in its centres and its remoter provinces. To study the characteristics and the modus operandi of Adam's earliest commissions is therefore to study some of the traits that neoclassical interior decoration would express as a whole.

We shall see that among the many ingredients that contributed to the elaboration of this language, three proved to be essential. Firstly, the presence of the skilled draughtsman Agostino Brunias, one of Adam's earliest employees, who emerges as instrumental in helping forge the new style; secondly, the *Admiranda Romanarum Antiquitatum* (1693), a celebrated publication on antique Roman reliefs that provided many of the decorative motifs incorporated by Adam; and finally a resourceful collection of drawings after the antique within easy reach of London – the Topham collection, preserved then, as now, at Eton College Library.[4]

The earliest convincing incarnations of Adam's new 'true style' – at least in terms of interior decoration – were two small rooms that no longer survive: Lady Caroline's Dressing Room at Kedleston Hall (Fig 4.1), and a small 'closet' (or cabinet) on the first floor of one of the angular towers at Syon House, both designed in early 1760 and very similar in size and decorative character.

The essential characteristics of the decorative language they expressed had already been conceived five years earlier. After his arrival in Rome in late February 1755, Adam became progressively aware that his future success rested on an easily available repertoire of forms and motifs that he could market as faithfully antique and clearly distinct from what the previous generation of British architects had offered.[5] His early correspondence from Rome reveals an obsessive preoccupation in assembling books, drawings and plaster casts from the antique, as well as originals, and months later the 'Antique mad' Robert could write to his sister Peggy that his lodgings at Casa Guarnieri were full of 'antique cornices, friezes, figures, bas-reliefs, vases, altars, so that I now have a room as full as it can stick from roof to the floors'.[6] At the same time he employed painters and draughtsmen to draw 'the fountains, the buildings, the statues' and everything that he considered to be of interest in or around Rome.[7] In the summer of 1755, while studying ornaments under Charles-Louis Clérisseau and Laurent Pécheux, and having just met Piranesi, Adam firmly decided to 'go to the bottom of things' and 'to outdo Chambers in figures, in bas-reliefs and in ornaments, which, with any tolerable degree of taste so as to apply them properly, make a building appear as different as night from day'.[8] This resolution perfectly sums up what would be the essential traits of Adam's originality in terms of interior decoration: the complex arrangement of painted panels, grisailles, plasterwork plaques and reliefs, marble tablets and grotesque ornaments, that made his style so clearly recognisable and distinct both from immediate predecessors such as Daniel Garrett, Matthew Brettingham and James Paine, and from his contemporary and rival William Chambers. It is true there had been several earlier individual attempts to create interiors based on antique decorative examples, notably by William Kent and by James 'Athenian' Stuart.[9] However, Adam's use of details derived from the repertory of Classical antiquity was conducted on a scale and with a consistency hitherto unknown.

Back in London in January 1758, Adam set out to make the best use of his vast collection of antique originals and printed, drawn or plaster-cast models. By February he had already rented a house in St James's Place and started to 'pay his respects to the great' and to invite visitors, as he now had 'a place for showing his things'.[10] A move shortly afterwards to larger premises in Lower Grosvenor Street gave him more scope;

Fig 4.1
Plan of the principal floor at Kedleston, by Robert Adam (from John Woolfe and James Gandon, Vitruvius Britannicus, *IV, London, 1767, plate 46). Lady Caroline's Dressing Room is indicated in red.*

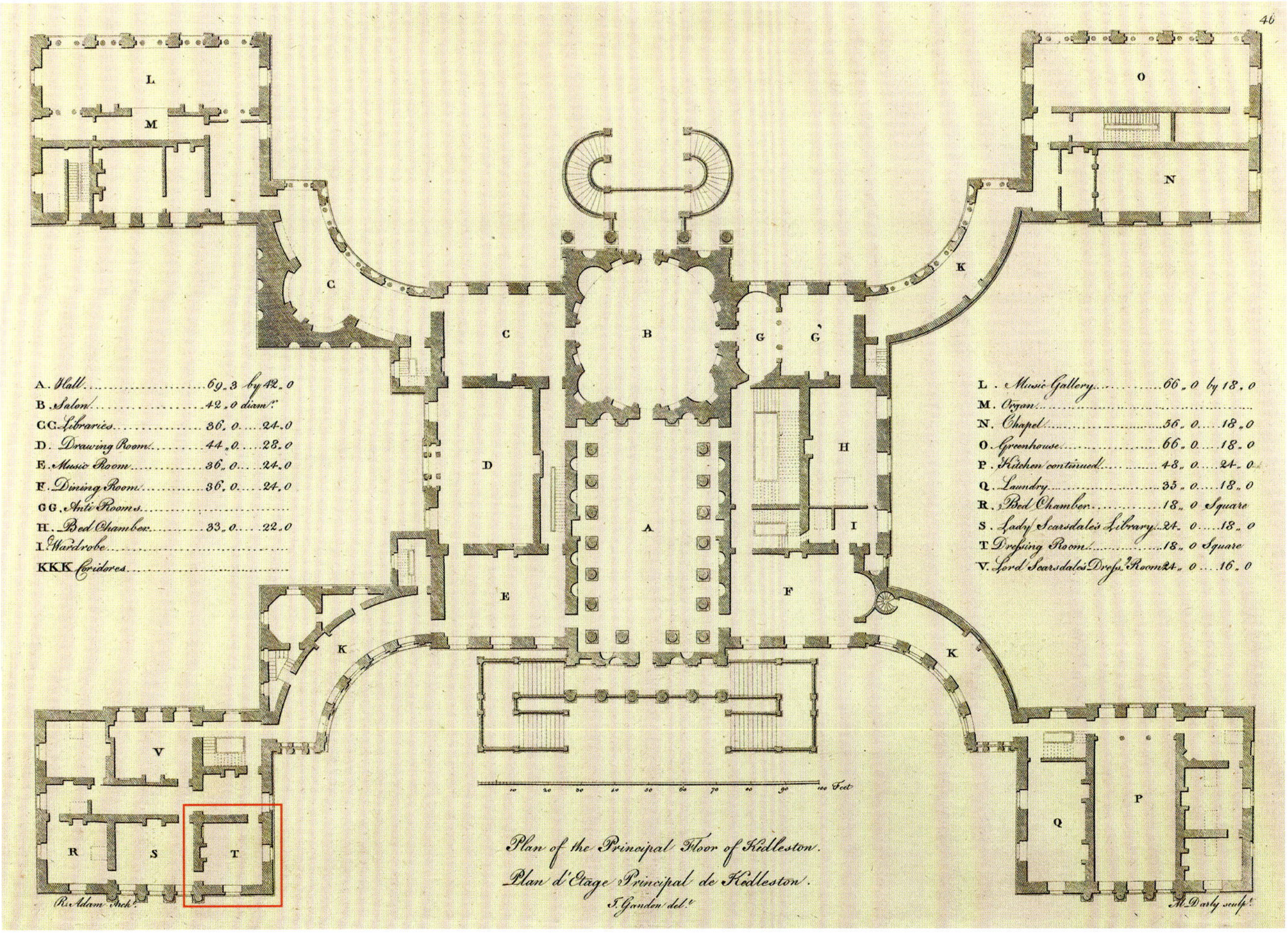

this apartment was clearly intended as a showcase to display his repertoire in order to gain new commissions, while at the same time promoting his related activity as a dealer – the full extent of which is revealed by Jonny Yarker in Chapter 2 of this volume. Adam's correspondence from London shows the crucial importance that he assigned to his drawings and how these started to attract his first clients.[11] In December 1758 he was introduced to Sir Nathaniel Curzon, newly embarked on building a grandiose house at Kedleston in Derbyshire to rival the Earl of Leicester's Holkham Hall, then still under construction.[12] Just a month before meeting Adam, Curzon had given the commission of his new house to the 60-year-old Matthew Brettingham, and now, seeing Robert's drawings at the architect's house, he was 'struck all of a heap with wonder and amaze' and 'every new drawing he saw made him grieve at his previous engagement with Brettingham'.[13] Robert soon went to Curzon's house, proposed alterations to Brettingham's designs, took the opportunity to comment negatively on previous designs submitted by 'Athenian' Stuart, and was put in charge of all the grounds on the spot.[14] A year and a half later, in April 1760, he was given a free hand on the whole house, a responsibility that until then had fallen upon James Paine, as Brettingham had left before the end of 1759. Adam could now finally concentrate on the interiors and apply his new language on a monumental scale.[15] By May he was producing various solutions for the plan and the garden façade, drawing wall elevations for several rooms, and he had been given full responsibility for decorating a small space in the north-west corner of the family pavilion: Lady Caroline's Dressing Room.[16] This, the first room that Adam executed at Kedleston and one of the first he ever designed, was only 18ft (5.5m) square, unconstrained by Curzon's collection of paintings and statues, and therefore a space where Adam could experiment with his new 'true style of antique decoration' in total freedom.

Fig 4.2
Adam office, Agostino Brunias, Design for a Window Wall for Lady Caroline's Dressing Room at Kedleston Hall, *probably as executed, pencil, pen and watercolour on paper, 1760.*

Adam, Brunias and Lady Caroline's Dressing Room at Kedleston Hall (1760–1)

The correspondence recorded in the Appendix to this essay reveals the different phases of the commission: in May 1760 Curzon sent a section of the room to Robert Adam, who by July could already write to his brother James that he 'had done a painted room' for Curzon, 'quite in a new taste'; a month later the overall design, 'the work of great labor [*sic*] and study', had been completed (Appendix, nos 1, 3–5). A preparatory drawing for the ceiling survives at Sir John Soane's Museum in London, while three finished elevations and a ceiling, most probably as executed, are still kept at Kedleston (Figs 4.2–4.4, *see* Fig 4.7).[17] All are here attributed to the hand of the Italian Agostino Brunias, one of the two draughtsmen, with the Flemish Laurent-Benoît Dewez, whom Adam had brought back from Rome to London to work in his office. While Dewez was mainly in charge of line drawings, Brunias had been trained as a painter and was responsible for 'ornaments, for landscape and figures and other things of that nature'.[18] Dewez fled the Adam office for Brussels at the end of 1758, but Brunias remained as Adam's master draughtsman from 1758 until 1764, when he in turn left in search of better employment.[19]

The vast majority of Adam's early decorative schemes preserved at Sir John Soane's Museum can be assigned to Brunias's hand, and his role in the formulation of the early Adam style must have been paramount.[20] As we shall see, he was at the same time in charge of copying drawings after the antique to provide a repertory of 'ornaments' to be employed in Adam's decorative schemes. Along with the drawings for the 'closet' at Syon – which were also produced by Brunias and are discussed on p 114 – these early drawings for Lady Caroline's Dressing Room constitute the first complete ornamental scheme ever produced by the Adam office to make extensive use

Fig 4.3
Adam office, Agostino Brunias, Design for a Chimney Wall for Lady Caroline's Dressing Room at Kedleston Hall, *probably as executed, pencil, pen and watercolour on paper, 1760.*

Fig 4.4
Adam office, Agostino Brunias, Design for a Wall for Lady Caroline's Dressing Room at Kedleston Hall, *probably as executed, pencil, pen and watercolour on paper, 1760.*

of grotesque decoration: not surprisingly, Adam intended to publish them to disseminate the novelty of his innovative solutions (Appendix, no 4). They are also the earliest Adam office drawings to survive in colour, which was used to evoke to its fullest the chromatic vivacity of the ornaments derived from the antique.

Brunias was also put in charge of the room's physical execution, as the decorations were to be painted in London using the *détrempe* technique, or distemper, presumably on rolls of paper (Appendix, no 4). The painted rolls were then sent and installed at Kedleston only after May 1761 (Appendix, no 9), while the mouldings and frieze were executed in relief by local craftsmen – the latter based on models provided by the plasterer Joseph Rose, who was to remain one of Adam's most important long-term craftsmen and collaborators (Appendix, nos 2–3).[21] The chimney piece of 'statuary marble, part gilt' was probably executed by Michael Henry Spang, a prolific Danish sculptor who had already provided several chimney pieces for Kedleston (Appendix, no 8).[22]

Fig 4.5
Agostino Brunias, Decorative Paintings in the Manner of Roman Fresco for Lady Caroline's Dressing Room at Kedleston Hall, *tempera paint on canvas, 1760.*

The mirror shown in the chimney wall elevation seems to have been executed (*see* Fig 4.3), as a 'looking glass in two plates 4 feet by 2 feet 6 inches gilt frame' is listed in an inventory of the room compiled in 1804.[23] The rectangular and oval figurative paintings on the walls and the circular one at the centre of the ceiling (*see* Figs 4.2, 4.4 and 4.7) were surely also executed by Brunias, who attempted to reproduce on canvas the same *détrempe* technique that he had used on paper for the other wall and ceiling decorations (Fig 4.5). This water-based medium unfortunately deteriorated quickly over the years to the point that in 1807 the room was dismantled. The Adam paintings survived in their frames at Kedleston until the 1960s, when they were removed and stored before being sold to the Victoria and Albert Museum in 1975.[24] The two best preserved, still in their original pinewood gilt frames, are today on display in the British Galleries of the museum, while the rest are kept in storage. The frames were executed by Sefferin Alken – the most famous ornamental carver of the time, who often worked for Robert Adam, as well as for William Chambers and Henry Flitcroft – and they were clearly judged too projecting when installed in the small room (Appendix, nos 9–10).[25] Finally, the room was furnished with two neoclassical giltwood window seats, as shown in the elevation for the window wall (*see* Fig 4.2), and a pair of tables with scagliola armorial slabs, all surviving today in the house.[26]

Adam's and Brunias's antique sources for the room

What resources could Adam, driven by his ambition to create a 'new taste', draw upon for such a commission? What antique 'figures', 'bas-reliefs' and 'ornaments' were available to him at that date, to help formulate a 'true style of antique decoration'? For the 'figures' – the framed figurative paintings – a source was readily to hand. The surviving canvases at the Victoria and Albert Museum all depict pairs of women in classical dresses in a landscape, symmetrically arranged around vases, altars, columns and baskets of fruit, in a manner very close to Brunias's surviving wall elevations, suggesting that the drawings show what was actually executed (*see* Fig 4.5). It has been proposed that these figures represent 'vestals, virgin priestesses of the Ancient Roman religion, highly regarded and respected for their devotion and purity' – possibly because four of them seem to be involved in taking care of the sacred fire.[27] Nevertheless, they should probably be considered as a more general reference to ancient Roman women caught in daily activities – a subject well suited to a lady's dressing room.

In fact, Adam and Brunias derived most of their figures from plates in the *Admiranda Romanarum Antiquitatum* (1693), a reference publication that enjoyed immense success among grand tourists, antiquarians and artists throughout the 18th century and that was omnipresent in the period's most important libraries (Fig 4.6).[28] The *Admiranda* had been produced by the greatest art historian and antiquarian of

Fig 4.6
Giovanni Pietro Bellori and Pietro Santi Bartoli, Admiranda Romanarum Antiquitatum, *Rome, Giovanni Giacomo & Domenico de Rossi, 1693, plates 37–39, 41, 'Palladis Artes' (works of Minerva).*

Fig 4.7
Adam office, Agostino Brunias, Design for a Ceiling for Lady Caroline's Dressing Room at Kedleston Hall, *probably as executed, pencil, pen and watercolour on paper, 1760.*

the 17th century, Giovanni Pietro Bellori, and illustrated by the antiquarian and engraver Pietro Santi Bartoli, and its collection of ancient artefacts constituted a constant source of inspiration for Adam.[29] Compositions derived from images in the book are ubiquitous in his country and urban houses.[30] Not surprisingly, the *Admiranda* was also used extensively for the later decoration of various main rooms at Kedleston. In this case, the elegant figures used in the dressing room, taken from Bartoli's plates illustrating the 'Palladis Artes' (works of Minerva) from the frieze of the Temple of Minerva in the Forum of Nerva in Rome, were to resurface in several successive Adam interiors, such as Syon House in Middlesex, Home House and Kenwood House in London, Newby Hall in Yorkshire, and even on the frontispiece of the first volume of his *Works in Architecture*.

While in many cases Adam retained Bellori's iconographic interpretation of Bartoli's plates in the *Admiranda* to create decorative cycles with a consistent subject, often the antique images were used purely for ornamental purposes. At Kedleston, Adam and Brunias cut and pasted only a few of the 'Palladis Artes' as convenient examples of dignified draped women, even mixing them from different plates and reversing the direction of some: inevitably Bellori's original explanation of them as acolytes of Minerva engrossed in the works presided over by the goddess was lost in the process (*see* Figs 4.5–4.6). Despite their very modest quality and decorative nature – Brunias was by no means a great figurative painter – the Kedleston canvases play a rather significant role in 18th-century art, as they are among the earliest examples of neoclassical painting, contemporary to Anton Raphael Mengs's *Parnassus* at Villa Albani in Rome (1760–1), or to the first classically inspired works of Gavin Hamilton or Joseph-Marie Vien.[31]

If Adam could rely on the publications and drawings that he had brought back from Rome for the 'figures' and the 'bas-reliefs' of his interiors, what he lacked at the beginning of his career were sources for ancient 'ornaments' that could be used to embellish walls, ceilings and carpets in order to give an antique consistency to the whole. As this author has explained in another essay, there were very few surviving antique 'grotesque' decorations in Rome by the middle of the 18th century, and publications illustrating them were even scarcer.[32] For ancient ceiling decoration, Adam and other architects of his generation, such as Chambers, relied heavily on the few that had been printed in antiquarian publications since the late 17th century. The Kedleston ceiling is in fact derived from one in the 'Bains des Augustes a Rome', reproduced in the celebrated *Antiquité expliquée et représentée en figures* published by the abbot Bernard de Montfaucon, copies of which were in both Adam's and Curzon's libraries (Figs 4.7 and 4.8).[33] This must surely be the reason why the first guide to Kedleston printed in 1769 described the decoration of the room as derived from the Baths of Diocletian,[34] while most probably the ceiling published by Montfaucon comes from one of the chambers of the nymphaeum in the Domus Transitoria on the Palatine Hill, known in the 18th century as the 'Baths of Livia' or 'Baths of Augustus'.[35]

However, the inspiration for the decoration of the rest of the room came from a different source. Before elaborating the design for Lady Caroline's Dressing Room, Adam must have discovered one of the largest English collections of drawings after the antique, which had belonged to the connoisseur and collector Richard Topham (1670–1730) and had been located at Eton College Library since 1736.[36] The Topham collection contained a vast number of drawings of ancient ceilings and wall elevations, all executed by Pietro Santi Bartoli's son, Francesco Bartoli. This easily accessible, colourful and detailed resource on ancient grotesque decoration was exactly what Adam needed for his 'ornaments' and was later plundered by other architects such as James Wyatt and Charles Cameron. Many surviving copies of these Bartoli drawings, in the hands of Agostino Brunias and later of Giuseppe Manocchi (who joined the Adam office in 1765), are today preserved among the Adam drawings at Sir John Soane's Museum and in an album at the Getty Research Institute in Los Angeles (the latter in the hand of Brunias only).[37] Furthermore, in the RIBA collection in London is a miscellaneous volume almost certainly once in Adam's possession that contains further coloured copies on tracing paper, attributed here to the hand of Agostino Brunias.[38] While the Brunias copies at the Soane and at the Getty were probably executed later – as their paper bears watermarks possibly dating from the late 1760s – those at the RIBA, although they are drawn on tracing paper and therefore not bearing any watermark, were most likely realised around 1760, as they show details all reused for the decoration of Lady Caroline's Dressing Room and the Syon House 'closet'.[39]

All these copies testify that Adam sent some of his most skilled collaborators to Eton to make drawings that could be used as a source for patterns by his office. In the ensuing years, these drawings became an integral aspect of Adam's interior designs, and there are direct references to them in the surviving Adam family correspondence.[40] They prompted a rapid change in Adam's decorative language: colour became increasingly important and a flatter, yet more delicate and complex decoration replaced the simpler and thicker forms of previous ceilings and wall elevations. Thanks to the Topham collection, Adam and his collaborators could finally complete their system of reference to the antique and introduce 'true' grotesque decoration in England.

The three surviving wall elevations for Kedleston show that Adam and Brunias drew most heavily for inspiration on three original Francesco Bartoli drawings in the Topham collection – all depicting details probably derived from the Domus Transitoria on the Palatine Hill.[41] The framed oval panels visible in Fig 4.4, the grotesque decoration around them and the garlands displayed on the upper part of the wall all came from a drawing illustrating the same ceiling from the 'Baths of Augustus' published by Montfaucon (Fig 4.9), copies of which survive in Brunias's hand both in the RIBA and the Getty series.[42] This drawing may also have served Adam as inspiration for the ceiling, along with the Montfaucon engraving.[43] But the frieze with white griffins and tritons on a blue background, visible on all the elevations and eventually modelled by Rose (*see* Figs 4.2–4.4), the rectangular figurative panels framed by grotesque decoration and the general colour scheme all derive instead from a large drawing depicting a complex ceiling, whose copy by Brunias at the

Fig 4.8 (left)
'Peinture a fresque d'une chambre des Bains des Augustes a Rome' (from Bernard de Montfaucon, L'antiquité expliquée et représentée en figures, *15 vols, Paris, Florentin Delaulne [*et al*], 1719–24, supplement III (1724), plate 59).*

Fig 4.9 (below)
Francesco Bartoli, Drawing of an Antique Ceiling from the Palatine, *pencil, pen, watercolour and body colour on paper. Inscribed in pen on recto, lower left, within the red border, 'Fran[cis]cus Bartolus. Fecit. Anno 1721–', and on the lower centre; inscribed in pen on verso, lower right: 'Volta del palazzo di Tito'.*

RIBA must have been used as a source (Figs 4.10 and 4.11).[44] The gilded floral decoration on a blue background in the dado and the general decorative pattern of the wall must have come from a colourful drawing depicting a large lunette (Fig 4.12), a copy of which appears among the drawings at the RIBA (Fig 4.13).[45] Finally, the figurative paintings shown in the Bartoli drawings (*see* Figs 4.9, 4.10 and 4.12) clearly served as inspiration for the arrangement of the female figures in the panels for the room (*see* Fig 4.5).[46]

On the basis of this series of well-known engravings and drawings, Adam and Brunias therefore conceived a decoration that was radically new for the early 1760s, both in England and on the Continent. Aside from the frames, mouldings and frieze, all the other details must have been painted, certainly to evoke the traditional flatness of grotesque decoration, something to be privileged in such a tiny space. This was a feature that was not always retained in Adam's later commissions. One wonders whether he took inspiration in this sense from the painted ceilings by William Hannan in the

Fig 4.10 (above)
Francesco Bartoli, Drawing of an Antique Ceiling from the Palatine, *pencil, pen, watercolour and body colour on paper, 521 × 521 mm, 1721, previously inscribed on verso, now pasted on canvas, 'Una volta trovata nel Monte Palatino, 1721'.*

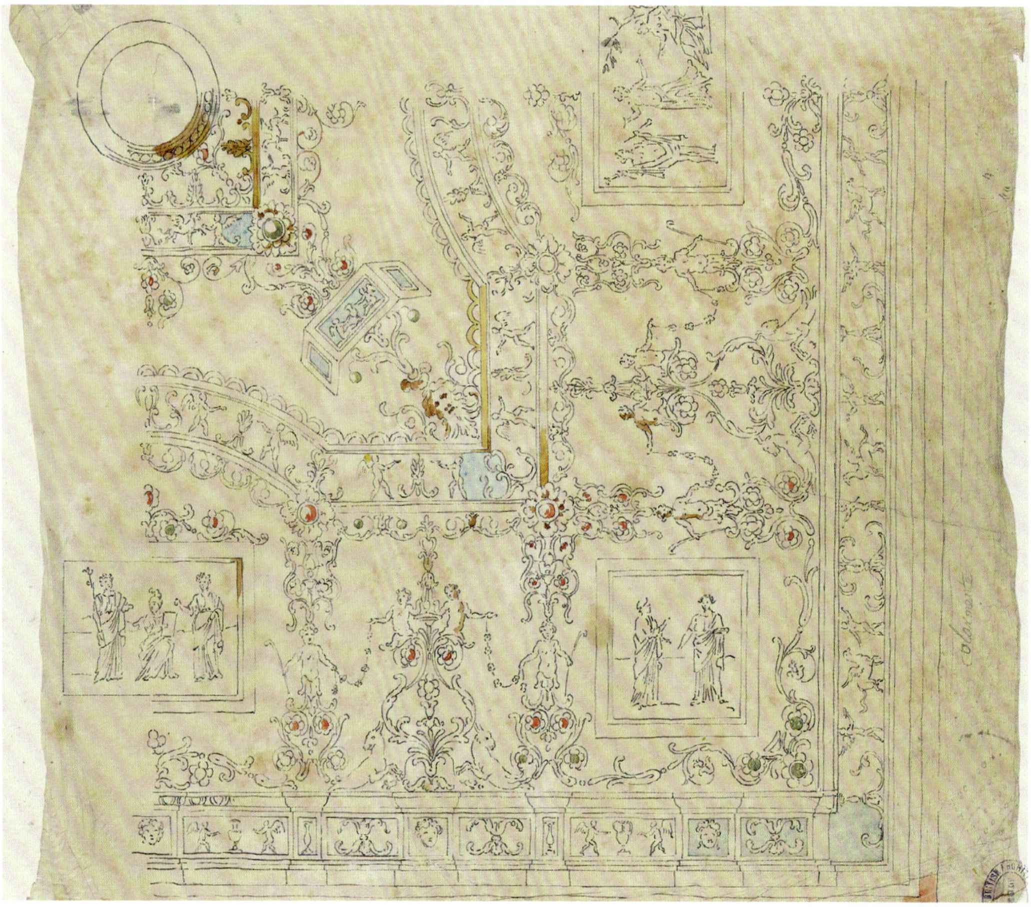

Fig 4.11 (right)
Agostino Brunias, Section of an Antique Ceiling from the Palatine, after Francesco Bartoli, *pen, watercolour and body colour on paper, c 1760, inscribed in pen on recto, centre 'Monte Palatino'.*

Fig 4.12
Francesco Bartoli, Drawing of an Antique Lunette from the Palatine, *pencil, pen, watercolour and body colour on paper, 1721, previously inscribed on verso, now pasted on canvas, 'Una lunetta di volta arcata trovata sul Monte Palatino, 1721'.*

Fig 4.13
Agostino Brunias, Drawing of an Antique Lunette from the Palatine, *from Francesco Bartoli, pen and watercolour on tracing paper, c 1760, inscribed in pen on recto, lower right 'Monte Palatino'.*

Red Drawing Room and Tapestry Room at West Wycombe Park, Buckinghamshire, both faithful copies of Francesco Bartoli's drawings at Eton and probably executed before Lady Caroline's Dressing Room at Kedleston.[47]

The Syon House 'closet' (1760–1)

Decorated at exactly the same time as the room at Kedleston was a small closet on the first floor of one of the angular towers at Syon House, the seat of the 2nd Earl and Countess of Northumberland in Middlesex (Figs 4.14–4.16). In the 1760s Syon was radically transformed by Adam into one of his most monumental and successful interiors, while the Northumberlands – elevated to the dukedom in 1766 – would become his most faithful patrons, commissioning work at Alnwick Castle and Northumberland House, their other estates in Northumberland and London.[48] The closet, actually a small cabinet, was evidently Adam's first experiment at Syon, and work on it must have started in the first half of 1760.[49] It was 'fitted, painted and ornamented after ye antique; being the first essay in that Taste executed in Engl[an]d w[hic]h has since so universally prevailed in this Kingdom', as Thomas Percy, the celebrated man of letters and tutor of the Northumberlands' sons, reported years later.[50] The 'closet' has long gone, possibly destroyed during 19th-century refurbishment works on the house, but Sir John Soane's Museum still preserves a preliminary sketch in Adam's hand and two wall elevations, here attributed to Agostino Brunias (*see* Figs 4.14 and 4.15); furthermore, there are two colourful elevations in the Brunias album at the Getty Research Institute (*see* Fig 4.16).[51] This small space shared many similarities with Lady Caroline's Dressing Room at Kedleston: from the character of the decoration, which was probably painted in the same *détrempe* technique, to the window and wall seats. Not surprisingly, it was also based on a set of Francesco Bartoli drawings in the Topham collection (Figs 4.17–4.19), copies of which survive among the RIBA drawings (Fig 4.20).[52]

To achieve an original and inventive scheme of decoration, Adam and Brunias dismissed the architectural structure shown in the Bartoli wall elevations (*see* Figs 4.17 and 4.18) but retained and freely reinterpreted many of the fanciful decorative details. This can be seen throughout the surviving Adam drawings: from the initial idea as laid out on the sketch, where the details are assembled freely for the first time (*see* Fig 4.14), to the finished coloured elevations (*see* Fig 4.16). For instance, the two herms shown in the large panels within architectural aedicules in Figs 4.15 and 4.16 derive from the central figure drawn by Bartoli in Fig 4.18, but with elements (such as the peacock feathers, foliage and the pedestals on which the herms rest) derived from decorative elements encircling the main central figures in Fig 4.17. The same is true of the more elongated vertical panels that appear in all the elevations (*see* Figs 4.15 and 4.16), where the figures and the scrolling foliage derive from Fig 4.17, but with decorative elements borrowed from the central panel in Fig 4.18. Furthermore, the dogs flanking an amphora in panels on the dado in the elevations are derived from the central frieze

Fig 4.14
Robert Adam, Preparatory Sketch for the Window Wall Elevations of the Closet at the Top of One of the Towers at Syon House, Middlesex, *probably as executed with modification, pencil on paper, c 1760.*

Fig 4.15 (above)
Adam office, Agostino Brunias, Designs for the Wall Elevations of the Closet at the Top of One of the Towers at Syon House, Middlesex, *probably as executed, pencil, pen and watercolour on paper,* c *1760, the right-hand drawing inscribed 'Sion' both in pencil on recto, lower left, and in pen on verso, top right.*

Fig 4.16 (left)
Agostino Brunias, Designs for the Wall Elevations of the Closet at the Top of One of the Towers at Syon House, Middlesex, *probably as executed, pencil, pen, watercolour and gouache on paper,* c *1760.*

panel in Fig 4.18, while the central panel above the window in Figs 4.15 and 4.16, with winged centaurs and Erotes, is almost a literal copy of Fig 4.19. These drawings show very clearly how Adam and Brunias made free and inventive use of the Bartoli originals, as a source of inspiration rather than a constrictive model. Brunias's coloured elevations (*see* Fig 4.16) most probably record what was executed, and suggest that the colours of the original drawings by Francesco Bartoli were retained for most of the details. As at Kedleston, the execution of the cabinet was most likely given to Agostino Brunias, who presumably then recorded the whole scheme in his volume at the Getty, therefore joining together in the same pages the 'antique' sources of the new decorative language and one of the first examples of its practical execution. Elevations of the Syon cabinet were later published by Michelangelo Pergolesi, another decorator who

Fig. 4.17
Francesco Bartoli, Drawing of an Ancient Wall Elevation from the 'Baths of Constantine', *pencil, pen, watercolour and body colour on paper, 1711, inscribed in pen on recto, lower centre 'in aedibus Rospiliosi, e Balneis Constantini M. Repert. 1711'; inscribed in pen on verso, lower left 'No. 2'.*

Fig 4.18
Francesco Bartoli, Drawing of an Ancient Wall Elevation, *pencil, pen, watercolour and body colour on paper, c 1710–25, inscribed in pen on verso, lower right 'Fran[ces]co Bartoli Fece' and 'del Palazzo di Augusto'.*

Fig 4.19
Francesco Bartoli, Drawing of an Ancient Ornamental Panel from the 'Sette Sale', *pen, watercolour and body colour on paper, c 1720–5, inscribed in pen on recto, within the red border, lower left and right 'Fran[ces] co Bartoli', and in pen on verso, lower left 'un ornamento del Palazzo di Tito'.*

Fig 4.20
Agostino Brunias, Drawing of an Ancient Wall Elevation from the 'Baths of Constantine', *from Francesco Bartoli, pen and watercolour on tracing paper, c 1760, inscribed in pen on recto, lower right 'Nelle Terme di Costantino al Palazzo Rospigliosi'.*

worked for Adam at Syon, as plates 64 and 70 in his *Designs for Various Ornaments* – a collection of plates published serially in London between 1777 and 1801, which served as a pattern book for neoclassical decorators.[53]

Conclusion

In these two small rooms, unconstrained by previous designs, displays of artworks or functional preoccupations, Adam and Brunias were able to develop for the first time a style of decoration that would spread all over Europe in the following decades. Most importantly, they established here a modus operandi based on the principle of quotation from antique sources that would be refined in subsequent years and adopted by most architects of Adam's generation. Adam would have provided the initial sketch of compositions, as for the cabinet at Syon (*see* Fig 4.14), while Brunias would have developed these early ideas into a final scheme with the help of the Bartoli drawings in the Topham collection. This extremely painterly manner of interior design soon mutated into a more 'sculptural' approach, and the wall paintings – though at times retained in small spaces – were usually replaced in later Adam schemes by an arrangement of plasterwork reliefs and plaques, marble tablets or painted panels, often in grisailles, to be inserted into the bare wall. Adam and Brunias turned increasingly to antiquarian illustrations for inspiration, especially those by Bellori and Bartoli, thereby establishing a sophisticated twofold system of reference to Roman antiquity: to the originals that could be experienced in Rome, and to the antiquarian publications that were common knowledge among Adam's patrons and their guests, and which were often physically contained within the architecture. Robert Adam managed to keep faith with his early determination to 'outdo Chambers in figures, in bas-reliefs and in ornaments', and his new language of decoration, fully expressed for the first time in Lady Caroline's Dressing Room at Kedleston and the cabinet at Syon, would become the key to his outstanding success.

Acknowledgements

For help and suggestions I am indebted to Eileen Harris, Olivia Horsfall Turner, Ashleigh Murray and Frances Sands. Special thanks are also due to Ann Jones for her help with correspondence and documents in the Kedleston Hall Archives.

Appendix: Documents related to Lady Caroline's Dressing Room at Kedleston Hall, May 1760 to July 1761

1 **Robert Adam to Lord Curzon, 7 May 1760**

'I have received the section of Lady Carolines Dressing room and shall get the Drawings made for it and send them and instructions to Mr Swan when done.'[54]

2 **Robert Adam to Lord Curzon, London, 6 June 1760**

'In the meantime I send you inclosed a Drawing for the base and surbase mouldings of Lady Carolines Dressing room to full size. The Ornament of the Mouldings must be carved, but the ornament in the frieze of the Surbase may be painted – so that they may proceed with these mouldings in the meantime. I should wish to have the Drawing of the Chimney Piece of that Dressing room that it might go into the section provided you have fixed on it, or if you incline that I should design one for it.'[55]

3 **Robert Adam to Lord Curzon, London, 16 July 1760**

'The sections for the Lady Carolines Dressing Room are well advanced and painted, so is also the Ceiling. Mr Rose has modelled and sent down the Frieze at large for this man [the executioner] to execute from. I am very hopeful you will approve of these drawings, as they have been the work of great labor and study, and I flatter myself they will be agreeable to Her Ladyship.'[56]

4 **Robert Adam to James Adam, London, 24 July 1760**

'When I was at Kedleston Sir N. spoke to me one day about publishing the Designs of His House, which I told him I certainly would do so soon as all was done & Executed. I have done a painted room for him, which is quite in a new taste, & I have Brunias now employed in painting in size to learn that method as Oyl Colours will by no means answer. They call that manner of painting in French, à la detrempe, and I think he succeeds wonderfully with it. But I tremble for B[runia]s for I do not think he has Constitution for this Countrey. I therefore whish you would cast about for some clever one in that way or some boy with good dispositions to Ornament & Figures, & keep him either at work with you or Send Him to England & I could put him under Bruny, as he complains much of his Eyes & I find grudges much drawing of ornament so small that it blinds him, so that I am convinced he would do every thing he could to learn one in [fol 3v] order to relieve himself. But I would not chuse one altogether Ignorant. Let me know what sort of Lad your Scholar is, & how he draws & what George is doing, I suppose C[lérisseau]– will keep them constantly busy. From what I have said you see that Supposing one could get the Best things they do engraved by degrees it would produce a Works insensibly of great value & great Fame. You may perhaps object that this is publishing all your Manner your Ornaments & Study's. That I grant is true, But at the same time I doubt that if you do not publish them yourself some other will which is worse. Sir Nats. Fronts & Sections may make near 10 plates including the painted room & if you publish detail of organ, Silver plate, Carpets &c many more. Then there is the Bridge, the Ruins, The Towers, The pheasant house (a new Contrivance), a new Gateway to his avenue and a new Temple. Then there is Mr Lascelles's plans & fronts. General Blands Room & Green House & Temple. I think I mentioned to you that the Green House turns out in the Execution Handsomer than it is possible to conceive & surprized myself as infinitely prettyer & more antique like in the reality than in the Drawing. Then there is my Lord Northumberland, my Lord Coventry, in Short many others that it is needless to repeat which would make no bad publication already, & if the plates could be gone on with I should think it no bad Scheme. But english Engraving is not to be ventured upon So that I only leave these things to your reflexion.'[57]

5 Robert Adam to Lord Curzon, London, 5 August 1760

'I have now finished the Design for Lady Caroline's Dressing Room which I must send in a small box by the wagon as rolling would destroy the drawings entirely.'[58]

6 Robert Adam to Lord Curzon, London, 11 August 1760

'I was informed that you was to be in town in a few days which made me delay sending the Sections of Lady Carolines Dressing room as I hoped to have the Honour of showing them to you here. Your answer will determine me.'[59]

7 Robert Adam to Lord Curzon, London, 23 April 1761

'I beg you'd be so good as desire Mr Wyatt [the clerk of the works] to Send me the Exact Sizes of the panels of the Window Shutters of the Painted Dressing room and also of the Soffite pannel within the Mouldings as it will be necessary to have Ornaments painted for them to match the whole compleat.'[60]

8 Robert Adam to Lord Curzon, London, 8 May 1761

'I received the Honour of Your Lordship Letter of the 5th Ins.t which by being much out of Town in little excursions I am not able fully to answer, particularly with respect to the Chimney piece doing by Spang for Lady Scarsdales dressing room. But I intend to see him tomorrow morning and to push him to get it compleated. Since your Lordship went out of Town I have made great progress in the paintings, and I think Twelve or 14 days will finish the whole.'[61]

9 Robert Adam to Lord Curzon, London, 15 May 1761

'I have now got all my affairs so settled that I would sett out for Kedleston directly. But as the Painted Room wont be finished for 8 days, I shall regulate myself according to Your Lordships desire. If I sett off directly it will be impossible to see the Painted Room begun, and going on whilst I have the Honour of being with Your Lordship, nor would I carry down my Italian with me to assist. And if I delay my setting out till the paintings are ready, I am affraid it will throw my time too near that of Your Lordship coming to the Birthday, which might prove inconvenient. I could go to Kedleston directly after the Birthday if that would answer equally well for Your Lordship and then both the painting and the frames making by Alken would be finished. But I leave it entirely to Your Lordships choice ... Whatever time you please to fix my coming to Kedleston, if it is after the paintings are finished, it would be necessary to appoint the people who are to cut out and paste up the ornaments that they might be present, so as to put them on the method of doing it right.'[62]

10 Robert Adam to Lord Curzon, London, 9 July 1761

'I have received both Your Lordship and Mr Wyatt's letters concerning the Picture Frames in the painted room, and am very sorry at the projections being too great. But if one compares them with ordinary picture frames they are not near so much raised, as even those for very small pictures, and particularly the oval frames which I made extremely thin. To be sure it would be the best way to sink them in the walls, if it was not for destroying the paintings by the dust and dirt it would occasion in the room by cutting away the plates, and therefore the least which is to be chosen which is to reduce the mouldings according to Mr Wyatts sketch, or rather to avoid the appearance of two astragal in this manner [a sketch of the section of the frame is drawn on the bottom-right corner of the page]. As for the circular frame in the centre of the ceiling, I think the best method for it would be to make it in two separate pieces and to let the inner part be only a small astragal with beads, and the outter ring of some other flatt moulding, and to cut the remaining ornament which goes between them by itself and fix it to the ceiling, and to paint the ground of a colour like the ground of the freeze or surbase there [a sketch follows of a section of the ceiling and frames with the inscriptions 'ceiling', 'astragal', 'ceiling', 'moulding', 'ceiling'; below the middle section of the ceiling is another inscription: 'upon this ground to put an ornament cutt in wood and fixd up, to be gilt and to have a purple ground']. This is all that occurred to me as a remedy unless, as Your Lordship proposes, you was to make the square frames answer your other pictures, and have new ones done wither [with a] plane, or if you have no pictures that answer exactly to them you might perhaps get somethings painted for them. I have not done the above sketches to any scale, not having any copy of the mouldings which were given to the joiner and carver.'[63]

Notes

1 On Adam's grand tour, Fleming 1962 is still invaluable, especially 109–244. *See also* Ingamells 1997, 5–8, and Tait 2008, especially 45–123. On the early days of the Adam London office, *see* Fleming 1962, 159–60, 168–9. On the formation and sources of his style, *see* Stillman 1966, *passim*, esp 10–12, 15–26, 31–40, 82–3, no 75; Harris 2001, esp 1–17.

2 Adam and Adam 1773–1822, vol I, 3 (note A).

3 Ibid, 3, 5 (note C) and 16.

4 *See* Aymonino *et al* 2013.

5 Fleming 1962, 109–244; Stillman 1966, 31–40; Tait 2008, 45–123.

6 Adam refers to himself as 'Antique mad' in a letter from Rome to his sister Peggy (Margaret), dated 4 March 1755 (NRS, GD18/4766, fol 1v). On Robert's collection at Casa Guarnieri, *see* NRS, GD18/4797, fol 1v.

7 NRS, GD18/4766, fol 1v.

8 NRS, GD18/4777, fol 1r.

9 *See* for example the Presence Chamber at Kensington Palace (1724) and the parlour at Rousham, Oxfordshire (1738), both by William Kent (Croft-Murray 1962–70, vol II, 234–5 and fig 41; Sicca 1986, 137, 142–4; Cornforth 2004, 137–88 and figs 174 and 175; Weber 2013, 101–2, 123–4, 135–6, 173–9). Or the interiors at Spencer House, London; Wimbledon House, Surrey; Nuneham Park, Oxfordshire, or those unexecuted for Kedleston Hall – all designed by James 'Athenian' Stuart in 1758–9 (Bryant 2006; Hewlings 2006). Adam explicitly acknowledged his debt to Kent and especially to Stuart for having introduced 'grotesque paintings' to England. See Adam and Adam 1773–1822, vol I, 5 and notes C and E.

10 NRS, GD18/4847, fol 1v.

11 *See* for instance ibid, fol 2r; NRS, GD18/4853, fol 1v; GD18/4854, fols 1r–2v.

12 The literature on Robert Adam's work at Kedleston Hall is vast. *See* esp Harris 1987 and Harris 2001, 19–39.

13 NRS, GD18/4854, fol 1r. *See also* Harris 2001, 19.

14 NRS, GD18/4854, fols 1v–2r.

15 *See* Harris 2001, 21–2.

16 Harris 1987, 52–4, nos 37–9; Harris 2001, 21–3. *See also* the Appendix to this essay. Lady Caroline's Dressing Room was later known also as the Painted Breakfast Room (not to be confused with the 'Painted Breakfasting Room' planned for the west end of the south front in 1768 but never executed) and is referred to as such in various publications. It is also referred to as Lady Scarsdale's Dressing Room.

17 For the preparatory drawing in Sir John Soane's Museum (SM, Adam vol 11/44), *see* Aymonino *et al* 2013, 28, cat no 18. For the elevations and ceiling kept at Kedleston, *see* Harris 1987, 52–4, nos 37–9; Bristow 1996, 82–3, fig 82; Harris 2001, 22–3, figs 20 and 22. In addition to these, at Kedleston is preserved also an outline sketch in pen of the chimney wall with measurements, almost certainly in the hand of Robert Adam (NT 109903).

18 NRS, GD18/4811, fols 1r–1v. On Brunias, *see* Fleming 1962; Huth 1962; Croft-Murray 1962–70, vol II, 177; Stillman 1966, 41–2; Kriz 2008, esp 37–69; Bagneris 2010; Aymonino *et al* 2013.

19 NRS, GD18/4854, fol 2v (for Dewez). I am indebted to Frances Sands for the date of Brunias's departure from the Adam office.

20 *See* for instance the drawings attributed to him by Frances Sands in the Sir John Soane's Museum's Collection Online Database. See also Appendix, no 4.

21 This is most likely Joseph Rose senior rather than junior, who would have been around 15 years old in 1760 (*see* Beard 1975, 237–44; Harris 2001, *passim*; Murray 2012).

22 Kedleston 1769, 23. For Michael Henry Spang, *see* Roscoe *et al* 2009, 1161–2.

23 National Trust, Scarsdale collection, Kedleston Hall (henceforth KH), L2–1/1, fol 7.

24 Hardy 1978, 199–201; Harris 1987, 52–4; Harris 2001, 23. *See also* the curatorial file in the Victoria and Albert Museum Archive (MA/1/S649).

25 For Sefferin Alken, see Beard 1981, 241–2; Roscoe *et al* 2009, 11–14. On his role at Kedleston, *see also* Hardy 1978, 201; Harris 2001, 23.

26 *See* Hardy 1978, 200–201, figs 6, 8; Harris 2001, 23.

27 *See* the Victoria and Albert Museum collection database entry for the paintings at http://collections.vam.ac.uk/item/O53065/painted-panel [accessed 30 June 2017].

28 Two copies of the book are preserved today in the library at Kedleston (NT 3042398 and NT 3042399), though we cannot be sure whether they were acquired by Lord Curzon or one of his descendants.

29 On Bellori and Bartoli's publications, *see* Petrucci, A 1964 'Bartoli, Pietro Santi', in *Dizionario biografico*

degli Italiani, VI. Rome: Istituto della Enciclopedia Italiana, 586–8; Pomponi, M 1992 'Alcune precisazioni sulla vita e la produzione artistica di Pietro Santi Bartoli'. *Storia dell'Arte* **75**, 195–225; Borea, E and Gasparri, C (eds) 2000 *L' idea del bello: Viaggio per Roma nel Seicento con Giovan Pietro Bellori* (exhibition catalogue, 2 vols). Rome: De Luca, esp vol I, 25–38, 113–20, 141–51, and vol II, 589–604, 625–72.

30 A comprehensive study on the use of antiquarian publications by Adam and in general by artists and architects in the 18th century is still lacking. But *see* Beard 1978, 7; Bristow 1996, 78–91; Harris 2001, 4; Aymonino 2010b; and Aymonino *et al* 2013, especially 16–21.

31 The modest quality of the decoration was already noted by the 1st Duchess of Northumberland, who, visiting Kedleston in August 1766, described the room as 'painted after the antique but sadly executed'. Archives of the Duke of Northumberland, Alnwick Castle (henceforth AC), DNP: MS 121/16, fol 54r.

32 *See* Aymonino *et al* 2013, especially 22–3.

33 *See* Harris 2001, 22, figs 20 and 21. For the book's presence in Adam's library, *see* the sale of the Adam brothers' property at Christie's, 20–22 May 1818, 2nd day of sale, 21 May, lot 144, published in Bolton 1922, vol II, appendix C, 331–2, and also in Watkin 1972, 189. For Curzon's copy, *see* the 1765 inventory of his library, KH, KCL1.

34 Kedleston 1769, 23: 'The Family-Pavilion. Breakfast Room. Eighteen feet square, finished with fresco paintings and antique ornaments, after the Baths of Dioclesian [*sic*]. Chimney-piece, Statuary Marble, part gilt.'

35 Aymonino *et al* 2013, 25, cat 15 (M Modolo); Gentile Ortona, E and Modolo, M 2016 *Caylus e la riscoperta della pittura antica: Attraverso gli acquarelli di Pietro Santi Bartoli per Luigi XIV. Genesi del primo libro di storia dell'arte a colori*. Rome: De Luca, 53–4, no 3.

36 Aymonino *et al* 2013, with previous bibliography.

37 Ibid, cats 16, 17, 29, 30, 37–40. Apart from the copies listed here, an additional copy of a Bartoli drawing in the hand of Brunias was sold at Sotheby's, London, 15 July 1993, lot 48, with the wrong attribution (since withdrawn) that it constituted a 'Design for the Painted Breakfast Room at Kedleston'.

38 RIBA, Drawings Collection, vol 54: Album of copy drawings, mainly from the published sources by Pietro Santi Bartoli, Francesco Bartoli and others, Adam, Robert, 1728–1792, office of. See the RIBA Drawings Collection catalogue, 1968–89, vol 1 (A), 17–18 (*sub* Adam, Robert, Office of). The RIBA also holds a volume with line copies of the Francesco Bartoli ceilings, which nonetheless cannot be assigned with certainty to the office of Robert Adam: RIBA, Drawings Collection, vols SB4–SB9, Copies of Drawings by F Bartoli, made by the latter in *c* 1721, of antique ceilings. *See* the RIBA catalogue, 1968–89, vol 2 (B), 58 (*sub* Bartoli, Francesco).

39 The Getty volume (Los Angeles, Getty Research Institute, 2002.M.22*) bears the watermark of J Whatman and of I Villedary; *see* Edward, H 1950 *Watermarks, Mainly of the 17th and 18th Centuries*. Hilversum: Paper Publications Society, 105–6, nos 1809–10, 1846, 1849, 1856. In Aymonino *et al* 2013, the Getty rather than the RIBA volume is considered, possibly erroneously, as the source for Adam and Brunias's earliest decorative schemes.

40 *See* Aymonino *et al* 2013, 23 and note 29 (additionally to the sources mentioned there, *see also* NRS, GD18/4946, fol 1r, where James Adam mentions 'all the profusion of Eaton [*sic*] and Santo Bartoli'). Adam's knowledge of the Eton drawings has been referred to already in passing, especially by Bristow 1996, 83, 89–90; Connor Bulman 2001, 346; and Harris 2001, 4.

41 I am indebted to Mirco Modolo for the reference to the Domus Transitoria. *See also* the inscriptions on the drawings illustrated here as Figs 4.10 and 4.12.

42 Aymonino *et al* 2013, 25, no 15. For Brunias's copy at the Getty, *see* ibid, 25, no 16, and for that at the RIBA, *see* RIBA, Drawings Collection, vol 54, fol 37. The RIBA copy resembles more closely what was realised on the walls of Lady Caroline's Dressing Room than Montfaucon's print, suggesting that this was Adam and Brunias's source. The original Bartoli drawing was later copied also by Giuseppe Manocchi – *see* Aymonino *et al* 2013, 28, no 17.

43 As suggested in Aymonino *et al* 2013, 25–6, nos 16 and 18 and note 43.

44 *See* Ashby 1914, 60, Eton VI, no 15; Polignac and Raspi Serra 1998, 62–3, no 39 (L Connor and F de Polignac, who incorrectly report that the drawing was copied by Clérisseau and published by Montfaucon); Aymonino *et al* 2013, 49 n 43. A copy by Brunias of this drawing also survives in the Getty album (Getty Research Institute, Los Angeles, 2002.M.22*, no 11) and by Manocchi at Sir John Soane's Museum (Bristow 1996, 82–3, fig 85; Rowan 2003, 22, cat 13; Curzi 2010, 201, fig 3). It was later engraved in reverse in Cameron 1772, plate 54. The decorative pattern was also used by Adam and Brunias in 1761–2 as inspiration for the ceiling of the Dining Room at Kedleston (*see* Harris 2001, 35, fig 47), which will be the subject of an article by the present author.

45 *See* Ashby 1914, 60, Eton VI, no 13. It was later engraved in reverse in Cameron 1772, plate 56.

46 Another Bartoli drawing at Eton shows the same gilded floral decoration on a blue background and other similar motifs and arrangement of figures. *See* Aymonino *et al* 2013, 43, no 41. A copy of this drawing appears among the RIBA drawings (RIBA, Drawings Collection, vol 54, fol 61). This drawing was copied by Manocchi – for which *see* SM, Adam vol 26/93, and Christie's, London, 16 July 2010, lot 46 (wrongly attributed to Brunias) – and also by Charles-Louis Clérisseau, *see* Chevtchenko *et al* 1995, 100–1, no 13. It was later engraved in reverse for Cameron 1772, plate 55.

47 Aymonino *et al* 2013, 40, 42, figs 4 and 5.

48 For Adam's work at Syon and other Northumberland estates, *see* Harris 2001, 64–103, and Aymonino forthcoming.

49 The first mention of work done for 'Lord Northumberland' appears in the letter of 24 July 1760 in which Robert Adam also discusses the new 'painted room' at Kedleston (see Appendix, no 4, fol 3v).

50 This comment appears in Thomas Percy's handwritten annotation (*c* 1777) to the duke's copy of Dodsley and Dodsley 1761, VI, 7: AC, DNP MS 93/A/14.
51 Aymonino *et al* 2013, 38–9, nos 37–40.
52 Ibid, 36–8, nos 34–6. A partial copy of the drawing illustrated in Fig 4.19 also survives at the RIBA (*see* RIBA, Drawings Collection, vol 54, fol 63).
53 Pergolesi's original drawings for the published plate are preserved in the Pierpont Morgan Library, New York. *See* Harris 1971, 156–7, plate 113.
54 KH, L1–2/5, fol 1r. The original spelling has been retained in the documents in the Appendix.
55 KH, L1–2/8, fols 1r–1v.
56 KH, L1–2/10, fol 1r.
57 NRS, GD18/4866, fols 3r–3v.
58 KH, L1–2/12, fol 1r.
59 KH, L1–2/13, fols 1v–2r.
60 KH, L1–2/18, fols 1r–1v.
61 KH, L1–2/19, fol 1r.
62 KH, L1–2/20, fols 1v–2r.
63 KH, L1–2/21, fol 1v.

5

Robert Adam's scenographic interiors

Miranda Hausberg

Robert Adam conceived of his domestic interior designs scenographically, using the techniques of the theatre to make them and the language of the theatre to describe them. In the first volume of *The Works in Architecture of Robert and James Adam*, published in 1773, Adam remarked that in his renovations at Syon House (from 1762) the 'inequality of levels [of the old floors] has been managed in such a manner as to increase the scenery and add to the movement, so that an apparent defect has been converted into a real beauty'. In his discussion of the same house's majestic anteroom, Adam boasted that six columns of 'verd antique marble ... form the room and heighten the scenery'. And when he described his plans for Syon's intended 'great circular saloon', which was never executed (*see* Fig 5.5), Adam attested that this room's 'scenery, like the decorations of a theatre, apparently increases the extent, and leaves room for the imagination to play'.[1]

For Adam, 'scenery' referred to the architectural forms that had been manipulated to make spaces appear larger and more complex than they were. He made explicit this meaning in *The Works*, informing his readers that the technique derived from traditional practices in the construction of theatre sets, which often incorporated illusionistic effects to make the shallow, simple space of the stage appear deep and complicated. The illusionistic treatment of space was an important aspect of Adam's scenographic interiors, but it was only one element of his vast and sophisticated scenic repertoire.

This chapter examines six of the key 'scenic' qualities of Adam interiors, with the intention of revealing the power and specific functions of this little-studied aspect of his work. It also identifies the salient points of intersection between architecture and theatre in this era, the so-called 'Golden Age' of British theatre, and shows how the term 'scenographic' should stand alongside 'picturesque' as a closely related and equally significant hallmark of Robert Adam's interior designs.[2] A final section explores the scenographic aspects of Adam's most theatrical inventions: temporary structures built for specific celebratory occasions and intended to be used only for one night.

But before looking in detail at Adam's interiors, a consideration of three contextualising issues is instructive: an elucidation of the term 'scenographic'; an explanation of the intimate relationship between the scenographic and picturesque in 18th-century Britain; and an outline of the state of the art of British theatre during the age of Adam.

Scenography

To apply the term 'scenographic' to particular types of interior space – such as those spectacularly executed by Adam and, later, Sir John Soane – is a recent trend.[3] In the 18th and 19th centuries, the word denoted the representation of three-dimensional objects in perspective and was associated almost exclusively with draughtsmanship and painting. According to the *Gentleman's Magazine* of 1743, for example, 'scenography is the representation of a place drawn according to its appearance in perspective, and is such a view of it as a painter would give'.[4] In his dictionary, Samuel Johnson similarly defined scenography succinctly as 'the art of perspective'.[5] In this sense, scenography has always been strongly associated with architectural drawing; and the closely related term 'scenery', as we have seen above, was applied directly to architectural interiors and most commonly denoted something approximate to 'beautiful appearance'.[6]

In the 20th century the use of the term broadened to refer also to the design and creation of theatre scenery, and to the design of performance environments, including lighting, costumes and special effects. Eventually 'scenographic' took on an expansive abstract connotation that embraced all the visual, experiential and spatial aspects of performance, as well as audience reception and engagement. In the 20th century the term was also applied to architectural monuments: John Summerson considered John Nash's composition of Regent Street in London (of 1809–26), for example, to be an exercise in scenographic composition, since his building façades were designed to be seen in perspective and experienced as a series of episodes.[7] Despite the increasing use of the term in architectural writing, however, the ways in which architecture can behave scenographically have yet to be analysed fully.

The scenographic and picturesque

Adam lived and worked in a culture saturated with ideas about scenography and the picturesque, and he held a nearly lifelong devotion to both. 'Picturesque' was a term originally used in landscape gardening, before its importation to other arts, where it alluded to all of the compositional aspects borrowed by gardeners from certain kinds of landscape painting, notably the works of Claude Lorrain, Nicolas Poussin and Salvator Rosa. The closely related terms 'picturesque' and 'scenographic' concern not merely the making of visual phenomena but also their reception by viewers. Both are associated with the belief that the purpose of art is to raise feelings and excite the imagination, and to act as the connective tissue between sensory experience and emotions. Fundamentally each concept is grounded on the relation of pictorial form to visual experience, and on understanding the means of heightening emotional and sensory states, rather than the intellectual.

By mid-century, and during Adam's most productive decade in the 1760s, both concepts were pervasive, though latent. In the 1770s, at a time when Adam's art and architecture were becoming increasingly aligned with the picturesque, the ideas and theories concerning that genre, which British writers and artists would explore and clarify towards the end of the 18th century, were then still nascent and irreconcilably intermingled.[8] At this time a precise terminology and a clear distinction between the picturesque and 'scenery' (as it applied to architectural interiors) did not exist, and Adam used both terms freely in his publications when discussing composition, taste, perception and aesthetics. Both in his architecture and in his writings, Adam demonstrated an awareness of the scenographic and the picturesque, which often functioned visually in identical ways.

British theatre in the age of Adam

Robert Adam lived during the so-called 'Golden Age of Theatre' in Britain, between around 1730 and 1830, when the fine and theatre arts were more closely intertwined than ever before or since.[9] This was also an age of popular theatre, when a high proportion of the population attended performances, as had been the case in 5th-century Athens and Elizabethan London. The expansion of theatre-going and theatre-building has never been matched: in 1740 the Theatre Royal, Drury Lane, was the sole established playhouse in London, but by 1805 over 280 places of regular theatrical entertainment existed. The Irish playwright and biographer Arthur Murphy in his *Life of David Garrick* (1801) mused that 'theatre engrossed the minds of men to such a degree ... that there existed in England a fourth estate, King, Lords and Commons, and Drury Lane Playhouse'.[10]

While theatre was thus becoming more closely allied to the fine arts and simultaneously providing gripping entertainment for the British masses, it underwent a gradual but dramatic stylistic shift under the guiding hand of one of the most prominent British actors and theatre managers of the 18th century, David Garrick (1717–79). Garrick's career spanned the transition in the history of British theatre from an insistently aural art, centred on showcasing an actor's rhetorical skill (represented by the sonorous James Quin), to a primarily visual one.[11] This shift demanded the transformation of the playhouse from a space reminiscent of an aristocratic residence to one in which every aspect of architecture and scenography was designed to transport the spectator's imagination to the 'other worlds' that the spectacle of performance sought to conjure. Before this transformation, early and mid-century Georgian theatres recalled grand domestic interiors in terms of their decorative motifs, lighting and the presence of picturesque landscape scenes (used as scenic backdrops in theatres). Moreover, like the great houses of the era, they were also sites for the complex dynamics of spectatorship and highly choreographed social rituals. It was not uncommon for members of the audience to move freely during the performance, to socialise, and to express their opinions stridently, and they were often seated on the stage itself.[12]

The domestic character of the theatre and the close relationship between actors and their audience began to be seen as damaging to new ideas about theatrical representation. Garrick campaigned tirelessly to remove the audience from the stage, a tradition he treated comically in the composition of his first dramatic work *Lethe, or Aesop in the Shades* (performed at Drury Lane on 15 April 1740). In *Lethe* the Fine Gentleman is made to say: 'I stand upon the stage, talk loud, and stare about, which confounds the actors, and disturbs the audience.' In 1762, more than 20 years later, Garrick was finally able to banish spectators from the Drury Lane stage.[13]

Although Garrick's crusade to make the theatre more scenic and illusionistic began as early as the 1750s, the transformation was not complete until well after his death in 1779. It was not until the early 1840s that the 18th century's place of performance, where the audience and actors shared the same architectural volume, light and condition of reality, would be entirely replaced. The new scenic theatres in Britain had divided interiors with discrete architectural spaces for spectators and actors, now separated by a decorated proscenium through which a fixed and silent audience watched the performers on the brightly lit stage from the dimly lit auditorium.[14] Adam witnessed this transition, seeing his own bright decorations for the auditorium at Drury Lane (Fig 5.1) 'all put out by painting over them' in 1783, less than a decade after their making. At that time, the house scenic artists Thomas Greenwood and William Capon covered Adam's work with sombre, darker tones that focused attention on the proscenium stage.

Adam, himself an amateur performer (as a singer), was thoroughly familiar with all aspects

Fig 5.1
Benedetto Pastorini, Drury Lane Theatre interior, 1775, as it appeared from the stage. From Sir John Soane's Museum's special coloured edition of The Works in Architecture of Robert and James Adam *(Adam and Adam 1773–1822), vol II, 1778, plate VII.*

of the theatre, and he must have learned a considerable amount about stagecraft and theatre design from his long working relationship with Garrick.[15] Garrick in 1772 purchased a residence in the Adam brothers' Adelphi development, and later commissioned Robert to make alterations to his Thames-side villa at Hampton, including the remodelling of the portico. Then in 1775 Garrick hired Adam to redecorate and renovate Drury Lane, the 100-year-old father of English theatres. A lifelong theatre-goer, both while on his grand tour and in London, Adam supposedly walked to Drury Lane each Wednesday evening to watch Garrick perform; late in life his brother James seems to have used his cousin Willie's influence with R B Sheridan to secure a box there as a 'favour'.[16] Also significant in this regard was the Adam brothers' friendship and working relationship with Giovanni Battista Piranesi, who was a stage designer as well as an artist, and a maker of hundreds of scenographic prints.

Adam and scenographic illusion

To understand Robert Adam as a scenographic architect, it is useful to begin with his own description of this mode of composition, to 'apparently increase the extent' of rooms. As others have noted, like a set designer, Adam was skilled at illusionistically enlarging or otherwise altering the perceived character of rooms. This characteristic of Adam interiors was accomplished by a remarkable number of shrewd and varied methods: through what Adam called 'piercings', or openings between rooms (as in the proposed rotunda for Syon, *see* Fig 5.5); through the insertion of shallow recesses, niches and apses (as in the dining room at Kedleston Hall, of *c* 1760–5); through illusionistic ceiling and floor patterns (as in the Long Gallery at Syon, which appears wider than its actual dimensions; *see* Fig 5.13); through the use of large mirrors

Fig 5.2
The library, Kenwood House, of 1767–70.

(as in the library at Kenwood House, of 1767–70, which seems wider and taller than it is in reality, Fig 5.2); through the placement of columns (as in the Syon anteroom, which seems square, although it is rectangular); through the elongation of the proportions of architectural orders; through the avoidance of 'ponderous compartment ceilings' and other weighty and projecting forms; and through the application of light colours to walls and ceilings.[17]

In all of these strategies for manipulating appearances, a leitmotif is the continual play with plane and recession, whereby precise depths remain imperceptible to the distant viewer. Sometimes Adam made recesses, such as niches, apses and alcoves, seem untruthfully shallow. He achieved this flattening, illusionistic effect by avoiding the use of framing elements, and by muting shadows through the application of light-coloured paint and intricate decorative patterns. Conversely, Adam at times also created an illusion of depth by using darker shadows and dark-toned wall decorations, and by incorporating architectural forms that articulated intricately organized space, as in the library at Kenwood.

At Syon, Adam played even more complex *trompe l'œil* games. In the dining room, for example, he inserted elements that made the space behind the column screen seem, at the same time, both untruthfully shallow and deep, thus creating a moment of striking spatial ambiguity (Fig 5.3). Here the patterned decoration in the apse and the continuation of the wainscoting from the adjacent walls worked together to make the space appear shallow, but conversely the door frame and shadows in the niches made the space seem deep. Similarly, throughout the house, Adam created the illusion of making the passages between rooms invisible. This *trompe l'œil* effect worked to 'add to the scenery' by illusionistically enlarging single rooms, to establish unity among

Fig 5.3
Syon House, dining room, 1762–9. The dark red-brown paint in the niches is a later addition.

rooms of radically varied characters, and to create what Adam called 'movement'. For example, in the hall (Fig 5.4; *see also* Fig 2.9), a doorway behind the column screen, when open and seen from directly in front, frames a niche in the distant back wall of the adjoining imperial anteroom (later filled with a statue), thus creating the illusion that the wall continues where an opening (a doorway) exists in reality. The greatly contrasting colours of the two rooms, and the intrusion of part of the entablature from the anteroom's own screen of columns, disrupt but do not fully break the illusion of a continuous wall.

One of Adam's most spectacular scenographic designs, the unbuilt rotunda at Syon, would have been a masterpiece of theatrical spatial manipulation (Fig 5.5). 'The form' of the rotunda, Adam tells us, 'is new and singular; it is a circle within a circle, the smaller opening into a larger, by eight piercings, adorned with columns, and terminated with niches and statues'.[18] In the centre of this dramatic space, the inhabitant would have seen the statue-filled niches of the outer ring framed seamlessly by the openings of the inner ring, creating the illusion that the inner ring was pierced only four times, rather than eight, and that the outer ring did not exist.

Adam's mastery of scenographic illusion was undoubtedly inspired and shaped by the innovations in Italian baroque scenography made by the Bolognese architect and painter Ferdinando Galli-Bibiena (1657–1743), and popularised by his book, *L'architettura civile* ('Civil architecture', 1711).[19] His style was brought to culmination in a visual art not directly associated with theatre in the engravings of Adam's friend and adviser

Fig 5.4
The east end of the entrance hall at Syon House, with an angled view into the anteroom beyond, described by Robert Adam as showing 'also the steps, rising to the anti-room, and the additional scenery occasioned by that circumstance'. When seen straight on, a niche in the anteroom (now filled with a statue) occupies the opening, giving the illusion of a continuous wall surface. Engraving by Thomas White from The Works in Architecture of Robert and James Adam, *vol I, no I, 1773, plate VI.*

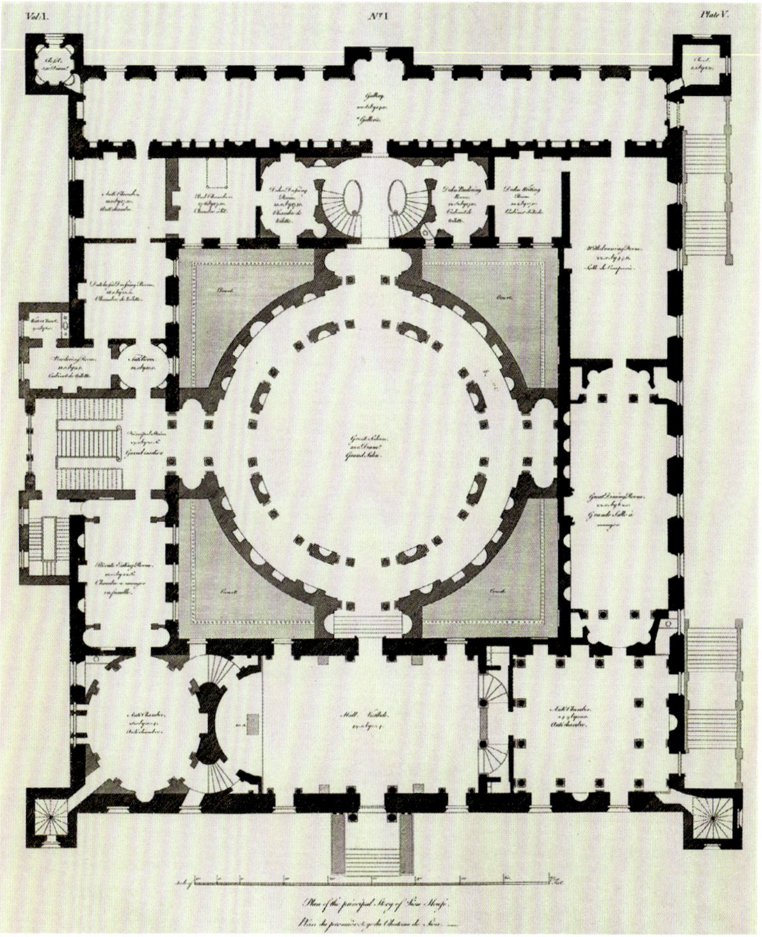

Fig 5.5
Plan of Syon House. From The Works in Architecture of Robert and James Adam, *vol I, no I, 1773, plate V.*

Piranesi, who drew heavily in the composition of his prints from stage designers of the time, and in turn inspired them.

The making of these spatial illusions was often accompanied by the artful mimicry of materials. Adam's well-known *trompe l'œil* effects included the imitation bronze plaque above the fireplace in the anteroom at Syon and the fictive Etruscan vases at Osterley (1761–80). Although he played with *trompe l'œil* illusions to induce both spatial and material deceptions, Adam never drifted into the realms of the fantastic or delusional, which were investigated by Piranesi and Adam's teacher the French architectural draughtsman Charles-Louis Clérisseau. Both Piranesi and Clérisseau frequently shifted points of orientation in their prints and drawings, and created alternate, imaginary worlds. In contrast, Adam played within the realm of the rational and created games for the senses that kept viewers actively engaged with the architectural interior and their own experience.

One building where Adam was publicly acclaimed for his scenographic skills was, fittingly, his renovation and decoration of the Theatre Royal, Drury Lane, in 1775. One reviewer enthused: 'At first view I was a good deal surprised to find that by some means or other the ingenious Artists had contrived to give an Appearance of greater Magnitude to the House. I knew it was not rebuilt but only repaired, and consequently that there could be no additional space within the old Walls and Roof.'[20] To achieve this effect of enlargement, Adam drew upon techniques he had employed in domestic interiors. As shown in the well-known engraving of Drury Lane by the Italian printmaker Benedetto Pastorini (*see* Fig 5.1), and attested by contemporary descriptions, Adam substituted slender piers for the weighty ones that Christopher Wren had used to support the two balconies, raised the ceiling by 12ft (3.66m), made the forestage shallower and deepened the galleries.[21] Significantly, Adam's structural renovations at Drury Lane would set the pattern for all other theatres in Britain.[22] In addition to these structural changes, Adam also made radical decorative alterations that transformed the auditorium's character: he painted the walls and seating boxes pale green and pale pink, and inlaid the pilasters, which decorated the slender piers, with coloured glass to create an effect described as 'leste [lustrous] and brilliant'.[23] The rhythms of the intercolumniation and decorations worked together to create the appearance of greater magnitude and spaciousness, as contemporaries noticed. Pastorini's engraving, which Adam commissioned for publication in *The Works*, exaggerates this scenic quality, as the human figures are far too small and the proportions of the house are amplified.

Adam and scenographic sculpture

The profusion of figurative sculpture, much of it gesturing theatrically, is a second scenographic element of the Adam interior. Like many aristocrats, Adam's patrons collected sculpture and displayed it in their houses, and sometimes had pieces made for particular spaces, or acquired especially on their behalf by the Adams.[24] This was part of the larger culture of exhibition in Britain, which showcased the wide variety of decorative objects brought home along the rapidly expanding travel and trade routes that linked Britain and the rest of the world.

Within this culture, statuary had a distinctive theatrical connotation, since comparisons of sculpture and actors were commonplace – particularly analogies between the 'body language' of actors and the *Apollo Belvedere* (*c* 120–140 AD) and the *Laocoön* (*c* 27 BC – AD 68).[25] These ancient works were popular not only because of their historical significance and visceral visual appeal, but also because they played starring roles in contemporary aesthetic debate. Syon was linked to both, as copies of both statues were originally intended for the entrance hall there; sadly, however, the *Laocoön* replica was never executed.[26]

In contemporary thinking, actors and statues were considered equivalent objects of aesthetic contemplation that could similarly lead the mind to a 'pleasing train of ideas'.[27] The successful tragic actor was expected to embody the Classical ideal of ancient sculpture; the legendary British actors Sarah Siddons and John Kemble were both said to have deliberately modelled their static poses assumed during monologues on Classical statuary (Fig 5.6). These theatrical associations, which viewers would have made with the Classical sculpture on display in domestic interiors, were given heightened significance by the era's quasi-philosophical notion that actors were exemplars of controlled behaviour. The statues would thus have fuelled the contemporary debate about 'proper' conduct, particularly in domestic settings, and would have also stirred musings about the conspicuous status of the actor and the burgeoning culture of celebrity. In this 'age of the actor', talented players received far more attention than the plays themselves, and books, pamphlets, critiques, prints and paintings of the time reveal the public's ruthless and fetishistic scrutiny of actors' public personae and private lives.[28]

Fig 5.6
Mrs Sarah Siddons in the character of Euphrasia (below), from Arthur Murray's tragedy The Grecian Daughter, *engraved by James Caldwall after the oil portrait by William Hamilton. The* Apollo Belvedere *(below right), engraving of 1617 by Hendrick Goltzius after the Roman 2nd-century AD copy of a lost Greek bronze, now in the Vatican Museum.*

Scenographic framing

The prominence of framing devices is a third scenographic aspect of Adam's interiors. Like the theatre, with its proscenium, and the picturesque landscape gardens that were inspired by carefully framed painting compositions, Adam's scenographic interiors were artificially constructed scenes that benefitted from framing. Although Adam drew heavily on the Palladian tradition of compartmentalised architectural ornament and the French rococo convention of dividing the wall into panels, he went beyond these precedents to employ emphatic frames for dramatic effect, to unify his interiors, and to assert the primacy of architecture in rooms that were filled with paintings and sculptures from the patron's collection (Fig 5.7). In his preoccupation with the architectural frame, Adam was also compelled to engage with the period's enthusiasm for arranging and displaying large numbers of framed paintings within architectural interiors.

The strong proscenium motif was Adam's most celebrated scenographic framing device, and its usual form was the column screen, exemplified in his iconic library at Kenwood.[29] Hyperbolically accentuated door jambs and fireplace mantels, niches and apses, mirrors, and framed panels on walls and ceilings also act as proscenia. Significantly, at precisely the time that Adam was designing these interiors, Garrick was campaigning to give new importance to the proscenium arch in the British theatre, where it had long functioned as a mere decorative device, set in the background of the thrust stages.[30] Garrick insisted that henceforth the proscenium should act as a fixed boundary between audience and stage – between the viewer and the viewed.

Adam also played with the tension between framed easel paintings and the framed compartments of his architectural wall decorations. Both systems directed attention to the rectangular neoclassical architecture of the room – the room's form – rather than the pictorial representations they enframed. In this frame-heavy design matrix, the framed easel painting became an instrument of Adam's comprehensive spatial scheme, providing eruptive glimpses of fictive depth that counterpointed the optical flatness of the surrounding wall decorations. This dynamic scenographic interplay exploited painting's illusionistic potential and intensified the viewer's awareness of the basic paradox of pictures: to appreciate the artistry of their spatial illusion requires recognition of their flatness. Adam's framing games also lured the viewer into a state of heightened awareness of the act of looking, and of the nature and limits of both sight (the sensory information taken in by the eye) and vision (the relay of visual information to the brain and the interpretation of images captured by the eye). He achieved this primarily through the play between the perception of depth and flatness, whether real or imagined.

Architecture as landscape

Two further elements in Adam's architecture combined the scenographic and the picturesque. The first was the bringing of picturesque scenery indoors, which Adam sometimes accomplished by enlivening his interiors with dramatic picturesque landscape *capricci*, such as the four painted panels in the music room at Harewood House (*c* 1765–71; *see* Fig 0.10). This conceit had a long tradition in the theatre, particularly in Italian and French opera. An even more illusionistic use of painted landscape panels was seen in the dining room at Osterley. Here, Adam centred picturesque paintings of ancient ruins by the Venetian artist Antonio Zucchi (1726–95) on each end wall. The large size, crisp detail, sharply angled views, and chiaroscuro treatment made these paintings more potent than the views of real scenery glimpsed through the windows, offering a heightened scenographic effect.[31]

Fig 5.7
The hall at Osterley Park, designed in 1767.

Adam also brought natural scenery indoors by manipulating views of the surrounding landscape through the canny placement of windows and mirrors.[32] At Gosford House (1790–1800), for example, the triplet Venetian windows on the west elevation of the main block command a panorama of the city of Edinburgh and the Firth of Forth. At Kenwood, Adam's strategy was different: he set mirrors in square recesses on the north wall of the library that duplicate by reflection the panorama of the city of London and the River Thames, seen through the windows on the opposite side.

Yet, Adam did not merely import representations and glimpses of picturesque landscapes into his houses; he also conceived of his interiors much like landscape gardens, and sought to create an experience in walking through them that was akin to walking through nature. This is a fifth scenographic aspect of Adam's interiors. In order to achieve such an effect, the architect must rouse in the visitor associations and ideas comparable with those generated by the variety and picturesque qualities of natural scenery. Adam did this by constructing a consecution of diverse exterior and interior spaces – different 'scenes', cast in varying interplays of natural light and shadow. Adam also configured some interiors to mimic caves in their spatial organisation, although the details were man-made; the sculpture gallery at Newby Hall (1767–74), which resembles a system of caves, is a prime example (Fig 5.8). These innovations paralleled trends in contemporary British stagecraft, which reached an apex in Philip James de Loutherbourg's *Eidophusikon* (opened February 1781), a miniature mechanical theatre that sought to recreate 'natural phenomena' – the sounds, light, weather and scenery of the natural world.[33]

Significantly, Adam also believed that experiencing a house and a landscape were roughly

Fig 5.8
Newby Hall sculpture gallery, 1767–74.

equivalent. The commonalities between architecture and gardens had been developed in the writings of his contemporaries the French theorist Marc-Antoine Laugier (1713–69) and Adam's friend the Scottish philosopher Henry Home, Lord Kames (1696–1782), but they did not discuss the particular relationship between interior design and landscape that Adam's work embodied.[34] Adam's designs are dramatic and scenographic because they intensify the sensory and mental experiences of the visitor or viewer when walking through an architectural interior, much like when moving through a garden, but it is the spectator who controls the method and organisation of this viewing, and the pace at which information is processed. Although in both cases the artist or architect determines the parameters and content, the spectator's experience of either house or garden is subjective, contemplative and associative, centred on identifying relationships among objects and natural phenomena, and between these elements and one's own worldly experience.

Room for imagination

The sixth and final aspect of Adam's work to be considered here is his creation of spaces that, in his own words, 'leave room for the imagination to play'. In this, he invoked concepts of active viewership that were central to the 18th-century discussion of both landscape design and theatre.

Adam activated the imagination primarily through what he famously called 'movement' – exemplified, for him, in the south façade of Kedleston Hall (*see* Fig 0.13).[35] Although movement is conventionally taken to refer to the creation of irregular, or perhaps even picturesque compositions, it also entailed a picturesque mode of perception that energised the role of the spectator. That is to say, Adam's idea of 'movement' referred to the movement of the viewer's eye, and of their feet, and ultimately of their imagination, galvanised by sensory and mental responses to arrangements of diverse forms.[36]

Adam induces movement of the eye by diversifying the contours of rooms and objects, creating a play of light and shadow, varying the heights and sizes of rooms, emphasising the variety of architectural forms and their picturesque decoration, and making shallow spaces appear deep (*see* Fig 5.3). Importantly, the gymnasium for the eye that Adam created in his decorative schemes was intended to exercise the senses and emotions rather than the intellect.

Adam prompted movement of the feet chiefly by establishing tantalising sight lines between spectacularly and variously decorated rooms – a visual idiom directly inspired by stage-set design. Movement was also stimulated by varying room heights, and by making intimate pockets of space that invited the inhabitant to move to or through them. In this scenographic interior the empowered spectator participated in a new mode of perception that is tied to the movement of the body through space.[37]

Adam's movement of the mind, his 'play of the imagination', is more poetic than the movement of the eye and of the feet. It might be best understood as a corollary of the pervasive 18th-century interest in the 'pleasures of the imagination', as explained by Joseph Addison, among others. Addison posited that two kinds of pleasures arise from sight: 'those that proceed from objects before our eyes and those which flow from the ideas of visible objects when the objects are not actually before the eye, but are called up into our memories'.[38] In consonance with that widely held view, Adam deployed forms and iconographic systems to activate both the mind and body, soliciting both intrinsic and associational or relative responses from visitors. And while a principal function of his work was to evoke memories of the grand tour, undertaken by well-to-do visitors, Adam also cleverly disposed his decorative works, like stage sets, to convey a more general sense of history – one that transcended a particular moment, event or period – primarily through an eclectic approach to design and a preference for landscape over narrative scenes.

Adam's scenographic temporary structures

Robert Adam's temporary structures, designed and built for elaborate royal and aristocratic parties and celebrations, were his most scenographic creations. At least five such 'monuments' have been attributed to him: a bridge and an architectural screen holding three large 'transparencies' (oversized glowing coloured images, created by painting a calico or linen canvas soaked in varnish and setting it before lamps), made for the garden at Buckingham House for the 25th birthday of King George III on 4 June 1763 (*see* Fig 0.16); a bridge and pavilion erected in the open central court at Syon House for a 'Grand *fête*' held in honour of the King of Denmark in October 1768; and the spectacular pavilion built for one of the most extravagant parties of the 18th century – the *fête champêtre* of Lord Stanley (Edward Smith Stanley, later 12th Earl of Derby), staged at The Oaks, near Carshalton in Surrey, on 9 June 1774 (Figs 5.9 and 5.10).[39] Adam also designed the various interior settings at No 3 (later 26) Grosvenor Square, London, including a temporary glass dome over the staircase (an installation which required the partial removal of the roof), when that house was significantly altered for the extravagant ball and supper hosted there by Lord Stanley on 31 March 1773.[40]

Each of these ephemeral creations provided a dramatic setting for a social 'performance', enlivened with dancing, singing, instrumental music and choreographed rituals for costumed 'actors' – namely the host and guests, who dressed in special clothing for the event. Lord Stanley's celebrations were particularly theatrical in this regard. A notice in the *Morning Chronicle* of 31 March 1773 reported that his festivity at Grosvenor Square would boast a 'grand Quadrille, consisting of eight, habited in the dress of the four seasons'. In this quixotic dance, Lord Stanley personified Spring, dressed in white 'spangl'd all over with little Tufts of real Violets'. At the same party, tea was served by 'Vestal Virgins', young women garbed in white lutestring (a glossy silk fabric) with blue ribbons.[41] At his lavish *fête champêtre* the playful host 'was dressed like [the Flemish painter Peter Paul] Reubens, and Lady Betty Hamilton like Reubens' wife', while many guests and all performers wore pastoral costumes.[42]

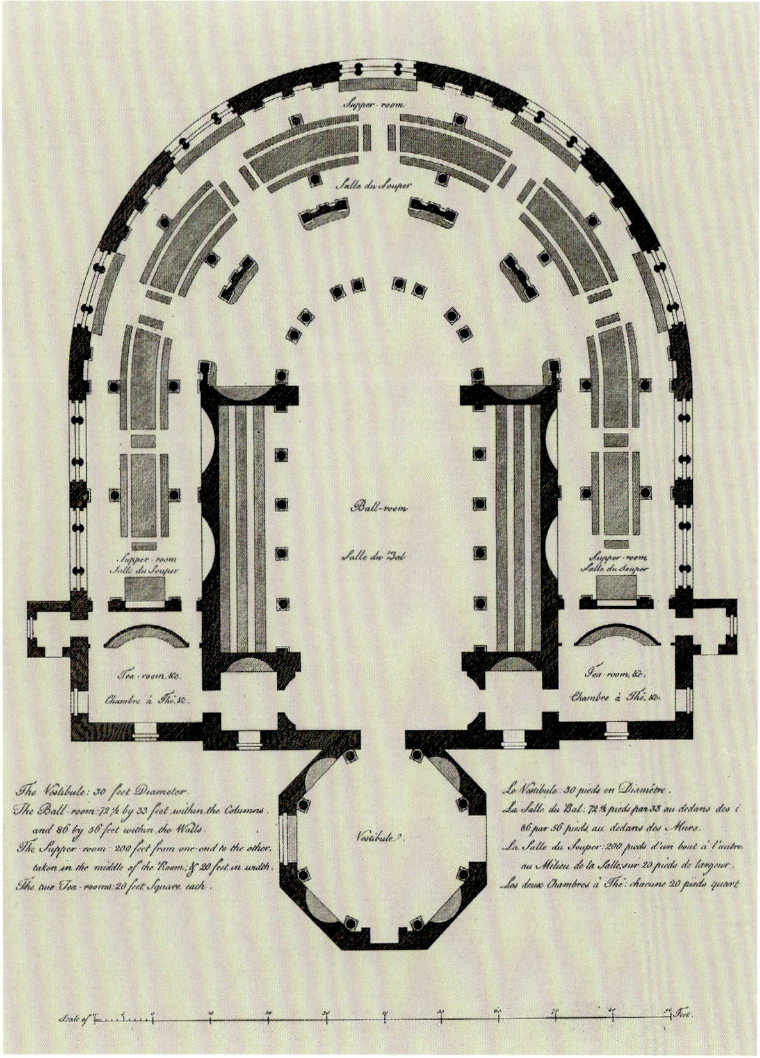

Fig 5.9
Plan of the pavilion for the Earl of Derby's fête champêtre *at The Oaks, Surrey, designed by Robert Adam, 1774. From* The Works in Architecture of Robert and James Adam, *vol III (1822), plate XX.*

In the conception and creation of these astonishingly lavish and complex projects, Adam played the roles of theatre manager and stage designer to create structures that embodied the qualities of theatrical architecture. His designs, however, were ahead of contemporary practices in the art of theatre and looked forward to those of the next century. Like 19th-century stage sets, each element of Adam's ephemeral entertainment spaces was carefully crafted to play a role in shaping a theatrical setting for a carefully planned programme of choreographed, sequential actions. These temporary structures were encompassing, integrated, total environments, whose intensity surpassed even that of Adam's intricate and unified domestic interiors. They were designed to shape mood strongly and to exhilarate the participants through the use of flamboyant architectural decorations, extravagant illuminations and dramatic changes in setting that recalled scene changes at a theatre performance.

As in the theatre, Adam's temporary structures brought to the fore complex issues of representation and spectatorship. Objects on a theatre stage and on Adam's 'stages' functioned both symbolically and practically, creating spaces that on the one hand realised illusions (as a remove from practical reality) and on the other resisted being categorised as illusions because of their functionality, which rooted both actor and viewer in immediate experiences. Thus a table in each setting was both a table and the symbol of a table. (In contrast, objects within Adam's scenographic domestic spaces merely approached the symbolic and never quite attained the level of illusion and detachment achieved by those in his party pavilions and in successful late 18th- and 19th-century theatre sets.) Also, like the theatre sets of the next century, Adam's temporary

Figs 5.10
Antonio Zucchi, the pavilion erected for the Earl of Derby's fête champêtre *at The Oaks, Surrey, 1774: the ballroom (left) and the supper room (below).*

structures were spaces of both symbolic and aesthetic representation. Many of the objects he brought into his settings were authentic artworks, intended to induce sensory pleasure simply through their beauty. Thus, as on a stage, a table in one of Adam's temporary monuments was not only a table and a symbol of a table, but also an autonomous work of art.

With regard to spectatorship, Adam's temporary structures, like later theatres, were both absorptive and theatrical settings, where audience and actors alike oscillated between awareness of a beholder's gaze (theatricality) and obliviousness of an audience due to engrossment in activities other than spectatorship (absorption).[43] This oscillation occurred because inhabitants in these scenographic environments acted simultaneously as self-conscious spectacles on view for the pleasure of others, and as participants in meticulously planned entertainments, such as dining, dancing and firework-viewing, which captivated them and temporarily suspended self-awareness.

Among these projects, the one most directly connected to the theatre was Adam's pavilion for the *fête champêtre* at The Oaks (*see* Figs 5.9–5.10). As Michael Burden has shown, this astonishing structure inspired de Loutherbourg's sets for the staging at Garrick's Drury

Lane of *The Maid of the Oaks* – an expanded version of the 'sylvan masque' that General John Burgoyne (Lord Stanley's uncle) had composed especially for the *fête*. Moreover, in designing the pavilion, Adam used a theatrical device – raising curtains to reveal a new scene – to delight and to surprise Lord Stanley's guests. After dancing minuets and cotillions at length in the ballroom, revellers were notified by an explosion of rockets of a shift in programme from dancing to dining; to their amazement, the six crimson ballroom curtains (which they had assumed were window curtains) were then drawn up to reveal the enormous dining room that encircled the dancing space.[44] Thus, Adam's Oaks pavilion also enlivened the guests' experience through dramatic, sudden expansions of space. In addition to the theatrical curtain-raising in the ballroom, Adam's plan of his giant entertainment space (*see* Fig 5.9) shows that upon entry into the relatively confined octagonal vestibule, one would have viewed the ballroom at an angle

Fig 5.11
The marble hall, Kedleston Hall. Begun to designs by Robert Adam of c 1760–4; completed 1774–7 to designs by his former draughtsman and protégé George Richardson.

Fig 5.12
Charles-Louis Clérisseau, Sketch of Painting of the Western Wall of the 'Ruined Room' in the Trinità dei Monti, c *1760.*

that initially concealed its great size. Lastly, in creating such a grand private space, Adam encountered the same problems faced by those who designed public auditoriums, in having to facilitate the intimate experience of watching a play for hundreds of people. The Oaks pavilion was of staggering dimensions and included the largest rooms that Adam ever designed or decorated: a U-shaped supper room of approximately 4,000sq ft (372sq m) and a ballroom of nearly 4,200sq ft (390sq m). This must have been a thrilling venture for an architect who always yearned for large-scale public commissions, and who had once considered vying for the job to replan the city of Lisbon after its destruction in 1755 by an earthquake and the ensuing tsunami and fires.

Conclusion

Adam's domestic interiors, like contemporary British landscape architecture and theatres, were designed to operate in a world of active viewership in which scenographic and picturesque elements were of great importance. Although these elements can be recognised in the interiors of earlier British architects, such as William Kent and James Stuart, it was Adam who first applied them to every branch of architectural design. In his experimentation with the scenographic and picturesque, Adam contributed to the powerful contemporary idea that a room was not merely a form and its contents, but the synthesis of space and the experiences and memories of both its inhabitant and architect. For Adam, the 'scenery' of a room was inseparable from the inhabitants' sensory and emotional experiences; it was not complete without an active being inside it.

Two final thoughts about Adam's scenographic interiors have the potential to initiate a broader discussion of these spectacular creations. Firstly, his experimentation was a component of the 18th century's fascination with theatrical, romantic and archaeologically inspired spaces, which were dedicated to the stimulation of pleasure and desire. In this context Adam's interiors might productively be compared with Clérisseau's decoration of a monastic cell (*c* 1760) in Santissima Trinità dei Monti, perched above the Spanish Steps in Rome, in which he painted all four walls and the ceiling to create the illusion of a ruin (Figs 5.11 and 5.12).

Secondly, Adam's scenographic interiors, although created for private clients, prefigured the forms of the new public architecture that his generation of designers was beginning to create for the revolutionary society of the late 18th century. The scenographic adventure of the intended Syon rotunda would soon be enlarged and

Fig 5.13 (above)
The Long Gallery at Syon House, 1762–9.

Fig 5.14 (right)
François Jean Delannoy, Galerie Vivienne, Paris, 1823.

eclipsed by the illusions conjured up in the newly invented 'panoramas', introduced in Edinburgh in 1788 by Robert Barker (1739–1806).[45] The display of aristocratic sculpture collections that Adam had orchestrated (as at Syon) were already in the process of being replicated in the first public museums, like the Museo Pio Clementino at the Vatican (founded 1771). And insofar as his richly complicated interiors induced modern eyes, feet and minds to wander productively, they laid a foundation for the visual culture of the next century – including the 'phantasmagoria' that the German philosopher and critic Walter Benjamin (1892–1940) would see in the 19th-century shopping arcades of Paris (Figs 5.13 and 5.14). In short, even as Robert Adam's scenographic architecture encouraged viewers to step from the audience onto the stage and become actors in the drama of their own lives, it also provided architectural models for the public realm that these modern actors would inhabit.

Notes

1 Adam and Adam 1773–1822, vol I, no I, 'Plans of Sion House' / 'Explanation of the Plates'. While his brother James was Robert Adam's partner and made significant contributions to the work of the architectural partnership, it has become customary to credit Robert with responsibility for the design and oversight of most of their projects.

2 A contrasting view of the intersection between architecture and theatre in the 18th century is provided by the work of the French architect Germain Boffrand (1667–1754). In his *Livre d'architecture* (1745), Boffrand presented architecture quite literally as theatre, arguing that buildings operate as stage-settings and as actors, announcing their character to the audience through their composition and ornament. This idea became a starting point for the architectural aesthetics of character. Boffrand's book is also significant because its design in many ways prefigures the advanced layout of the Adam brothers' own *Works in Architecture*. Both books begin with theoretical discussions before giving brief descriptions of the works, followed by plates with explanatory captions.

3 *See* Kondo 2012, 70ff ('Scenographic appeal').

4 *Gentleman's Magazine* **13**, 1743, 584.

5 Johnson, S 1755–6 *A Dictionary of the English Language*, 2 edn, vol II. The definition remained constant in later editions of Johnson's dictionary at least until the end of the 18th century.

6 Use of the term 'scenery' in reference to architectural interiors is also found in the letters of Adam's contemporary, the author and politician Horace Walpole. The latter, for example, described a staircase in William Kent's miniature palazzo for Lady Isabella Finch on the west side of Berkeley Square as 'as beautiful a piece of scenery, and considering the space, of art, as can be imagined' (cited in Cornforth 2004, 177 n 65). Similarly, Walpole, in consideration of the great Elizabethan prodigy house at Burghley, exclaimed, 'Burleigh. A noble Pile! The Inner Court is beautiful Scenery' (cited in Toynbee, P (ed) 1928 *Horace Walpole's Journals of Visits to Country Seats*. Walpole Society **16**, 58). I would like to thank Jeremy Musson for bringing the latter quote to my attention.

7 *See* Summerson, J 1980 *Life and Work of John Nash*. London: George Allen & Unwin.

8 The picturesque in modern times is largely the result of the reorganisation of ideas set forth by a group of British artists, landscape designers and writers – most prominently William Gilpin, Sir Uvedale Price, Richard Payne Knight and Humphrey Repton. A useful watershed date for the establishment of the idea of the picturesque in artistic culture is 1795, as the first publications of the latter three theorists cluster around that date, when public controversy over the meaning and value of the picturesque was at its height. In art historical scholarship, the period 1780–1830 is often considered the apotheosis of the picturesque. What happened earlier, therefore, is typically framed as pre-history, and what happened later remains a story of the development and impact of the concept. Today the picturesque is considered to constitute an ensemble of concepts that draw not only on 18th-century materials but also on those of the 19th and 20th centuries, where it has been applied very broadly across cultural theory.

9 This era in theatre is difficult to study, in large part because no British theatres survive in their 18th-century states and are generally not captured well in contemporary images. Moreover, although most English landscape and genre painters of the period — including many Royal Academicians — were involved in decorating either stage or theatre, none of their work of this kind survives, and their names were not advertised on theatre bills until the arrival in London in 1771 of the painter and set designer Philip James de Loutherbourg (1740–1812). *See* Stone, G W 1962 *The London Stage. Part 4: 1747–1776*, vol I, *A Critical Introduction*. Carbondale, Illinois: Southern Illinois University Press, cxvii ff; also Mackintosh, I 1975 *Georgian Playhouse: Actors, Artists, Audiences and Architecture*, 1730–1830 (exhibition catalogue). London: Arts Council of Great Britain, preface XII, 'The splendor of the scenes'. In spite of these challenges, however, we do know enough to examine the 'scenographic' nature of Adam's work.

10 Mackintosh, *Georgian Playhouse*, introduction, and preface to part 1.

11 For more on the shift from aural to visual theatre, *see* Nicoll, A 1980 *Garrick Stage: Theatres and Audience in the Eighteenth Century*. Athens: University of Georgia Press; Manchester: Manchester University

Press. *See also* Price, C 1973 *Theatre in the Age of Garrick*. Totowa, NJ: Rowman and Littlefield; Leacroft, R 1973 *The Development of the English Playhouse*. London: Eyre Methuen; and Mackintosh, *Georgian Playhouse*, ch 10.

12 Davis, J 2007 'Spectatorship', in Moody, J and O'Quinn, D (eds) 2007 *Cambridge Companion to British Theatre, 1730–1830*. London: Cambridge University Press, 58.

13 Ibid.

14 A significant feature of the Georgian stage is that the auditorium remained almost as brightly lit as the stage. The introduction of gas lighting in London between 1817 and 1827 enabled the auditorium lights to be dimmed for the first time. Chandeliers suspended over the stage – a convention since the opening of the theatres in 1660 – were abolished in the mid-18th century, as they were a deterrent to illusion and obscured a complete view of the stage. Inspired by continental lighting practices, Garrick insisted on doing away with them. They were replaced with lamps fastened to strips of tin installed out of sight of the audience on either side of the stage. *See* Hogan, C B 1968 *The London Stage. Part 5: 1776–1800*, vol I, *A Critical Introduction*. Carbondale, IL: Southern Illinois University Press, lxv–lxvii. Concerning 'picturesque' scenery, it is notable that many Royal Academicians began their careers in the field of landscape as scene painters for one of the London theatres, and that the grander playhouses, such as Drury Lane and Covent Garden, paid scene painters liberally. A 1750s source states that 'All these [stage-set scenes] should be done by a Master, if such can be procured; otherwise, they should be as simple and unaffected as possible, to avoid offending a judicious Eye.' *See* Baugh, C 2007 'Scenography and technology', in Moody, J and O'Quinn, D (eds) 2007 *Cambridge Companion to British Theatre, 1730–1830*. London: Cambridge University Press, 44.

15 On Adam's relationship with Garrick, *see* Lees-Milne 1947; Fleming 1962; and Murphy, A 1801 *Life of David Garrick*. London, 2 vols.

16 Fleming 1962, 250; BAM, 1454/4/11/6.

17 In his preface to the first number of *The Works* (1773), Adam enumerated the innovative forms he had introduced into modern interiors, all of which create the illusion of increasing a room's dimensions: 'The massive entablature, the ponderous compartment ceiling, the tabernacle frame, almost the only species of ornament formerly known, in this country, are now universally exploded, and in their place, we have adopted a beautiful variety of light mouldings, gracefully formed, delicately enriched and arranged with propriety and skill. We have introduced a great diversity of ceilings, freezes [*sic*], and decorated pilasters, and have added grace and beauty to the whole, by a mixture of grotesque stucco, and painted ornaments, together with the flowing rainçeau, with its fanciful figures and winding foliage.' Adam and Adam 1773–1882, vol I, no I, preface.

18 Ibid, 'Explanation of the plates'.

19 Among Galli-Bibiena's contributions were the introduction of angled perspective (*scena per angolo*) and the enlargement of the scale of architectural features to make only a small portion of a building visible to the audience (rather than the entire building, as was the previous tradition). Angled perspective is a convention that organises perspective designs around two or more vanishing points, often placing the architectural features in the middle of the painting with the vistas out to the sides. The implications of both innovations are highly significant, in part because they invite the spectator to complete the stage picture in his or her imagination.

20 *Public Advertiser*, 30 September 1775.

21 Mackintosh, *Georgian Playhouse*, 25.

22 *See* note 11 for relevant references.

23 Cited in Nicoll, *Garrick Stage*, 46.

24 For instance, for Adam's sculpture galleries at Croome Court and Newby Hall, *see* Harris 2001, 48–53, 215–21. Also, for sculpture collecting in 18th-century Britain generally, *see* Guilding 2014.

25 On the relationship between actor and sculpture in late 18th-century Britain, *see* West, S 1991 *Image of the Actor: Verbal and Visual Representation in the Age of Garrick and Kemble*. New York: St Martin's, 119–22.

26 *See* Lessing, G E 1766 *An Essay on the Limits of Painting and Poetry*, for a discussion on the Laocoön, and Winckelmann, J J 1764 *History of the Art of Antiquity*, for a poetical celebration of the *Apollo Belvedere*. For a brief discussion of the *Laocoön* replica intended for Syon House, *see* Harris 2001, 66.

27 The first chapter of Lord Kames' *Elements of Criticism* (Edinburgh, 1762), 'Perceptions and ideas in a train', made popular the concept of a 'pleasing train of ideas' and inspired the pervasive use of the phrase, subsequently found throughout aesthetic discourse of the 18th and early 19th centuries.

28 *See* Thompson, P 2007 'Acting and actors from Garrick to Kean', in Moody and O'Quinn, *Cambridge Companion to British Theatre*, 3; also West, *Image of the Actor*, 122.

29 For Adam's use of column screens as proscenia, *see* Harris 2001, where they are discussed for their 'picturesque and scenic' effect.

30 *See* Baugh, 'Scenography and technology'.

31 *See* Kondo 2012, 71.

32 Ibid, 72.

33 De Loutherbourg left Drury Lane during the winter of 1780–81 to open a miniature theatre (sometimes referred to as an early panorama) called the *Eidophusikon* in Lisle Street, just off Leicester Square. The stage was 8ft by 6ft (2.44 by 1.83m). He was inspired to construct this theatre in order to create visual effects that could not be achieved at Drury Lane. *The Morning Herald* of 12 March 1782 advertised the second programme of scenes of the *Eidophusikon*: 'sun rising in the fog of an Italian sea port, the cataract of Niagara, Storm at Sea, Setting of Sun after a rainy day, with a view of the Castle, town, and Cliffs of Dover, the rising of the Moon, with a Water-Spout, exhibiting the effects of Three different Lights'. The effect of the *Eidophusikon* was significant and was remembered by scenic artists and painters for several decades after its presentations in the 1780s.

34 Marc-Antoine Laugier, in the fifth chapter of his *Essai sur l'architecture* (Paris, 1753), wrote of using a garden as a model for city planning. Lord Kames, in vol III of his *Elements of Criticism*, discussed general similarities in a chapter entitled 'Gardening and architecture'. Notably, Kames also wrote in a chapter in vol I (on 'Emotions and passions') that 'furniture increaseth in appearance the size of a small room, for the same reason that divisions increase in appearance the size of a garden'. *See* Kames 1762, vol I, 216.

35 Adam and Adam 1773–1822, vol I, no I, preface. This footnote from *The Works* is reminiscent of the opening paragraphs of a more extended section on movement entitled 'Of the Elevation & the movement', from the draft of an unfinished essay by James Adam on architectural theory of 1762 (with additions of 1772) in NRS, GD18/4954. This is transcribed in Fleming 1962, 315–19.

36 I would like to acknowledge and thank David Leatherbarrow for the artful phrasing 'movement of eye, foot, and mind'.

37 The new kind of observer that emerged in the 18th century is explored extensively by Oleksijczuk, D B 2011 *The First Panoramas: Visions of British Imperialism* (Minneapolis: University of Minnesota Press) and Crary, J 1990 *Techniques of the Observer: On Vision and Modernity in the Nineteenth Century* (Cambridge, MA: MIT Press). Oleksijczuk argues that the panorama enabled two modes of perception: 'a static mode in which the viewer perceives depth by focusing on one perspective point, and a mobile, haptic mode of visuality, in which the observer confronts the objective qualities of the panorama's multiperspectival construction' (70). She asserts further that these meanings were important for a new conceptualisation of the self. Crary asserts that, in the 1820s and 1830s, vision became inseparable from the mental and physical make-up of the human body, and, therefore, subjective rather than objective, as Classical models had suggested. This shift, however, occurs decades earlier that Crary argues it does.

38 Addison, J 1712 'On the pleasure of the imagination'. *Spectator*, **411**, 21 June, 473.

39 Adam designed at least four other temporary structures: a 'Building which May be contrived to Answer for the Fireworks at a General Peace', 1759 (SM, Adam vol 9/108); an 'Illumination for Sir Charles Frederick', *c* 1761 (SM, Adam vol 10/105, 141; vol 49/66); a casino, 1764 (SM, Adam vol 7/200); and an 'Illumination for the Earl of Hopetoun', at his town house in Cavendish Square, 1789 (SM, Adam vol 49/65). It is not known with certainty if these monuments were built, except for the illumination for the Earl of Hopetoun, which was reported widely in the London press. For Adam's temporary structures, *see* Burden, M 1993 'Robert Adam, de Loutherbourg and the sets for *The Maid of the Oaks*', in Worsley 1993, 65–9; Harris 2001, 67, 279–82; and Doderer-Winkler, M 2013, *Magnificent Entertainments: Temporary Architecture for Georgian Festivals*. New Haven and London: Yale University Press for the Paul Mellon Centre for Studies in British Art, 18–25, 34–7, 53–74, 120–4.

40 The diary of the Duchess of Northumberland tells us that 'The Top of the Staircase was taken off & a Cupola carried to a vast height & beautifully illuminated with a profusion of Lamps of almost every colour in the Rainbow & appeared like a Fairy palace'. Cited in Doderer-Winkler, *Magnificent Entertainments*, 60; *see also* Greig, J (ed) 1926 *The Diaries of a Duchess: Extracts from the Diaries of the First Duchess of Northumberland, Elizabeth Seymour Percy, 1771–1776*. London: Hodder & Stoughton, 36. The same diary also tells us that 'The Ball Room had one of the Side Walls taken down & a large Orchestra & 2 Niches carried into the Garden' (ibid).

41 Cited in Harris 2001, 279.

42 Cited in Doderer-Winkler, *Magnificent Entertainments*, 65. *See also* Hall, A (ed) 1862 *Autobiography and Correspondence of Mary Granville, Mrs. Delany*. London: R Bentley, vol II, 1 (Mrs Delany to Mrs Port, June 1774).

43 Here, the terms 'theatricality' and 'absorption' are imported from Michael Fried's study of French painting of the second half of the 18th century, *Absorption and Theatricality: Painting and Beholder in the Age of Diderot* (University of California Press, 1980; repr University of Chicago Press, 1988).

44 For descriptions of this effect, *see* Doderer-Winkler, *Magnificent Entertainments*, 68; *Gentleman's Magazine* **44**, June 1774, 264; *also* Mrs Delany to Mrs Port, June 1774, in Hall, *Autobiography and Correspondence of Mary Granville*, vol II, 2.

45 The medium of panorama was first patented on 17 June 1787 by Robert Barker, an Irish artist living in Edinburgh. In his first exhibitions, in Leicester Square in London, Barker's design was built by the Scottish architect Robert Mitchell. From a central raised platform one could view 360-degree representations of a scene on a scale of 1:1. He called his invention 'La nature a coup d'oeil', which he translated in his London publications as 'nature at a glance'. In 1791 Barker changed its name to panorama, derived from the Greek terms *pan* (everything, the whole world, all) and *horama* (a vision). This neologism described both the specially designed rotunda and the circular painting it housed. The patent specifications make clear that the spectator was to be completely surrounded by the massive, continuous picture and the goal was to give the impression that they were 'really on the very spot' rather than in an architectural interior. To create this effect, Barker specified that the platform existed to prevent viewers from getting too close to the surface of the drawing or painting; a roofing component hid the picture's top edge, and skylights were placed directly above the picture to bathe it in natural light, leaving the platform in relative darkness.

6

Design by correspondence: Robert Adam and Headfort House

Conor Lucey

On 17 September 2009, following a five-year restoration project grant-aided by the World Monuments Fund, the eating parlour at Headfort House in Kells, County Meath – the centrepiece of the sole surviving suite of Adam-designed interiors in Ireland – was unveiled to the public (Fig 6.1).[1] The results polarised academic opinion. While some considered the reinstatement of the room's original and surprisingly complex 'green chiaroscuro' colour scheme to be an accurate recreation of Robert Adam's intentions, others argued that it was instead a

Fig 6.1
The restored eating parlour at Headfort House, designed by Robert Adam in 1771–5.

local reinterpretation, even corruption, of the architect's vision, deviating from the surviving designs and from what generations of Adam scholars had established regarding the architect's approach to colour.[2] The focus of the dispute rests on the non-architectural colouring of the relief plasterwork, including the alternating painted pattern of dentils and bead mouldings, and the delineation of plastic form with fine signwriting lines (Fig 6.2). Given its painstaking complexity, the extraordinary amount of time required to reinstate it and the fact that it represents one of the few Adam interiors subjected to sustained scientific analysis, this unique and idiosyncratic colour scheme remains a source of fascination and contention.[3] In her address to the audience assembled at Headfort House for the official unveiling, Eileen Harris acknowledged the difference of opinion among her peers, adding: 'I yearn for the debate to be resumed and resolved.'[4] With a special focus on the eating parlour, the following discussion officially resumes that debate – though its full resolution may yet prove elusive.

This chapter explores the arguments for and against Robert Adam's involvement in the realisation of the interior decoration at Headfort House. Despite the discrepancy between the designs and the restored scheme, it is suggested here that the surviving corpus of Adam drawings for the house – those issued to the client and the office copies – are the key to understanding the decoration of its formal spaces as a complex, collaborative process. Eileen Harris tended to downplay the role of the drawings in her assessment of the restored decorations at Headfort – here I want to emphasise their primacy. The relationship between architect and patron is of course a well-rehearsed narrative in histories of 18th-century architecture, as is the emergence of the architectural profession during that period and its increasing design 'authority' over the building process.[5] Adam is a representative figure in both categories, and the so-called 'developed surface interior' mode of drawing that he used at Headfort and elsewhere was a pivotal instrument in his attainment of a coordinated architectural effect; it was one of his most significant innovations. But as Adam scholars have consistently shown, the success of his practice relied on, and was facilitated by, large teams of draughtsmen and building tradesmen, and projects undertaken at a remove from London necessarily involved individuals beyond his ken. Just as architecture may be characterised as 'more like a highly organised collaborative business than a fine art', so Barry Bergdoll has described Adam's 'signature' style as 'a corporate creation'.[6] The extraordinary accomplishment of the team of decorators employed at Headfort House between 2005 and 2009, supervised by the renowned conservator Richard Ireland, is that they have restored the original 18th-century colour scheme of its state rooms and formal circulation spaces.[7] Its significance is absolute. But the question remains: did Adam supervise this project?

Fig 6.2
Details of the restored painted decoration in the eating parlour at Headfort House.

Design by correspondence

While Ireland's long-standing cultural, economic and political relationship with Great Britain gave rise to a shared architectural sensibility, design there in the early Georgian period was largely the province of architects resident in Ireland. From mid-century, however, and led by the example of James Caulfeild, 1st Earl of Charlemont (who in the 1750s–60s engaged William Chambers to design Charlemont House and the Casino at Marino), Irish patrons increasingly turned to British architects for houses and interiors in the neoclassical idiom. As well as Chambers, Robert Adam, James 'Athenian' Stuart and James Wyatt enjoyed sustained patronage from Ireland's nobility and gentry from early in their careers, leading Maurice Craig to brand the second half of the 18th century in Ireland as 'a period of domination by English architects'.[8] Nonetheless, and often in spite of the scale of the works or the standing of the client, it seems that English architects were reluctant to travel: Chambers never crossed the Irish Sea, and Wyatt, despite an extensive country-house practice in Ireland, is believed to have visited only once, in 1785, to advise on progress at Slane Castle, County Meath.[9] Instead, drawings were sent by post to Ireland from the architects' London offices, facilitated by a regular mail packet service between Holyhead in Anglesey and Ringsend in Dublin, and the undertaking of building works therefore became subject to different circumstances and outcomes. The execution of Chambers' designs for the Earl of Charlemont, for example, was supervised by the accomplished sculptor and stonemason Simon Vierpyl, and these rank high among his most important and admired works.[10] On the other hand, the same architect's designs for Lord Loftus at Rathfarnham Castle, County Dublin, were not overseen by an approved deputy, a situation that occasioned a protracted and often strained correspondence between architect and client (and interior finishes that John Harris has dismissed as 'provincial'). Chambers himself held a similar view, as is confirmed by a letter he wrote to Lord Loftus from London in January 1771: 'I am apt to think your Lordships workmen are rather careless in perusing the designs & seldom read what is written upon them which Occasions their Difficulties.'[11]

Here it is important to note, though, that design by correspondence and its associated complications was not simply confined to Ireland; letters between architects, clients and tradesmen reveal the myriad problems attendant on building by proxy in the British Isles generally, leading Howard Colvin to describe it as a practice where 'the scope for misunderstanding and muddle was enormous'.[12] Moreover, many of these buildings in Ireland enjoy positions of considerable importance within the opera of their respective authors. John Martin Robinson's recent monograph on James Wyatt is generous in its praise for Castle Coole in County Fermanagh (begun 1790), describing it unreservedly as 'a house of unsurpassed classical splendour', although the architect himself probably never saw it, and Chambers' Casino at Marino, outside Dublin (begun 1759), has long been admired as a paragon of European neoclassical taste.[13] Headfort House, by way of contrast, does not hold an equivalent status in Adam's extensively catalogued *oeuvre*.

Headfort House: the Adam work in context

The long and complex architectural history of Headfort House was first described by Christopher Hussey in a series of articles for *Country Life* in 1936, and later revised by John Harris in the catalogue for the exhibition *Headfort House and Robert Adam* held at the RIBA in 1973.[14] The house was conceived as early as 1750 by Sir Thomas Taylour of Kells, 2nd Bt (d 1757), and designs were elicited from a number of architects, including Richard Castle, the foremost designer of country houses in early Georgian Ireland, and Sir William Chambers, who was paid 40 guineas for a plan and elevation in 1765. By this date Taylour had been succeeded by his son, also Thomas (1724–95), whose rapid advancement in the peerage from Baron Headfort (1760) to Viscount Headfort (1762) and then Earl of Bective (1766) evidently stimulated progress of the building works. Bricks, stone and slates 'for the intended house' were being acquired between 1758 and 1762, and the shell of the house was completed by 1769–70, when the scaffolding was taken down. Completed to designs by the Irish architect George Semple, Headfort House is a representative example of the 18th-century dichotomy, particularly pronounced in this instance, between reticent exterior and elaborate interior (Fig 6.3). Here, Semple's austere elevation, famously disparaged in 1787 by Charles Manners, 4th Duke of Rutland, as 'a long range of tasteless building', conceals Adam's exceptional suite of reception rooms.[15] Lord Bective's decision to appoint Adam was apparently inspired by his father-in-law, the Rt Hon Hercules Rowley, who had commissioned interiors for his Dublin town house from Adam in 1765.[16]

The drawings supplied by Adam to Lord Bective, now at the Yale Center for British Art, document the evolution of the decoration at Headfort House.[17] The earliest are monochromatic presentation designs and working drawings for the entrance hall and eating parlour, dated 1771 (Fig 6.4).[18] Closely related to this first group, and dated January 1772, are monochromatic designs and working drawings for the stair hall.[19] These were followed, in November 1772, by polychromatic ceiling designs for the saloon and the 'Lady's Room' (including four sheets of full-scale working drawings for the latter). The pace of the decorating works at Headfort House was evidently protracted, and three years later, in December 1775, Adam provided a new set of drawings for the eating parlour. In the interim, it appears that Bective had changed his mind about this room, specifically the form of the ceiling. Commissioned to produce a new design appropriate to a coved rather than a barrel-vaulted ceiling (as in the original scheme of 1771), Adam responded with a single elevation that addressed the new wall height that resulted from this adjustment (Figs 6.5 and 6.6). A comparison between the two elevations confirms how little of the original 1771 decorative scheme was in fact altered in the 'revised' 1775 scheme, indicating that Bective did not request new designs for the room per se.[20]

What conclusions can be drawn from this? Firstly, and most importantly, it seems evident that Lord Bective did not commission Adam to design a parade of reception rooms from the outset, and so at this stage did not yet conceive of the ground-floor circuit as a compositional entity.[21] It seems more likely that, being satisfied with Adam's designs of 1771–2 for the entrance hall, eating parlour and stair hall, Bective then proceeded to commission further designs for the remaining reception rooms, from late in 1772. His need was not necessarily for an architect, but for impressive and modish schemes of neoclassical interior decoration.

Fig 6.3
Headfort House: the garden front.

Design for finishing the Chimney side of the Eating Parlor. for The Right Hon.ble The Earl of Bective.
Scale of
10
20
30 feet.

dark green
Light green
Picture
Ruin
Picture
Ruin

To this sequence of events may be added information from two separate collections of Headfort family papers, deposited at the National Library of Ireland in Dublin and at the Senate House Library in the University of London. From these manuscript sources it can be established that the Dublin plasterer James McCullagh (d 1795) was paid for the decorative plasterwork in the hall and saloon (a formal Palladian axis) in November and December 1774.[22] Ten years later, in 1784, a local craftsman from the town of Kells, George Booker, was paid for 90 days' work 'Painting the Great Parlor', which suggests that the plasterwork decoration in the eating parlour had been completed in the intervening years (it is neither itemised nor described in the account books and is presumed also to be by McCullagh). Significantly, this is the only reference to the painting of the eating parlour from these sources.[23] However, it is now clear from the complexity of the painted decoration revealed during the recent restoration that this work could not have been completed within the period of Booker's engagement: Richard Ireland has confirmed that the repainting of this room under his guidance employed an average of eight specialist craftsmen working 55-hour weeks for a period of six months.[24] Further glimpses of painting activity for Bective recorded in a receipt book covering the years 1783–95 are equally tantalising but inconclusive. Here, references to 'Paints for my Parlor', 'Paints for Headfort' and 'Paid McCullagh for Painting' throughout the course of 1784–7 provide evidence of decorating work but remain ambiguous: a large payment of £159 4s 9½d in December 1788 for 'Painting by Callaghan' at the 'House', for example, is likely to refer to work at Lord Bective's recently acquired Dublin town house in Rutland (now Parnell) Square.[25] Further references to the purchase of oil and paint abound in these manuscript sources, but the nature of the tasks, or the location of the works, is indeterminate.[26] What is clear from this slight documentary evidence, though, is the fact that Bective was scrupulous in his choice of craftsmen. By 1774 James McCullagh had also completed plasterwork decoration in a neoclassical style at Powerscourt House in Dublin, for Richard Wingfield, 3rd Viscount Powerscourt, and 'Callaghan' probably refers to Cornelius Callaghan, a glazier, painter and seller of maps and prints in Great Britain (now Parnell) Street in Dublin, and an agent for the London-based designer Placido Columbani's publications in Ireland.[27] Already professionally acquainted with

Fig 6.4 (opposite, top) Original 'Design for finishing the Chimney side of the Eating Parlor for The Right Honble, The Earl of Bective', 1771.

Fig 6.5 (opposite, bottom) Office record of revised 'Design for finishing the Chimney side of the Eating Parlor for The Right Honble, The Earl of Bective', 1775.

Fig 6.6 Revised 'Design for finishing the Chimney side of the Eating Parlor, for The Right Honble The Earl of Bective', 1775.

the modish Adam-style repertoire of urns, paterae and anthemia, McCullagh and Callaghan evidently satisfied Bective's demand for craftsmen with a demonstrable ability to 'read' the type of architectural drawings issued by the Adam practice, and the tacit knowledge of form to translate his visual shorthand from paper to plaster and paint.

Interestingly, Laura Jacobus notes that the 'developed surface interior' type of drawing so valued in the Adam office (as opposed to the true section) – also sometimes referred to as the 'laid-out interior' – originally gained currency in the work of 'those on the fringes of the profession', by which she means amateur architects and building craftsmen.[28] By mid-century, Isaac Ware had fully endorsed this drawing technique as an aid towards decorative unity, and craftsmen across the British Isles were by then already familiar with its graphic conventions.[29] In a provocative essay of 1989, Robin Evans' apposite characterisation of this drawing type as 'highly worked exhaustive individual room portraits' reiterates the representational focus on decorative effect rather than spatial volume.[30] But while noting how it became 'a basic mode of production' in the Adam office, Evans neglects to consider either its symbolic or its practical functions in 18th-century terms: namely, as a sign of the emerging authority of the professional architect with respect to the design of the coordinated interior (from architecture to upholstery), and as a means of dictating that authority through the medium of the drawing. At Newby Hall, according to Eileen Harris, Adam's designs were 'so detailed that large-scale working drawings for the craftsmen could be made directly from them'.[31] Given the scale of projects undertaken at a considerable remove from London – not least in Ireland – the developed surface interior was clearly valued by Adam for its visual clarity and legibility: it satisfied both aesthetic and utilitarian functions, as required by client and tradesman.[32]

Headfort House in Adam's *oeuvre*

The critical issue with the restoration of the eating parlour at Headfort House is the non-architectural colouring of beads, dentils and drops, and whether or not this 'unconventional' and, according to Eileen Harris, 'highly sophisticated' decorative effect was endorsed by Adam.[33] The reasoning for Adam's involvement is contingent on two established facts: the relative expense of the verdigris pigment used so liberally throughout the suite of formal reception rooms, and the intricate, labour-intensive 'picking in' and 'picking out' of surfaces and ornaments.[34] Given the levels of expense and intricacy, the argument continues, Lord Bective was unlikely to have ceded responsibility to anyone but Adam – or, perhaps, to one of his famous 'regiment of artificers'. Adam, in turn, given his holistic approach to design, is unlikely to have posted the drawings to Ireland without making provision for their proper implementation. Leaving aside the disparity between the drawings and the restored scheme – a topic we will return to presently – it is necessary to explore the likelihood of Adam's involvement by first considering the place of Headfort House within his wider *oeuvre*.

A long-acknowledged feature of Adam's Irish works is their narrow reach. Of his projects in Ireland, more than half were commissioned by members of the Taylour, Rowley and Upton families, all of whom were related through bonds of kinship.[35] More significantly, these were the only works actually executed in Ireland; the balance remained paper projects only.[36] Ireland was, however, clearly in Adam's sights during the early 1770s, and advertisements for the first two numbers of *The Works in Architecture* in British and Irish newspapers during the course of 1773–4 announced that it would contain 'Designs invented and executed by them in Great Britain and Ireland'.[37] Although this never in fact transpired, as advertising rhetoric it served a number of unambiguous marketing purposes, and Headfort was undoubtedly the catalyst.

On the other hand, and notwithstanding Lord Bective's status as Adam's foremost Irish client of the 1770s, we should not overstate the significance of Headfort House as a commission. The number of autographed and dated drawings produced by the Adam office during 1771–2 and 1775, for example, was considerable, including designs for large-scale building projects like Newby Hall and Osterley Park. For London alone, in fact, there are a large number of drawings for this same period, extending to the speculative housing schemes at Mansfield Street and Portland Place, as well as some of Adam's most distinguished suites of interiors as at Home House in Portman Square, and 20 St James's Square. And as Adam scholars well know, these years were dominated by financial crises, precipitated by the brothers overreaching London speculative developments at Portland Place and, especially, at the Adelphi – the latter described by Alistair Rowan as 'a considerable embarrassment to them for three years, at least, from 1772 to 1775'.[38]

More specifically, a wider review of the corpus of Adam drawings reveals numerous similarities between those provided for Lord Bective and those already sent to other clients, including designs for ceilings and chimney pieces for Harewood House in Yorkshire (1766–7), Lansdowne House in London (1766) and Saltram in Devon (1768).[39] While this was by no means an unusual practice – and it certainly was not restricted to Adam's Irish patrons – it rather suggests that the interior decoration of Headfort House was not an especially significant project for the practice. This is further confirmed by the office copy drawings for the 1775 eating parlour scheme: according to Stephen Astley, these are unusually perfunctory in terms of record keeping – an indication, or an acknowledgement perhaps, that the commission did not extend to Adam's personal supervision (*see* Fig 6.5).[40]

Project management

Alistair Rowan's synopsis of Adam's modus operandi describes three basic systems of professional engagement: the preparation of attractive presentation designs in response to a commission; the issuing of working drawings for use by the tradesmen on-site following approval of those designs; or the full supervision of the project from the outset, for which Adam charged a percentage of the total cost of the building works (and which typically involved an on-site deputy or proxy).[41] Writing in 1791, Adam explained: 'When I am employed to make designs only I am paid for my plan and I have done with the business.'[42] Was this the procedure followed at Headfort House? Architects were certainly by then familiar with the practice of working up designs from annotated plans or tabulated dimensions supplied by their patrons. In response to a letter from Viscount Loftus in 1770, regarding the decoration of rooms at Rathfarnham Castle, Sir William Chambers advised: 'If your Lordship will send me the exact dimensions of the Plan with the heights of the rooms & situation & dimentions of doors Windows &c &c and how you wish to have them fitted up I will immediately make the designs.'[43] That said, it is clear that Adam engaged a clerk of works at many of his most celebrated country houses. Eileen Harris notes that the 'carpenter turned architect' William Belwood worked under Adam at Harewood House, Nostell Priory, Newby Hall and Syon House; later, at Dalquharran Castle and Culzean Castle in Scotland, the stonemason Hugh Cairncross acted in this capacity.[44] But rather than speculate on the existence of an equivalent arrangement in Ireland, in the absence of any such documentary evidence it is more instructive to examine the ways in which similar aspects of architectural projects in England were often managed and realised. Reflecting on the decoration of the gallery at Syon House, Alistair Rowan described Adam as 'first and foremost an architect and one for whom the details of antique plasterwork, for all their novelty, take second place'.[45] Just how concerned with the minutiae of cornice and frieze – and by extension colour – were architects when faced with managing projects at a considerable remove from London? And to what extent was control ceded to those engaged on-site?

In fact, there is substantial documentary evidence to indicate that the architect's authority was not always absolute, and decisions made on locations remote from the office were customarily awarded to trusted deputies and building craftsmen. For the saloon at Ampthill Park, Bedfordshire, Sir William Chambers' ceiling design included instructions regarding the degree to which the ornaments and mouldings should project from the ceiling plane ('Ornament under the festoons to project in the highest parts 3/4 [inch]').[46] For the cornice of the billiard room, however, he was less prescriptive. Writing to John Rentham, clerk of works, on 28 May 1770, Chambers suggested: 'Let there be a plain ceiling & one of the same Cornice as in the Ground floor at Woburn Abbey – no matter which ... provided it be not like the Cornice put up in any of the other rooms at Ampthill.'[47] Evidently, as long as it was different from those already executed throughout the house, the design of the cornice in the billiard room at Ampthill was not a cause of particular concern or interest. (It should be noted too that Ampthill and Woburn Abbey were in progress at the same time, and Chambers engaged the same craftsmen and clerk of works at both sites.[48]) Adam on occasion also relinquished interest in such 'minor' details, particularly in deference to his client's wishes. In a letter of 1790 to Hugh Cairncross, clerk of works at Dalquharran Castle, he was positively cavalier about the decorative enrichment of the bookcases: 'In the section of the library I have drawn composition ornaments in the pilasters & frize of bookcases that can be put in at any future time if Mr Kennedy should incline to have them, or left plain if he chuses them so.'[49]

In her exposition on Adam's use of colour, Eileen Harris states: 'In addition to the finished ceiling designs made for the client and the office copies, Adam also supplied coloured working drawings for the use of the painter, who was normally an experienced London craftsman chosen by him.' As evidence, she cites a letter from Adam to Sir Edward Knatchbull, dated 10 February 1772, regarding the decoration at Mersham-le-Hatch in Kent: Adam wrote that he did 'not think any of the country hands could do [the painting work] as it ought to be'.[50] But read in context, it seems clear that the recommendation of a London painter in this instance was prompted as much by the particular demands of the client as by any model of best practice. Knatchbull had in fact posed a number of questions concerning the legibility of the scheme that Adam had proposed for the drawing-room ceiling, including an observation that 'the Colours are hardly strong enough to be seen and distinguished at a great Distance'.[51] With this in mind, Adam's comment about 'country hands' is clearly a riposte to an exacting patron who was notoriously tight with the purse strings: in order to get the work completed to his satisfaction, Knatchbull would have to pay for the privilege.[52]

While the skills of an Adam-endorsed artificer might well have been an attractive proposition in theory, circumstances and temperaments often dictated otherwise. In August 1780 Adam's estimate for painting two adjoining drawing rooms at Sir John Hussey Delaval's house in Hanover Square, London – including 'the ornamented Ceiling and Enriched Entablature, to be painted and pikt in accodg to the designs' – came to the princely sum of £122 18s.[53] Delaval demurred, and instead solicited the services of John Bromley, a 'coach, sign and house painter' of Little Queen Street, Lincoln's Inn Fields, who estimated the work, including the 'two ceilings in party colours as the Drawings', at only £58 – less than half of Adam's original quote. Cost, rather than Adam's personal supervision, was patently the deciding factor here, but the client nonetheless acquired an Adam colour scheme by employing a tradesman to work according to his designs.[54] But neither it seems did the on-site presence of a clerk of works necessarily guarantee fidelity to Adam's intentions. At Newby Hall, according to Eileen Harris, Adam's 'high opinion' of his clerk of works, William Belwood, was 'evidently shared' by the client William Weddell, 'who seemed prepared to entrust him with the execution of his more idiosyncratic wishes that might have been anathema to Adam'. This extended to the design of bookcases for the library, in light of Weddell's rejection of Adam's original proposal.[55] In fact, Adam's acute understanding of the vagaries of the design process was acknowledged in a very public manner: in the text accompanying the design of a harpsichord for the Empress of Russia, published in the first volume of *The Works in Architecture*, he revealed, 'This design was considerably altered by the person who executed the work.'[56]

The drawings as evidence

The divergence between the restored scheme and the 1775 drawings for the eating parlour, in vivid shades of green, represents the crux of the argument regarding the degree of Adam's alleged involvement in its original implementation. Designs limited to tints of green alone – sometimes referred to as 'green chiaroscuro' – were of course an established feature of Adam's repertoire by this date: paint historian Ian Bristow has described this palette as 'austere', and links it to 'the increasing simplicity of form' in Adam's ceiling designs of the early 1770s.[57] This is corroborated by Adam's own appraisal of his monochromatic schemes. In his description of the anteroom at Derby House, Grosvenor Square, of 1774, published in the second volume of *The Works in Architecture* (1779), Adam writes: 'The ornaments are all in stucco; and the grounds, both of the cieling and the side-walls, are all picked in, with different tints of green, which has a simple and elegant effect.'[58] The language, I think, is significant here: it is the grounds of the ceiling and wall planes that have been 'picked in', and this produces a 'simple and elegant effect'. This stands in marked contrast to the decorative effect restored in the eating parlour at Headfort House.

With the foregoing in mind it seems certain that the key to unlocking Adam's supervision (or lack of it) lies in a close reading of the various presentation and working drawings, now at the Yale Center for British Art, and the office copies at Sir John Soane's Museum. Here I am embracing the concept that 'the full significance of architectural imagery can best be recovered within the serial context of the group'.[59] Particularly instructive in this regard is a pair of polychrome drawings of 1771, one of which is clearly inscribed 'how to finish one end of The Parlour next the Salloon' (Fig 6.7).[60] Categorised as a 'preliminary study' by John Harris in 1973 and as a 'sketch' by Richard Ireland and John Gundry in an article published in the *Journal of Architectural Conservation* (2011), it begs the question: why would the Earl of Bective have been sent a preliminary or 'sketch' colour scheme? Ireland and Gundry, acknowledging that the drawing's textual annotations render it 'particularly interesting', are elsewhere unequivocal that 'it is not immediately clear how complex instructions concerning design or the use of colour were communicated to workers on site'.[61] However, in terms of graphic style and finish – being less precise and without ruled borders – it is certainly distinct from the finely rendered presentation drawings whose 'normal function' was, as Alistair Rowan notes, 'to capture the enthusiasm of a client by the skilful and attractive presentation of the proposals'.[62] So if not to convey instructions regarding 'how to finish' the decorated wall planes with coloured grounds of pink and green, what function did this drawing serve? Related designs for the ceiling of the

Fig 6.7
Working drawing for 'how to finish one end of The Parlour next the Salloon', 1771.

Fig 6.8
Coloured working drawing for ceiling of the eating parlour, 1771.

Fig 6.9
'Cieling for the Eating room for The Earl of Bective', 1771.

eating parlour – including a similar polychromatic guide (issued to Bective) and an annotated office copy – confirm that the polychromatic drawings Bective received in 1771 were no mere sketches (Figs 6.8 and 6.9). It becomes clear once these coloured drawings are viewed as a group that they provide an unambiguous guide for the intended decorative effect – to client and tradesman – without the expense of fully rendered presentation designs, the cost of which was prohibitive by this date.[63] They are unmistakeably examples of the 'coloured working drawings' referred to in Eileen Harris's account of Adam's method, being equivalent in form and execution to Robert Adam's annotated design for the ceiling of the dining room at Newby Hall, and to James Adam's annotated design for the library at Nostell Priory – drawings with colour instructions intended for use by the craftsmen engaged to undertake the works.[64]

Indeed, the survival of such a large number of working drawings for the interior decoration at Headfort House – 16 compared with 17 presentation designs – should come as no surprise.[65] Drawings of any type were a precious commodity throughout the 18th century, particularly those for commonly used mouldings and ornaments in the Classical idiom. At Sir John Soane's office, the sorting of drawings for cornices and architraves by type rather than by the client or location is a testament both to their iterative function and to their relative value.[66]

'Directions and caprices' that 'spoil the whole'

Extending this close reading of presentation and office-record drawings to a comparison between the drawings and the executed scheme in plaster and paint reveals the extent to which the interiors at Headfort House must be regarded as an interpretation of Adam's designs: an interpretation, moreover, that he would not have endorsed. While the moulded decoration *in situ* largely conforms to what Adam prescribed in his annotated designs and working drawings, there are a number of instances where it does not.[67] A good example is in the decoration of the stair hall: here Adam's frieze design of alternating ox and ram heads was merely simplified in execution; at ground-floor level, on the other hand, a corresponding design for the door entablatures 'in the bottom part of the Staircase' was entirely disregarded. The significance of this is easily overlooked. When the drawings are viewed together, it becomes clear that Adam designed the two friezes to complement one another at the different levels of this circulation space, employing identical motifs in alternate compositional modes (Figs 6.10 and 6.11). By ignoring this instruction, Bective and his workmen compromised the coordinated effect that Adam intended for this space. Moreover, while the restored green colouring of the ceiling in the stair hall and the green painting of its corresponding joinery might create a 'transitional balance' between the

Fig 6.10 (below, left) Working drawing for 'Entablature for the Staircase', 1772.

Fig 6.11 (below) Working drawing for 'Entablatures for the Doors in the bottom part of the Staircase', 1772.

colour palettes employed in the entrance hall (grey) and eating parlour (green), this contradicts Adam's preference for neutrally coloured halls and passages, as, for example, in the stair halls at Lansdowne House (1766) and Home House (1775), the schemes for which are known through drawings and confirmed by documentary evidence.[68] Nor, indeed, is there any precedent in Adam's *oeuvre* for a colour scheme shared between formal reception rooms and circulation spaces. Rather, there was a crescendo of ornament and a variable palette that created a 'Variety and gracefulness' with respect to the 'form, arrangement, and relief of apartments'.[69]

The colour scheme revealed by the restoration compromises Adam's intention in other significant ways. In his analysis of Adam's ceiling designs, Ian Bristow describes the use of tinted grounds as a means of emphasising the basic geometry of a composition (with smaller areas of richer colour for added piquancy). But at Headfort House the palette of green 'tints' for the ceiling of the eating parlour has been reduced in execution from three, as specified on the drawing, to only two.[70] While this does not interfere with the figure–ground relationship generally, it reduces the legibility of the geometry as Adam intended (Fig 6.12). Further disparities between intention and execution were found in the saloon: paint sample analysis revealed that the ceiling colour scheme had 'nothing in common' with Adam's design

Fig 6.12
'Design of a Cieling for the Eating Parlor for the Right Honourable The Earl of Bective', 1775.

of 1772, while the 'general treatment' of walls, entablature and overdoors followed Adam's original 1771 'preliminary study' drawing for the adjacent eating parlour.[71] (Evidently a 'preliminary study' functioned as a working drawing in this particular instance.)

More problematic still is the finicking painted detail of relief ornament for which there are no working drawings. One explanation for their absence is the suggestion that 'at the very least, comprehensive written or verbal instructions would be necessary to direct painters so meticulously' – but this is, I think, counter-intuitive.[72] Designs by Sir William Chambers for an unidentified room at Newby Hall, Yorkshire, clearly indicate how the complex 'picking out' of architectural mouldings and relief ornament was effectively communicated through the medium of large-scale working drawings – a method, moreover, that made sense when designing by correspondence.[73] These drawings invite comparison with the painted scheme revealed during the restoration of the Casino at Marino in the 1980s, where the colouring of the door entablatures in the saloon revealed 'a most delicate scheme of picking out in the white ground' that followed Chambers' specifications of 1769.[74] A comparison between the Casino's 'delicate scheme' and the painted decoration in the eating parlour at Headfort House serves to point up the difference between an application of colour that enhances architectural form, and one that (perhaps unintentionally) diminishes it.[75]

A related argument suggests a 'potential link' between the intricate painting of architectural elements at Headfort and the famous Etruscan Dressing Room at Osterley Park, created by Adam for Robert Childs in 1775.[76] But while it is certainly true that the painting of the three-dimensional plaster forms at Headfort creates 'a flat, abstract, decorative appearance', the scheme at Osterley is properly two-dimensional and the 'mouldings' are rendered in a consistent illusionistic manner.[77] By contrast, the colouring of three-dimensional architectural members at Headfort is not uniformly handled: the beads, dentils and guttae of frames and entablatures, for example, are not consistently 'flattened'. Instead, the individual forms are alternately coloured green and white, following a simple contrasting pattern of dark and light. The over-painting of three-dimensional relief ornament at Headfort is therefore entirely distinct from the carefully painted two-dimensional 'mouldings' at Osterley.[78]

Conclusion

Does the restored eating parlour scheme at Headfort House reveal what has been described as 'the inadequacy of scale and detail of [Adam's] design proposals to present anything but a general view of intended decoration', or does it rather emphasise the collaborative nature of the architectural process generally and the system of design by correspondence particularly?[79] Is it, in the moderating words of Christine Casey, an example of 'Robert Adam at his most abstract and uncanonical' or the work of 'an Irish painter-decorator of extraordinary ambition'?[80] In her overview of Adam's practice, Eileen Harris quotes Adam's friend and patron General Robert Clerk, who in 1765 decried any sort of interference with the architect's designs, suggesting that 'Such "directions and caprices" had a tendency ... to spoil the whole.'[81] But leaving aside its relative merits, it is important to recognise that the complex decorative effect originally created and now restored in the eating parlour at Headfort House was not casually undertaken; nor should it be seen as an example of Irish provincialism.[82]

Despite its visual clarity and high level of graphic detail, Adam's developed surface interior typology did not guarantee that a decorative scheme would be executed as illustrated. Architectural drawings are, of course, a means to an end rather than an end in themselves, and might be modified or repurposed without an architect's consent or knowledge.[83] Robin Evans characterised presentation drawings as 'projections of a plausible outcome for a set of instructions and proposals already defined elsewhere but not yet accomplished'.[84] This is how Adam's designs for Headfort should be understood: although clearly not a record of an executed scheme, they are an unambiguous record of what he intended.[85] Central to this thesis are the large quantity of surviving working drawings, none of which record the kind of complex 'picking out' of architectural mouldings and relief ornament revealed during the restoration. Notwithstanding this disparity, Eileen Harris has made bold claims on behalf of the reinstated scheme: 'I think I am justified in saying that – as far as colour is concerned – Adam's interiors at Headfort are the most authentic anywhere. They also turn out to be the most peculiar – quite unlike the Adam we know – or think we know.'[86] But this seems imprudent. Does one 'peculiar' scheme warrant an entire re-evaluation of Adam's approach to colour?

Alistair Rowan notes that 'working drawings from the Adam office are comparatively few'.[87] With an almost identical number of presentation designs and working drawings, the portfolio from Headfort House at the YCBA represents perhaps an opportunity to rethink how Adam managed projects through drawings, particularly those at a considerable remove from London.[88] But just as some questions might be put to rest, others take their place. Why do the drawings issued to Bective in 1775 lack the finish characteristic of Adam's presentation designs, being only partially rendered and coloured? While this was a standard graphic convention for office-record drawings, it was not the usual method for costly designs issued to clients.[89] Was it perhaps a way of suggesting a decorative scheme to a client without being too prescriptive?

In the decades since Headfort House was first brought to public attention in the pages of *Country Life*, there has been much speculation about the Earl of Bective's purse, specifically the fact that some of the decoration designed by Adam – including the articulation of the wall surfaces in the entrance hall and staircase – was not executed. This has traditionally been ascribed to the earl's restricted finances, a matter refuted by Loreto Calderón's recent appraisal of his accounts.[90] We might therefore credit the 'process of simplification and omission' of the proposed decoration – and the related tickling up in paint of its plaster ornaments and mouldings – to Lord Bective's taste. Indeed, given the number of architects commissioned for designs – beginning with Richard Castle in 1750 and concluding with Adam in 1771 – it seems clear that Headfort House was a project very much controlled and directed by the client. With that in mind, John Harris's appraisal of its interiors as 'less autocratically decorated' and 'more rational' than many of Adam's 'overpowering' schemes for English patrons in the 1770s should be ascribed, at least in part, to his client's intervention.[91] Perhaps it is time for the Earl of Bective to join the Earl of Cassilis at Culzean Castle, Lord Bute at Luton Hoo and Lord Scarsdale at Kedleston as patrons of Adam who, as Eileen Harris puts it, 'had strong ideas of their own'.[92]

Notes

1 For an overview see www.igs.ie/conservation/project/headfort [accessed 2 April 2018].

2 Preliminary findings, based on scientific analysis by conservator Richard Ireland, were first brought to public attention on 12 December 2007, when he and Eileen Harris presented an illustrated lecture at the Bank of Ireland, College Green, Dublin. Alistair Rowan was in attendance and was invited to contribute to the discussion that followed. As two of the foremost scholars of Adam's work, Harris and Rowan represent the opposing sides of the argument regarding Adam's direct involvement, by proxy or otherwise, in the painting of architectural form and ornament in the reception rooms at Headfort House. At the heart of the dispute lies a difference of opinion as to whether Adam authorised the scheme as restored (Harris) or not (Rowan). Rowan described the picking out of the architectural elements as 'utterly illiterate', whereas Harris found it 'utterly inconceivable' that anyone but Adam could have authorised it. These opposing views are recorded in 'Shades of Adam?', *Monumentum* (Spring 2008), 6–7.

3 Ireland 2008, vol I, 21. Prior to the investigation at Headfort, only the library ceiling at Osterley Park in Middlesex had been subject to scientific analysis, which established that it had originally been painted white and not coloured as in Adam's designs. The library ceiling at Kenwood was subsequently investigated: its gilded relief ornaments were proved not to have been part of the original decorative scheme. *See* Harris 2001, 167; Jenkins 2012, 71–2.

4 Harris, E 2009 unpublished lecture for the Irish Georgian Society at Headfort House (hereafter IGS lecture).

5 *See* Hanson 2003.

6 Saint, A 2013 'The conundrum of "by"', *in* Airs, M and Whyte, W (eds) *Architectural History after Colvin*. Donington: Shaun Tyas, 67; Bergdoll, B 2000 *European Architecture, 1750–1890*. Oxford: Oxford University Press, 37. For an overview of craftsmen who collaborated with Adam, *see* Stillman 1966, 41–50; Beard 1978, 1992.

7 At the time of writing (2016) the stair hall and eating parlour have been fully restored. The original colour schemes for the entrance hall, saloon and 'Chinese Room' have also been established but have yet to be fully reinstated.

8 Craig, M 1982 *The Architecture of Ireland*. London: Batsford, 244.

9 Robinson 2012, 103.

10 Designs for Irish clients featured regularly among those William Chambers exhibited at the Royal Academy and Society of Artists in the 1760s–70s. *See* list in Harris 1970, 175.

11 Harris 1970, 242; BL, Ms 41133, Sir William Chambers Letter Books, vol I, 26v.

12 Colvin, H 1991 'Architect and client', *in* Worsley, G (ed) *Georgian Architectural Practice: Papers Given*

at the Georgian Group Symposium 1991. London: The Georgian Group, 7.

13 Robinson 2012, 124: 'Even if he had designed nothing in England, Wyatt's Irish work in absentia, and especially Castle Coole, would have established him as one of the most accomplished architects of the late eighteenth century.' For Chambers' Casino, *see* Harris, J 1996 'A Franco-Roman triumph: The Casino at Marino', *in* Harris and Snodin 1996, 39.

14 Hussey 1936; Harris 1973. *See also* Cornforth 1973, and Calderón 2008, which introduced information from the Headfort MSS at Senate House, University of London. For Richard Castle's designs, *see* Griffin, D 1998 'A Richard Castle design for Headfort, Co Meath', in Manning, C (ed) *Dublin and Beyond the Pale: Studies in Honour of Patrick Healy*. Bray, Co Wicklow: Wordwell, 269–72; Harris 2007, 124–7.

15 Historical Manuscripts Commission 1888–1905 *The Manuscripts of His Grace The Duke of Rutland Preserved at Belvoir Castle*, 4 vols. London: HMSO, vol III, 419. The house was then 'painted white'.

16 The origin of Adam's commission remains obscure. A suggestion that Bective 'had a large house in London' where the decision might have been made is without foundation (Ireland 2008, vol I, 82). Bective owned several houses in Ireland – Prospect Lodge (later Ardgillan Castle) in Balbriggan, Co Dublin; a hunting lodge in Co Cavan; and a town house in Dublin (first in Smithfield, and after 1789 in Rutland Square). *See* Calderón 2008, 8. Adam's designs for Rowley were for his town house in Mary Street, Dublin (now demolished). One of these ceiling designs was later recycled for the Chinese Room at Headfort House – *see* Harris 1973.

17 Yale Center for British Art (hereafter YCBA), B1975.2.228–40; B1975.2.785–95; B1975.2.797–805. Richard Ireland states that Adam was responsible for the final spatial organisation of the eating parlour, fusing two smaller rooms into a single large one (Ireland 2008, vol I, 70). However, this is a misreading of Christopher Hussey's attribution to Adam in 1936 of a series of designs that have since been reattributed to the Irish architect George Semple. In fact, as Hussey noted, this 'important alteration' to the eating parlour is not represented in any designs, and he credited Bective with the decision (Hussey 1936, 303, 328). There is in fact no evidence to indicate that Adam was employed to do anything other than to provide designs for the decoration of the rooms. For the attribution to Semple, *see* Harris 1973; *also* Cornforth 1973, 948.

18 Designs for the entrance hall comprise a laid-out interior, a ceiling design and a working drawing for mouldings. The laid-out interior for the hall is undated, as is the office copy in Sir John Soane's Museum. In terms of finish it is identical to the dated ceiling design (1771), and to the designs for the eating parlour (also 1771) and staircase (January 1772). The office copy of the entrance-hall ceiling design is dated 12 February 1772, as is another fully rendered presentation design inscribed 'Ceiling for the Hall for The Earl of Bective' and appended 'not right' (SM, Adam vol 12/99). This is likely a 'correction' made by an individual in the Adam office and is not in Lord Bective's hand. Designs for the eating parlour comprise four elevations, a chimney-piece design and three working drawings for mouldings. I suggest here that the original presentation ceiling design, missing from the other presentation designs now at YCBA, is likely to be the highly finished 'office copy' design in the Soane Museum collection (Adam vol 12/96). This is formally presented in ink and watercolour with a ruled border, annotated with instructions ('N.B. This Cieling is drawn supposing the Arch extended or laid down flat') and signed 'Robt. Adam Archt. 1771', as are the other presentation drawings issued to Bective. Stephen Astley suggests that highly finished designs of this type in the Adam volumes are likely to be those returned to Adam by clients, as at Headfort House, where Lord Bective did not execute the original design of 1771 and sought a new one in 1775. Stephen Astley, 'The Adam office and Adam drawings', lecture, 'Robert Adam and His Brothers', Georgian Group Symposium, RIBA, London, 23–24 September 2015. Alistair Rowan demurs, and thinks it unlikely that unsatisfied clients returned designs to Adam (in conversation with the author 26 June 2016).

19 These comprise four elevations, a ceiling design, and five working drawings for plasterwork and door entablatures. Although the YCBA drawings are undated, there are office copies in Sir John Soane's Museum dated 24 January 1772 (SM, Adam vol 12/98; vol 50/60–3). Designs for friezes in the 'Great Room', 'Hall', 'Salon' and 'Lady's Room' exist as a single-folio office copy drawing and likely date from 1771–2 (SM, Adam vol 53/29). In each case, the designs tally with the friezes executed (bar the Lady's Room, which is no longer extant).

20 'A comparison of the drawings with the rooms as carried out shows that the wall decorations [proposed in the original scheme of 1771] were not much affected': Hussey 1936, 329.

21 The long gestation was not unusual in itself: the decoration of the interiors at Osterley Park, one of Adam's most celebrated schemes, was undertaken in several stages, with designs produced in 1763 (dining room and drawing room), 1766–7 (library and entrance hall) and 1772–6 (tapestry drawing room, state bedroom and Etruscan Dressing Room). Harris 2001, 157–79.

22 Two payments on 22 November and 19 December 1774, amounting to £200, confirm McCullagh's work in the hall and saloon. National Library of Ireland (hereafter NLI), Headfort papers, Ms 25,307(17), 116, 137. In the months prior to this there are references to the purchase and provision of materials, such as paints and oils. McCullagh was later (in 1774–6) paid a total of £161 14s 9½d for 'Plastering & Stuccoing of the West Wing of Offices &c.', NLI, Ms 25,307(18), 30. John Harris notes a payment to 'John McCulla' for the staircase plasterwork but this is not recorded in the return books and Harris does not provide a source. *See* J Harris 1973.

23 NLI, Headfort papers, Ms 25,307(26), 107. The underside of a marble-topped table once at Headfort, now in the National Museum of Ireland, bears the

inscription: 'John McAllester Painter began to work in Headfort House in the year 1780'. The only record of this individual in the Headfort papers is as a witness to a bill in April 1781. Ibid, Ms 25,307(22), 95.

24 Information provided by Richard Ireland (9 April 2009). It remains unclear what Booker might have been engaged to do during his 90 days' employment.

25 University of London (hereafter UL), Senate House, Headfort papers, 'Book entering Receipts for Bills Paid', 15v, 23r, 33r. Two payments 'for Painting' in February 1785 and June 1787 amounted to £25 4s 5d (ibid, 20r, 33r). The large payment for painting by Callaghan at Rutland Square is dated December 1788 (ibid, 45r).

26 Such references continue into the 1790s. A final payment of £27 7s 1½d to Callaghan 'for Painting, Glazing, &c' was recorded on 29 March 1793. UL, 'Book entering Receipts for Bills Paid', 78r.

27 For recent accounts of McCullagh's practice, *see* Lucey, C 2007 *The Stapleton Collection: Designs for the Irish Neoclassical Interior*. Tralee: Churchill House, and Lucey, C 2011 'Pattern books and pedagogies: Neoclassicism and the Dublin artisan', *in* Mulvin L (ed) *The Fusion of Neoclassical Principles*. Dublin: Wordwell, 131–50. Cornelius Callaghan is listed at 31 Great Britain (Parnell) Street in 1786–89. Placido Columbani's *A New Book of Ornaments* (London, 1775) and *Variety of Capitals, Freezes, and Corniches* (London, 1776) were available in Dublin from Callaghan's shop. *See Dublin Journal*, 10–13 April 1788.

28 *See* Jacobus 1998. Adam described these drawings as sections and evidently did not distinguish between this manner of representation and that of a true section: a bill of 1771 for drawings for Shelburne (later Lansdowne) House, for example, includes references to 'a section of four sides for the dining room' and 'a section of 4 sides of the Drawing Room'. Reproduced in Bolton 1922, vol II, 240.

29 Jacobus 1998, 155. 'The four sides of the room being laid down on paper, with the space or proportion of floor between, the figure represents at once to the eye the whole and its several parts: they are easily seen to be uniform or disagreeable; fancy can at pleasure raise them perpendicularly, and see the room in miniature'. Ware, I 1767 *A Complete Body of Architecture*. London, 478–9.

30 Evans, R 1989 'The developed surface: An enquiry into the brief life of an eighteenth-century drawing technique', *in* Evans 1997, 204. Charles Rice, in his overview of Evans' argument, also misses the point of this drawing type and arguably overstates its supposed problems with spatial legibility: 'Developed surface drawings do not partake of the doubled sense of the interior, where the interior as spatial volume is understood consciously, and is accompanied by representational techniques that double this spatial sense' (Rice 2007, 25).

31 Harris 2001, 219.

32 According to Evans, the purpose of an architectural drawing is 'to provide a complete determination in advance'. Evans, R 1986 'Translations from drawing to building', *in* Evans 1997, 156.

33 Harris, IGS lecture.

34 Ireland 2008, vol I, 83.

35 Adam's executed projects in Ireland were: interiors for Langford House, Dublin (1765), for the Rt Hon Hercules Rowley; Headfort House; remodelling of Castle Upton, Co Antrim (1783–8), for Clotworthy Upton, 1st Baron Templetown; and stables and a mausoleum at Castle Upton, Co Antrim (1788–9), for John Upton, 1st Viscount Templetown.

36 Adam's unbuilt projects in Ireland were: extension to Summerhill, Co Meath (*c* 1765), for the Rt Hon Hercules Rowley; statue pedestal for the Royal Exchange, Dublin (1770); gates for Loftus Hall, Co Wexford (1785), for the Rt Hon Charles Loftus; and a Gothic tower at Tollymore Park, Co Down (nd), for James Hamilton, 2nd Earl of Clanbrassil. Five drawings in the Adam volumes at Sir John Soane's Museum for a house (comprising three plans, one elevation and one section) dated 1775 are designated for 'James Alexander Esq'. Bolton identifies Alexander as 'Merchant, Stirling' (Bolton 1922, vol II, 60 'Index of clients'). But these drawings may relate to James Alexander, later 1st Earl of Caledon, who was in the market for a grand house at this time and later commissioned designs from James Wyatt in 1779 (SM, Adam vol 43/33–7).

37 The first number of *The Works* was announced in *St James's Chronicle*, or, *The British Evening Post*, 20 March 1773. Its first appearance in a Dublin newspaper was in the *Hibernian Journal*, 12–15 November 1773. The latter advertisement indicates that William Wilson of Dame Street was selling the first number in Dublin early on. His name is appended to the London newspaper advertisements throughout 1774–6, alongside booksellers in Paris, Amsterdam and Edinburgh. From the third number onwards, the tag line 'in Great Britain and Ireland' was dropped.

38 Rowan 1974, 678. For a comprehensive account of this, *see also* Rowan 2007.

39 Harris, IGS lecture; Harris 1973.

40 Information from Stephen Astley: only two of the office copy designs are coloured; the remainder contain textual notes only.

41 Rowan 1988, 19; Sanderson 1992, 84.

42 National Library of Scotland, MSS 19992–3 (Paterson–Adam correspondence), 13 April 1791, cited in Sanderson 1992, 84.

43 BL, Chambers Letter Books, vol I, 11r. Adam's presentation designs of 1771 for the eating parlour are annotated with room dimensions (YCBA, B1975.2.788–89, 793).

44 Harris 2001, 213, 321. The architect John Paterson (d 1832) also helped run Robert Adam's Edinburgh office and was employed as clerk of works for many of Adam's important late works in Scotland, until dismissed in 1791; he later became a leading proponent of the Adam castle style as an architect in his own right. His place as clerk of works was filled by other Adam assistants, notably Hugh and Alexander Cairncross, and Andrew Cairns. *See* Sanderson 1992, 82, 84, 86, 120, 122, 126; Sanderson 1982, 43.

45 Rowan 1988, 20.

46 Harris 1971, 51 (plate 37).

47 BL, Chambers Letter Books, vol I, 7v.

48 Harris 1971, 57.
49 NRS, Kennedy of Dalquharran MSS, GD27/7/323, no 1.
50 Harris 2001, 12, citing a letter from Adam to Sir Edward Knatchbull, 10 February 1772 (Maidstone, Kent Record Office, U951/A18).
51 Harris 2001, 13. Interestingly, King notes that the yellow in the design is 'unusually deep', a distinction that might be given to Knatchbull's influence. King 1991, 116 and plate 150.
52 Pevsner described Sir Edward as 'tight with his money'. Pevsner, N and Newman, J 2012 *Kent: West and The Weald – The Buildings of England*. New Haven and London: Yale University Press, 425–6.
53 Northumberland Record Office, Delaval of Seaton Delaval MSS, 2DE/23/2/1. Cited in Calderón 2008, 22–7.
54 Bromley's estimate for figurative centrepieces for these ceilings as shown on Adam's drawings was, at £57 15s, also less than half of Adam's specification for 'historical paintings' in these rooms (£129 6s). *See* Calderón 2008, 27–8, 38.
55 Harris 2001, 231. Harris also suggests (p 219) that Belwood 'adopted' Adam's style, and that 'Bypassing Adam (or his London office)' represented a saving for Weddell. Belwood also designed a screen for the upper landing and half landing: Harris describes them as a 'weak imitation of [William] Chambers's grand staircase at Gower House'. Belwood's bookcase design is equally dismissed as 'old-fashioned' (ibid).
56 Adam and Adam 1773–1822, vol I, no V, description of plate VIII.
57 Bristow 1996, 110–11. Bristow describes the use of 'three greens alone' in a design for General Fitzroy of 1772, a combination shared with the revised eating parlour design for Headfort House of 1775. Bristow also notes designs in shades of grey in 1774 (ibid, 112).
58 Adam and Adam 1773–1822, vol II, no I, description of plate II.
59 Blau, E and Kaufman, E 1989 'Introduction', *in* Blau, E and Kaufman, E (eds) *Architecture and Its Image: Four Centuries of Architectural Representation*. Montreal: Canadian Centre for Architecture/Centre Canadien d'Architecture, 13.
60 Although the coloured working drawing issued to Bective for the original eating parlour ceiling is undated, the identical annotated office copy is dated 1771. Therefore it seems clear that the monochromatic presentation drawings for the eating parlour are of the same date as the two coloured working drawings for the same room.
61 Harris 1973; Ireland and Gundry 2011, 63.
62 Rowan 1988, 19.
63 The difference in price between monochromatic and polychromatic drawings was surely a factor when commissioning designs from the Adam office. In 1765 Adam's pen-and-wash ceiling design for the bedroom at Coventry House, Piccadilly, was invoiced at 7 guineas; in 1766 a coloured ceiling design for the drawing room at Lansdowne House, Berkeley Square, was estimated at 12 guineas. *See* Harris 2001, 55, 119.
64 Harris 2001, 12. The drawings for Nostell Priory and Newby Hall are reproduced on pp 200 (plate 293) and 223 (plate 331). The correspondence between plasterer Joseph Rose and his client Sir Christopher Sykes, for work done at Sledmere House, Yorkshire, in 1791–2, refers to drawings being sent by 'packing case'. University of Hull Archives, Joseph Rose letters, DDSY/10/62. (Information from Colin Thom.) Portfolios were often used for sending presentation drawings to clients. Lever, J 1991 'The architect's office', *in* Worsley, G (ed) *Georgian Architectural Practice: Papers Given at a Georgian Group Symposium 1991*. London: The Georgian Group, 62. John Adam often sent working drawings folded up with his correspondence. (Information from Alistair Rowan.)
65 This contradicts the statement that 'Working drawings of as-executed work are in much shorter supply and it is assumed that they were lost or destroyed during the construction process': Ireland and Gundry 2011, 63. I am here counting the highly finished presentation design for the ceiling of the eating parlour of 1771 in Sir John Soane's Museum as one of the 17 presentation designs.
66 Ayres, J 1998 *Building the Georgian City*. New Haven and London: Yale University Press, 23.
67 Richard Ireland's suggestion that the existence of so many 'at large' detail construction drawings on linen-backed paper in the Headfort collection is because 'they were not approved and therefore not handed over to the tradesman and "consumed"' is contradicted by a careful examination of the original drawings (Ireland 2008, vol I, 83). There is in fact no consistency to the mounting of drawings: of the 17 working drawings at YCBA, only 6 are mounted on either canvas or linen; 2 of the presentation drawings are also mounted. Of the mounted working drawings, at least 2 (B1975.2.235, B1975.2.234) relate to work executed in the house.
68 Ireland 2008, vol I, 77. The designs for the stair hall at Headfort, issued in 1771–2, are monochromatic and contain no textual notes on colour. At Syon the hall was originally 'stucco'd & left white', according to Lady Shelburne, who visited in March 1765 (Harris 2001, 68). At Home House, the walls of the staircase were 'dead white, grey and fawn to look like stone; even the bannisters were painted white and grey' (ibid, 300). For Lansdowne House, *see* SM, Adam vol 39/66, reproduced in Harris 2001, 127 (plate 187). White ceilings remained popular throughout the 18th century, particularly for halls and staircases (Bristow 1996, 122). For the walls of halls and passages, Bristow notes the 'preferred' use of stone colours, although they were 'occasionally' painted in 'delicate tints' of green and grey (ibid, 154). The sole polychromatic designs for a staircase hall in Adam's *oeuvre* relate to the so-called 'Hell Staircase' at Burghley House, Lincolnshire, where in 1779 Adam designed a scheme clearly intended to complement Antonio Verrio's celebrated ceiling painting (created 1697–1700). David King, oddly enough, suggests otherwise. *See* King 2001, 175.
69 Adam and Adam 1773–1822, vol I, no I, description of plate V. Harris describes Adam's 'thoughtful application of different colours' as 'a finely tuned balance and great variety at the same time' (Harris 2001, 12).

70 Bristow 1996, 104–7. The presentation design clearly employs three tints of green. This is confirmed by the uncoloured office-copy drawing which specifies 'light green', 'second tint' and 'darkest tint'.

71 Ireland 2008, vol I, 16, 18, 49. The restored green colouring of the stair-hall ceiling also contradicts Adam's monochromatic design of 1771, and his preference for neutrally coloured circulation spaces. *See* note 68 above; *see also* Harris 1973.

72 Ireland 2008, vol I, 83.

73 Coloured working drawings for Newby Hall are attributed to William Chambers in Cornforth, J 1979 'Newby Hall, Yorkshire–III'. *Country Life* **165**, 21 June, 2006–9, figs 7 and 9. On Chambers' use of coloured relief ornament, *see* Bristow 1996, 141–2. Adam sent large-scale working drawings by road or sea from London. In a letter to his client George Baillie in 1778, Adam wrote that he would send drawings of the 'ornaments at large' for the decoration of the gallery at Mellerstain House, Berwickshire. *See* Sanderson 1992, 85.

74 Redmill, J and Bristow, I C 1984 'The Casino at Marino, Dublin'. *Transactions: Association for the Studies in the Conservation of Historic Buildings* **9**, 41. No drawings survive, but Chambers recommended a white entablature 'touched with blew'.

75 Ireland finds 'strong parallels in the disposition of picking out of alternating elements' between Chambers' coloured working drawings for Newby Hall and the articulation of painted ornament at Headfort, but this does not bear close scrutiny. The designs by Chambers clearly articulate the forms of the mouldings: there is no alternating pattern. Ireland 2008, vol I, 84.

76 Ireland and Gundry 2011, 77; Ireland 2008, vol I, 84.

77 Harris, IGS lecture; 'Shades of Adam', 6.

78 The argument is further weakened by comparing painted three-dimensional plaster ornaments (at Headfort) with a two-dimensional drawing (for Osterley). Ireland and Gundry 2011, 77, fig 18.

79 Ireland 2008, vol I, 82.

80 'Shades of Adam', 7.

81 Harris 2001, 4.

82 Elsewhere, correspondences between ceiling and door entablatures in terms of colour – pink in the saloon, violet in the Chinese Room – follow the established methods for coordinating the architectural elements within an interior.

83 The colour scheme that Adam proposed for the walls of the eating parlour in 1771 was in fact used for the saloon (as revealed during the scientific investigation). Ireland 2008, vol I, 49.

84 Evans, R 1989 'Architectural projection', *in* Blau and Kaufmann, *Architecture and Its Image*, 19.

85 Eileen Harris, noting the difficulty of drawings as 'conclusive evidence of what was finally executed', is nonetheless clear that colour annotated drawings or written instructions are a 'somewhat safer guide' in determining what was intended. Harris 2001, 11.

86 Harris, IGS lecture.

87 Rowan 1988, 19. With this in mind, it is interesting to note two record drawings for Langford House, Mary Street in Dublin, among the Adam volumes at Sir John Soane's Museum (SM, Adam vol 52/21–2).

88 Just as the organisation of the drawing office has received sustained scholarly attention, the different types of drawings and their functions demand further research. On various aspects of the Adam drawing office and building business, *see* Stillman 1966, 41–50; Rowan 1974; 1985, 10–12; 1993; 2007; and Tait 1993, 71–102.

89 Stillman notes that office copies were 'only partially filled in and coloured' (Stillman 1966, 44). Was cost a factor? For 'a Room in the Gothick Stile' at Hulne Abbey, Northumberland, Adam charged 17 guineas for 'a section of the four sides and a ceiling'. Ibid, 111–12 (citing Alnwick Castle Archives, U.I.46).

90 John Harris speculated about Bective's reduced finances in 1973, finding evidence in the protracted progress of building works, 'and the measures taken to economize in the decoration process of simplification and omission' (Harris 1973). Cornforth, following Harris, ascribed the 'considerable amount of cutting back' of decoration proposed by Adam to Bective's 'relative lack of means' (Cornforth 1973, 948). However, Loreto Calderón has more recently shown that Bective 'never had money problems' and that his account books indicate his 'careful financial management' (Calderón 2008, 7–8).

91 Harris 1973.

92 Harris 2001, 4.

7

A 'Classical Goth': Robert Adam's engagement with medieval architecture

Peter N Lindfield

Robert Adam's grand tour in 1754–8 had a profound impact upon his understanding of the forms and ornament of Classical architecture. After his return to Britain this became manifest in his country-house designs, such as the south façade of Kedleston Hall, Derbyshire (from *c* 1760; *see* Fig 0.13).[1] Beyond this elevation's dynamic advancement and recession of planes, the architecture paraphrases two notable monuments of ancient Rome: the Arch of Constantine and the Pantheon. Kedleston's interior, including its celebrated saloon, was equally responsive to antique forms, and the tour's impact upon Adam in shaping his eponymous neoclassical style is explored elsewhere in this volume.

The vast majority of Adam's country-house architecture – from the remodelling of existing homes like Osterley Park, Middlesex (1761–80), through to the creation of new structures such as Witham Park, Somerset (1762–70) – was neoclassical.[2] Indeed, of his realised work, only 3 of his 89 known country-estate commissions were overtly Gothic and all were south of the border: the interior of the Church of St Mary Magdalene, Croome d'Abitot, at Croome Court, Worcestershire (1763); the Round Room at Strawberry Hill, Twickenham (1766–7); and his work for the Duke of Northumberland at Alnwick – namely the interiors for Alnwick Castle (1769–80) and its associated Gothic eye-catcher, Brizlee Tower, Hulne Park (1777–83). In 1958 John Fleming suggested that the Gothic interiors of the so-called Lord's Tower and garden house at Hulne Priory were also by Adam, a claim later refuted by Alistair Rowan, though it should be pointed out that the chimney piece in the upper room of the garden house resonates with aspects of Adam's proposed chimney pieces for Alnwick Castle – in particular the blind panelling and clustered columns of his 1769 design for the drawing-room chimney piece.[3] To this short catalogue of finished Gothic works can be added the earlier Decorated-style west front of Yester Chapel, on the outskirts of Gifford, East Lothian, designed around 1753 by Robert in partnership with his older brother, John, and also unrealised proposals for Sir William Pulteney (1684–1764), 1st Earl of Bath, at Shrewsbury Castle, Shropshire, of *c* 1769, in a quasi-Perpendicular style of Gothic not too dissimilar from Adam's proposals for Alnwick's interiors of around the same date.[4]

A further 17 of Adam's known country houses were castellated, such as Culzean Castle, Ayrshire (1777–92), but these cannot be considered wholeheartedly Gothic, incorporating as they do forms and features, including Venetian windows, entablatures and pilasters, from Classical and other architecture.[5] Alistair Rowan has referred to this castle style as 'amongst the most original creations of eighteenth-century European architecture ... with a vocabulary and syntax of its own'.[6] The difference between Adam's Classical–Gothic hybrid structures and the Gothic Revival buildings of the earlier 18th century, including Horace Walpole's Strawberry Hill at Twickenham (1747–80) – a country villa intended by Walpole to represent 'the castle ... of my ancestors' – relates, in part, to the time and place in which Adam was working, and to his very personal synthesis of the styles that appealed to his romantic sensibilities.[7] Adam saw these castle-dwellings 'not as a Gothic house but in some sense perhaps as an ancient Caledonian fort', and infused also with his memories of the ancient Roman fortifications he had seen in Italy.[8] The handful of self-consciously Gothic commissions, sketches and preparatory designs considered here, albeit small in number, is extraordinarily varied, and is important for what it reveals about Adam's evolving engagement with the language of medieval architecture – an imaginative and largely overlooked undercurrent to his trademark Roman style.

Gothic design in Georgian Britain developed tremendously between around 1730 and 1840, and Adam's place within that tradition has not been assessed fully.[9] His experiments with Gothic date from the mid- to late 18th century and are thus sandwiched between the work of the earlier, whimsical Gothicists and their later, more antiquarian-minded successors. He postdates the Classical-Gothic of William Kent in the 1730s–40s, as seen at Esher Place, Surrey (*c* 1733), and York Minster (1740–41); also the designs presented by Batty Langley in *Ancient Architecture, Restored and Improved* (1741–2); and the highly decorative, whimsical rococo-Gothic promoted by the likes of William and John Halfpenny and Thomas Chippendale in their publications.[10] But, as this essay suggests, Adam's post-grand-tour Gothic is indebted, at least initially, to Kent's distinct interpretation and reuse of medieval forms, which falls short of the later and more rigorous work of architects deeply concerned with understanding medieval fabric, such as James Essex (1722–84).[11] Adam did not share Essex's antiquarian pretentions; instead he developed a highly personal style of Gothic that, at its creative zenith, manipulated medieval forms and fused them with or into Classical motifs. In finish, this neoclassical-Gothic also mirrored his more familiar Classical colour schemes in the tinted grounds of pastel colours that helped define the ornamental patterns.[12] Adam, in effect, pioneered neoclassical-Gothic design: a phase of Georgian Gothic architecture and furniture that has only recently been given serious attention.[13]

The Works in Architecture: Adam, Wyatt and Gothic

Given the minor role that Gothic design played in Adam's *oeuvre*, and the overwhelming preference for Classicism in late 18th-century Britain, it is not surprising to find him being critical of the style when the occasion suited. This is, of course, consistent with the received architectural wisdom of the period. Sir Christopher Wren's memoirs, *Parentalia*, had only recently been published (in 1750), in which Wren largely dismissed Gothic as a rude and unsophisticated style.[14] Writing in the first volume of *The Works in Architecture of Robert and James Adam* (1773), Robert Adam echoed the by then well-established criticism of pointed-arch architecture. He implied that Gothic was a low point in British architectural history, remedied by the sophisticated appropriation and deployment of Classical forms by post-medieval architects:

> We intended to have prefixed to our designs a dissertation concerning the rise and progress of architecture in Great Britain; and to have pointed out the various stages of its improvements from the time, that our ancestors, relinquishing the gothick style, began to aim at an imitation of the Grecian manner, until it attained that degree of perfection at which it has now arrived.[15]

But perhaps we should not take Adam's words in *The Works* too seriously. The book was a work of polemic, intended to bolster his and his brother James's positions as the foremost neoclassical architects of their day. In order to stress the pre-eminence of his new, pioneering 'revolution' in design, Adam also took great pains in the text to distance himself from the preceding British fashion for Palladianism, yet so much of his own architecture was firmly rooted in its forms and proportions.

Referring obliquely in the same publication to James Wyatt (1746–1813), the arriviste and highly fashionable architect then challenging the Adams' architectural hegemony, the brothers dismissed him as a copyist rather than an inventor – an imitator, not an original genius.[16] Their argument was that Wyatt had nothing new to offer 18th-century architectural design:

> We hope it will be thought no more than justice to ourselves, thus to ascertain the originality of our designs, and enable the world to discover, where they have been imitated with judgment, and where they have been servilely copied or misapplied. – An artist who feels in himself an inability of presenting to the public any thing from his own store of invention, has no title to be offended if an author is solicitous to vindicate himself to posterity from any imputation of plagiarism.[17]

The brothers were clearly troubled by the instantaneous success that accompanied Wyatt's arrival on the British architectural scene, courtesy of his spectacular Pantheon in Oxford Street, London, which opened to critical acclaim in January

1772. Even Walpole, that great supporter of Gothic architecture, had nothing but praise for Wyatt's Pantheon.[18]

Much like Adam, Wyatt is known primarily as a Classical architect, but he also took recourse to Gothic forms, with early Gothic-style works including the Great Hall at Beaudesert, Staffordshire (1772), the exterior and Entrance Hall of Sandleford Priory, Berkshire (1780–9), and Sheffield Place, Sussex (1775–87).[19] A comparison with Adam's Gothic is instructive. Unlike Adam, Wyatt's Gothic interventions brought him notoriety and criticism because he attempted something that Adam never did: the restoration of medieval monuments, especially at Westminster Abbey (from 1803) and the cathedrals of Litchfield (1787–93), Salisbury (1787–92), Hereford (1788–96), Durham (1795–1805) and Ely (1796–1802).[20] This isolated Wyatt and placed him in direct conflict with preservationists and antiquaries who opposed such 'improvements' to medieval fabric. John Carter (1748–1817), for example, a topographical draughtsman and staunch medievalist, labelled him 'the Destroyer' for his interventions.[21] But, contrary to this criticism, Wyatt's Gothic country houses, culminating in Fonthill Abbey, Wiltshire (1796–1812), and Ashridge Park, Hertfordshire (1808–13, completed by Jeffry Wyatt, later Sir Jeffry Wyatville, 1813–17), reveal a sophisticated understanding of medieval architecture.[22] Although distinctively Georgian, Wyatt's Gothic works from 1780 were clearly concerned with imitating true medieval forms. This cannot be further removed from the tenor of Adam's interest in the Gothic. Even at his most sober, Adam did not, for example, attempt to follow strictly the visual language and appearance of medieval architecture. Instead, he created Gothic buildings, furniture and interiors from the absolute position of a Georgian designer unhampered by antiquarian interest and restraint. Consequently, he was able to recreate and recast medieval Gothic in a decidedly fashionable 'Adamesque' mode – since described as 'bizarre and even barbarous'.[23]

As with his castle architecture, discussed elsewhere in this volume, Adam's Gothic was heavily indebted to the language of the Classical antique and late Georgian aesthetic fashion. This mixture of styles is something that Wyatt never really harnessed for his Gothic projects: while the Adams may have considered him a plagiarist who threatened their primacy, there was no attempt on Wyatt's part to imitate, improve upon or even approach Robert Adam's style of Gothic. Their respective output in this mode helps separate the two as architectural designers, though this crucial difference has received little sustained attention. The case studies considered below illustrate the distinct evolution of Adam's Gothic and show how his innovatory genius – a genius that he was so keen to promote and defend in *The Works* – was not confined solely to Classical and castle architecture but included also the medieval, and in a more pronounced manner than is generally acknowledged.[24]

Adam's Gothic background: early designs, sketches and musings

Despite Adam's implied criticism of medieval fabric in *The Works*, he obviously paid it serious attention. In the sale of the Adam architectural library by Mr Christie in 1818, one of the first lots was described as a large parcel 'of gothic Architecture and Antiquities'.[25] In addition to this cache, there is plentiful other evidence of an absorbing interest in the subject from early in his career. Regarding his trip south of the border to England in 1749–50, Adam remarked on his shifting interest from Gothic to Classicism: 'At London I first felt the Change of Taste grow on me from that I had contracted in Scotland'; his friend and brother-in-law the draughtsman John Clerk of Eldin agreed that it was during this trip that Adam 'first began to prune the exuberance of his fancy, & to correct his taste'.[26] This 'exuberant fancy' was expressed directly in Adam's Gothic capriccios of the period. Despite his increasing turn towards Classicism, reflected in a robustly Palladian proposal from May 1750 for 'Pavilions to Lord Pembroke's House at Wilton', Wiltshire, Adam's sketchbook from this trip contains eight imaginative and highly ornamental Gothic designs (Fig 7.1).[27]

It has been suggested repeatedly that Adam's early Gothic capriccios are indebted to the plates in the first Georgian pattern book on Gothic architecture: Batty Langley's *Ancient Architecture* (1741–2), of which there was a copy in his father's library at Blair Adam.[28] But in terms of their form, lively filigree ornament and architectural ethos, these 1749–50 sketches bear no relationship to Langley's work and its mostly linear, Classical structuring (albeit that Adam would later rewrite his Gothic in neoclassical terms). Instead, they are characterised by dilated

Fig 7.1
Two early Gothic fantasies from Robert Adam's 1749–50 English sketchbook: folio 17 (above right) and folio 19 (above).

and cusped Y-tracery windows, and galleries of cusps and pendants running against lines of enlarged crockets. Indeed, Adam's designs are significantly over-crocketed in comparison with typical medieval architecture and the work of other Georgian designers. These ornamental and imaginative predilections emerge strongly throughout the pocketbook, but especially so in the sketch on folio 19, where a hexagonal window is cusped to make a sexfoil, and the external angles are finished with finials, creating a hexagram (*see* Fig 7.1). Equally ornate are the gables on what appears to be a church in folio 17; these are cusped externally, similar to those on the tomb of Edmund Crouchback at Westminster Abbey, and bedecked with crockets (*see* Fig 7.1). Adam's fertile imagination overloads these Gothic designs with crockets, cusps and windows to create impossibly busy and decorative structures that are unrelated to England's medieval fabric. Adam stuck mostly to the ornamental micro-architectural vocabulary of Decorated Gothic, but he deployed these motifs in an imaginative manner never anticipated by early Georgian architects like William Kent or Batty Langley (Fig 7.2).

Other drawings in the Blair Adam collection show that this overtly ornamental and picturesque Gothic style of Adam's extended beyond the pages of his English tour sketchbook. A 1749 drawing of a folly, worked up with body colour and shadow, is of similar character, albeit with the addition of ogee-arched openings.[29] Another contemporary drawing that relates directly to the 1749–50 trip is his *Elevation of the Cross at Winchester,* of 1750 (Fig 7.3). It has been suggested that this inaccurately scaled depiction of the Winchester Cross is based on a 1741 engraving by George Vertue (Fig 7.4), published by the Society of Antiquaries of London in the first volume of their *Vetusta Monumenta* (1747).[30] Vertue's crudely depicted ogee gables and overabundant aedicules above the primary niche are not, however, reproduced in Adam's version. This is not the only difference: more significant, perhaps, is the angle from which the cross is depicted. Whereas Vertue's engraving renders the structure from a high viewpoint that introduces perspectival distortion, Adam's elevation is drawn orthographically, preserving the architectural details, and is thus akin to studies in early 18th-century pattern books, such as the various volumes of Colen Campbell's *Vitruvius Britannicus* (1715–25). Additionally,

Fig 7.2
Batty Langley, 'Four Examples of Arcades for Piazza's', from his Ancient Architecture: Restored and Improved *(1741–2), plate 29.*

the aedicules and buttressing, along with the finer details – pinnacles, finials and crockets – are all delineated by Adam with close attention to form and scale, more so, actually, than in Vertue's engraving.[31] These differences suggest that Vertue's work was almost certainly not Adam's model; he either viewed the structure in person (he was close to Winchester in 1750) or worked from another rendition. Though its source remains uncertain, the Winchester Cross drawing shows that Adam was no mere Gothic fantasist, but was willing when required to attempt a close interrogation of medieval architecture.[32] And the difference between this study and his other, more imaginative capriccios emphasises the significance in his work of the creative design process.

There are more such sketches of Gothic buildings, mostly dating from 1753, shortly before Adam's departure for the Continent, including designs for a tower and church.[33] These structures, much like those in his sketches of the late 1740s, are also said to have been influenced by Langley, but they, too, owe very little to the linear plates in *Ancient Architecture*, which lack the elaborate gables, finials and swags seen, for example, in Adam's Gothic folly.[34] Indeed, the underlying rectilinear form and mural mass of the tower and church are, instead, derived almost exactly from the plates in William and John Halfpenny's *Rural Architecture in the Gothic Taste* (1752). Plate 11 in that book, for example, *A Lodge or House of Retirement in the Gothic Taste*, depicts a country house with a

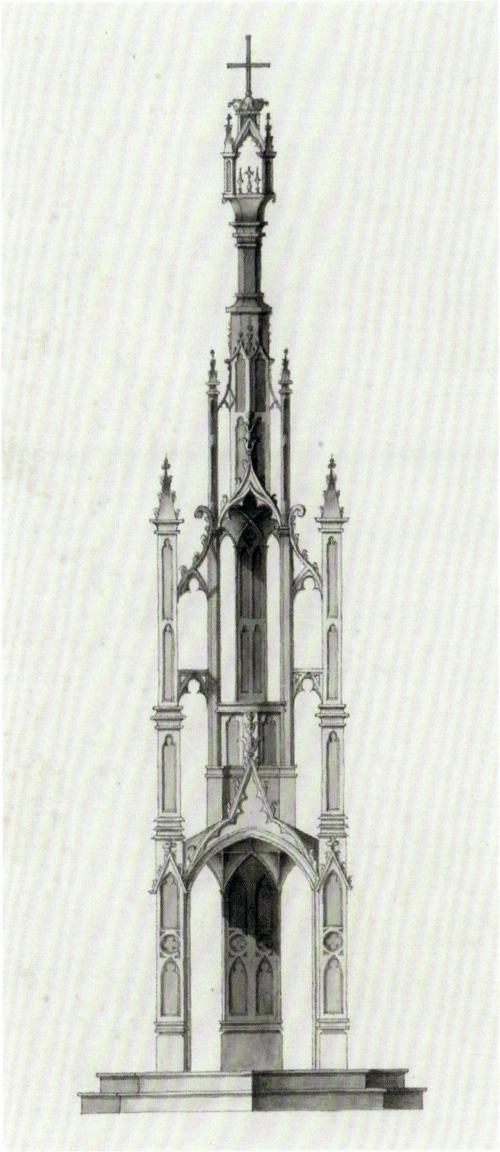

Fig 7.3 (far left)
Adam's Elevation of the Cross at Winchester, *1750.*

Fig 7.4 (left)
George Vertue, Winchester Cross, 1741, from Vetusta Monumenta *(1747), plate 61.*

large, quasi-Palladian accommodation block, which, along with its pavilions, is 'made' Gothic by the addition of lancet-shaped windows, ogee arches, crenellations, spires and crockets.[35] This technique – not seen in Adam's known 1749–50 sketches – was used by him again in an almost identical fashion for his 1753 church design, suggesting a familiarity with the Halfpennys' work. The tower's distinctively Scottish crown steeple, on the other hand, comes from a source closer to home – that on the tower of St Giles' Cathedral, Edinburgh – and its offset and attenuated buttressing imitates that in his drawing of the Winchester Cross.[36] One particularly distinctive feature of the church's design is its Gothic Venetian window. While Batty Langley did not include this motif in his pattern book, it accords exactly with the architectural ethos driving *Ancient Architecture*. Langley wished to rehabilitate the Gothic style and reinstate its respectability by applying its motifs to the underlying form and structure of Classical architecture.[37]

Adam's early, ornamental brand of Gothic, as seen in his 1749–50 capriccios, was realised in 1753 at Yester Chapel, East Lothian, for John Hay, 4th Marquess of Tweeddale. Here Adam and

Fig 7.5
One of Adam's Classical-Gothic capriccios from the Italian leg of his grand tour, c 1756.

his brother John refashioned the truncated collegiate church begun in 1421 (a letter from the Adam office of March 1753 records: 'We are busy with the drawings of the old church').[38] The brothers' interventions concentrated on the church's west front; the adjacent cusping and crocketing on the façade's gable, together with a cusped parapet, reproduce, almost exactly, a capriccio by Robert in his English tour sketchbook.[39] John Adam made similar sketches in his youth (as indeed did James Adam), and the ornate early Gothic, almost rococo musings of the three brothers are manifest in the west front of Yester Chapel. The drawings and the Yester commission establish beyond doubt the depth of Robert Adam's early interest in Gothic design. This fascination stayed with him and was to return in subsequent capriccios, especially his many picturesque renderings of castles within landscapes, such as *A Rocky Landscape with a Castle, c* 1780.[40]

When he arrived in Italy in 1755 for the crucial phase of his grand tour, Adam's interest in Gothic was, by all accounts, extremely obvious to his architectural tutor, Charles-Louis Clérisseau. The latter's assessment of Adam – 'very ignorant of Architecture when he came to me, except the Gothick: but I put him off that; and gave him some taste for the Antique' – seems surprising, given the young Scot's grounding in Classical and Palladian architecture.[41] His grand tour capriccios are predominantly of Classical structures, but a few possess Gothic details and on occasion relate to his 1749–50 sketches, such as one of a tower with orthogonal turrets (where the Y-tracery and windows resonate with those incorporated into designs in his 1749–50 sketchbook), and also a low Gothic folly, encrusted with crockets and cusping.[42] A number of sketches betray the influence of Italian Gothic cathedrals, such as Milan, Siena and the baptistery at Pisa: the latter especially informed some capriccios incorporating dominant rings of Gothic gables (Fig 7.5). Others also seamlessly hybridise Classical and Gothic elements, an artistic process that would prove influential for Adam's post-grand-tour Gothic architecture, especially at Alnwick.[43] Despite Clérisseau's best efforts, Adam clearly retained an interest in medieval buildings. His imaginative engagement with and playful recontextualisation of Gothic forms was hardly dinted by his Continental experiences; instead, he appears to have embraced and harnessed Italian Gothic and Classical styles in his capriccios; their fusion was ultimately reproduced in his work back in England in the 1760s.

Case studies

The early designs by Adam for Gothic structures discussed above, however, have very little connection with the character and ethos of his post-grand-tour Gothic architecture. Confined to an 18-year period, this later Gothic output nevertheless reveals an important and distinct evolution in ornament and syntax, which, like his castle architecture, is highly personal and dependent upon his admiration and understanding of Classical forms. Adam's three Gothic commissions considered here – the Church at Croome Court, Worcestershire (1763); the Round Room at Strawberry Hill, Twickenham (1766–7); and the interiors at Alnwick Castle, Northumberland (1769–80) – are each examined chronologically, revealing a nuanced and evolving style of neoclassical-Gothic.

Adam at Croome Court

Croome Court, Worcestershire, the family seat of George William, 6th Earl of Coventry, was a mid-17th-century double-pile house, improved by Lancelot 'Capability' Brown from 1750 according to designs by Sander Miller.[44] Miller's and Brown's interventions were Classical – the end bays of the original 1640s structure were taken down and rebuilt as corner towers in a Palladian style. Adam took over from Brown in 1760 and was responsible for Croome's interior spaces, particularly the gallery, as well as for the addition of suitable fittings, such as seat furniture and the library bookcases.[45] Beyond working at the house, Adam also made designs for fitting out the interior of the new Croome Church, built in the grounds of the house around 1758–63, probably according to a proposal by Brown, replacing an old village church that had stood inconveniently close by.[46] Whereas Adam's work at the house presented his characteristic brand of neoclassicism, not surprisingly his designs for the church were Gothic. This respected and reflected Brown's church exterior, which in turn had followed the medieval precedent for English ecclesiastical architecture and, simultaneously, the 18th-century fashion for Gothic structures in landscaped gardens.[47]

A number of Adam's unexecuted proposals survive for the church's internal furnishing programme.[48] This corpus demonstrates that, despite his facility for devising hypothetical Gothic structures, initially he turned to a printed source to guide the design of this project; there is, of course, a difference between playful sketches for the exteriors of imagined structures and formal proposals for the fitting up of an actual building. Significantly, Adam's two presentation designs for the church's interior reveal a fundamental shift in his Gothic style over a relatively short period of time.[49] What is certainly his first, and most straightforward, rendering of the church's side-aisle walls emphasises his reliance on a Gothic pattern book (Fig 7.6).[50] It is not, as is the case with his 1750s designs discussed earlier, related to the Halfpennys' *Rural Architecture in the Gothic Taste* but John Vardy's *Some Designs of Mr. Inigo Jones and Mr. Wm. Kent* (1744). A posthumous volume, bringing together the work of two of Britain's most prolific Classical architects, Vardy's *Some Designs* compiled and published a range of Kent's Gothic designs, from a candlestick through to the screens at Westminster Hall (1739, heightened 1755) and Gloucester Cathedral (1741), and York Minster's pulpit (1740–41).[51]

The panelling beneath the windows in Adam's design for Croome Church matches closely that incorporated by Kent in his York Minster pulpit – an ogee-headed lancet subdivided with a Y-shaped pendant at the head and a double cusped and finialled base – as well as paired shaft rings on the adjacent clustered columns, and a Gothic-pendant frieze (Fig 7.7). This panelling can also be found on the interior of Shobdon Church, Herefordshire, *c* 1755, a building either designed or at the very least directly influenced by Kent.[52] Similar panelling is also incorporated into the side piers of a highly unusual polychromatic cabinet for a friend of William Kent's, Henrietta Louisa Fermor, Countess of Pomfret, and probably executed around 1752–3 by the Westminster cabinet-maker William Hallett Senior.[53] Adam's first proposal for the aisle panelling at Croome Church therefore sits happily within a limited but notable corpus of Georgian architectural works directly referencing Kent's idiosyncratic Gothic designs. As I have argued elsewhere in regard to these two separate occurrences of Kent's motifs, this suggests that Adam may have been reluctant to develop Croome Church's interior without an external source to guide him.[54] This is understandable, especially given that his architectural practice up until this point had been almost entirely classically oriented.[55]

This reliance upon Kent's particular brand of Classical-Gothic was only temporary, however. Adam's second proposal for the church's side aisle, dated to 1761 and also unexecuted, similarly places Gothic panelling along the aisle walls, though their form in this instance bears no relationship to Kent's work (Fig 7.8).[56] Instead,

Fig 7.6 (below)
Robert Adam, 'Section of the Church at Croome for the Right Honble The Earl of Coventry', c 1761.

Fig 7.7 (below right)
William Kent, 'A Pulpit in the Cathedral at York', from John Vardy's Some Designs of Mr. Inigo Jones and Mr. Wm. Kent *(1744), plate 51.*

Adam's new proposal sets a rectilinear, gable-like tone that his later, executed Gothic work in the church and at Alnwick would follow, together with his unrealised and executed designs for the Classical-Gothic funerary monument at Milton Abbey, Dorset, for Lady Caroline Milton (d 1775).[57] This runs contrary to mainstream Georgian Gothic design that was speckled, often liberally, with Gothic architecture's pointed and ogee arches, as can be seen in the work of Kent at Esher Place and by Thomas Farnolls Pritchard at Croft Castle, Herefordshire (1765), and in cabinet-makers' designs of the 1750s and 1760s such as those by Chippendale in his *Director* (1754, 1755, 1762) and Ince and Mayhew in their *The Universal System of Household Furniture* (1762).

Adam's use of the gable in this way (as distinct from the shallow-pitched Classical pediment) is, nevertheless, consistent with medieval architecture, where it was used for portals, for gable ends, on tomb canopies and on buttressing.[58] But the gable had not previously been incorporated widely and as a primary decorative form in fashionable Georgian Gothic design. Adam, consequently, was offering the 6th Earl of Coventry a newly developed, distinctive interior for Croome Church. Adam deploys this motif widely: as hood mouldings over the windows, on the inter-window buttress-like blind panelling and on the sub-window 'choir-stall'-like seating. Contradicting this overtly linear phrasing of the interior's elevation is a profusion of scroll-like crockets, finials and cusping added to the gables, details that tie them into the Gothic scheme. The end result is an attenuated and filament-like arrangement of motifs ranged around the gable – a Gothic equivalent to Adam's neoclassical wall designs, for example as in the Painted Breakfast Room in the Family Pavilion at Kedleston Hall (1760).[59]

Adam's reconfiguration of Gothic forms according to his motif-based design aesthetic had the by-product of giving the side-aisle seating the appearance of the coronation throne repeated along the wall. These aisle seats were never executed, but a throne matching their design was made for Croome Church, and the pattern was later reused by Adam at Alnwick Castle, and also for a parson's chair in the Gothic chapel built at Audley End House, Essex, by John Hobcroft (or Hobcraft) and James Essex in 1770–72. Hobcroft was the carpenter at Croome Church, responsible for realising Adam's designs for the Gothic interior, including the triple-decker pulpit (now sadly separated from its boxes), throne and pews, and his familiarity with Adam's work there explains the reappearance of the throne's design at Audley End.[60] The pulpit (Fig 7.9) is based

Fig 7.8 (below left) Robert Adam's design for 'Finishing one of the Side Walls of the Church at Croome for the Right Honble The Earl of Coventry', 1761.

Fig 7.9 (below) A recent photograph of Adam's pulpit in Croome church.

on that by Kent at York Minster (*see* Fig 7.7) but with the addition of cusped and crocketed gables like those found in the inter-window panels of Adam's 1761 side-aisle design (*see* Fig 7.8).[61] The overall effect is delicate, attenuated and, with a large number of crockets, best characterised as 'prickly'. This is significantly different from the overwhelmingly curvilinear Gothic preferred by cabinetmakers and other Gothic architects at the time, including those in the circle of Horace Walpole.

Equally idiosyncratic are Adam's two unexecuted designs for the church's vaulting. Both demonstrate an awareness of medieval vaulting patterns, especially the fan vault and attendant pendants, as well as the rose window. In his simpler proposal for the ceiling, Adam places a pendant at the centre of each bay, and creates the impression of a rose window where these bays intersect (Fig 7.10, top). A more complex second proposal arranges four pointed arches with cusped-panel borders around a rose window in each of the ceiling's central bays (Fig 7.10, bottom); the lateral arches also include an unprecedented use of cusped Y-tracery in imitation of the side windows.[62] With its different arch head-heights, and the misalignment between the centre and flanking bays' fan-vault patterns, Adam's proposal is a flat, imaginative and purely decorative rearrangement of medieval forms, without conveying any real sense of a coherent Gothic structure. In both designs, Adam's use of Gothic forms is akin to his treatment of antique Roman motifs in his neoclassical proposals.

Ultimately these innovative designs, like so many of Adam's proposals for Croome Church, went largely unrealised. The ceiling's decoration was pared back to a simple, easily moulded and reproduced Gothic frieze of intersecting and cusped Y-tracery and quatrefoils running around the bays' perimeters (seen in Fig 7.10), together with a modest ornamental pendant (similar to the types seen in Fig 7.10) in the middle of each of the nave's central bays, and an oval of tracery (not a blind rose window) in the sanctuary vault. Of his contributions to the interior, the pulpit and throne are the most ambitious stylistically. Adam's Gothic would evolve in later commissions at Strawberry Hill and Alnwick Castle, but the design-based reconfiguration of medieval forms evident in his Croome Church proposals would remain at its heart.

Working for the 'high priest' of Georgian Gothic: Adam at Strawberry Hill

Adam's next and notable foray into Gothic was for Horace Walpole at his villa, or 'little Gothic castle' – Strawberry Hill in Twickenham. This links Adam directly with one of the most important and well-documented structures of the 18th-century Gothic Revival.[63] It is noteworthy that Walpole should seek the advice and involvement of a professional architect-designer – and the most fashionable Classical architect of the day at that – as for the most part Strawberry Hill had, up until this point, been designed by Walpole himself and a group of male amateur architect friends known as the 'Strawberry Committee'.[64] Adam worked on the Round Drawing Room, located at the end of the gallery (or state apartment), on the first floor, part of the major additions made to the house by Walpole in 1759–71.[65] He was not, however, given free rein to refit the apartment as he liked but was guided closely by Walpole. In September 1766 Walpole provided him with the specific source material necessary to determine and complete the room's two most important decorative elements, the ceiling and chimney piece – William Dugdale's *The History of St. Paul's Cathedral in London* (1658) and John Dart's *Westmonasterium, or, The History and Antiquities of the Abbey Church of St. Peters Westminster* (1742):[66]

> Mr Walpole has sent Mr Adam the two books, and hopes at his leisure he will think of the ceiling and chimney-piece. The ceiling is to be taken from the plate 165 of St. Paul's, the circular window. The chimney from the shrine of Edward the Confessor, at Westminster. The diameter of the room is 22 feet. The enclosed little end is for the bed, which Mr Walpole begs to have drawn out too.[67]

Walpole turned to Adam after attempting to design the chimney piece himself in direct imitation of the shrine of St Edward the Confessor in Westminster Abbey. In September 1768 he explained in a letter to the diplomat and art collector Sir William Hamilton:

> For this year past I have been projecting a chimney in imitation of the tomb of Edward the Confessor, and had partly given it up, on finding how enormously expensive it would be. Mr Adam had drawn me a design a little in that style, prettier it is true, and at half the price. I had actually agreed to have it executed in scagliola, but have just heard that the man complained he could not perform his compact for the money settled.[68]

Clearly, Walpole wished for the chimney piece to be made in imitation of the shrine of St Edward the Confessor, and, significantly, he recognised that Adam's proposal was only 'a little in that style'. Indeed, in *The Description of the Villa of Mr. Horace Walpole* (1784), Walpole wrote that the chimney piece was 'taken from the tomb of Edward the Confessor, improved by Mr. Adam'.[69] Adam effectively transformed the 1268 tomb base by 'Peter of Rome' into a neoclassical form enriched with anthemion, guilloche, husks and affected Corinthian-like capitals, all largely typical of his contemporary neoclassical work.

A series of five drawings, dated to 1766–7, trace Adam's incremental 'improvements' to the Shrine (Fig 7.11).[70] The initial designs reproduce the structural polychromy of the shrine's Cosmati mosaics, including the alternation of wide and narrow diamonds, and the spiral-turned columns at the orthogonals (*see* Fig 7.11, upper pair).[71] It is clear that, even from the start, Adam based the chimney piece on one of the shrine's trefoil-headed prayer niches, stretched laterally to form the grate opening. This softened out the niche's pointed arch, and Adam retained only a subtle nod to the design's Gothic heritage by preserving the cusps. The tomb of John of Eltham (1316–36), 1st Earl of Cornwall, was similarly stretched and reinterpreted (without losing its Gothic identity) by Richard Bentley for the library's chimney piece at Strawberry Hill the decade before, so Adam's reconfiguration of Walpole's source material was by no means unprecedented.[72] The Round Room's chimney-piece ornament, unlike its form, gradually developed away from the Westminster model: two detailed proposals for the scagliola decoration demonstrate how the Cosmati patterns were progressively modified – losing their looped extensions, connecting the diamond pattern with anthemion, and inserting typically Adamesque ornament throughout the entablature and frieze (*see* Fig 7.11, lower pair).

Perhaps to compensate for this reinvention of the Cosmati patterns, Adam introduced other Gothic motifs into what had become an overtly neoclassical scheme. The barley-twist columns had a certain Gothic currency, for they are found on Walpole's favourite medieval turned ebony furniture – what he called 'the true black blood' –

Fig 7.10 (opposite) The two ceiling designs by Adam for the Church at Croome, 1761, show a simpler proposal for the ceiling (top) and also a more complex one.

that he collected and displayed at Strawberry Hill.[73] Above these columns Adam incorporated narrow cusped Gothic panel motifs to replace the Cosmatesque diamond pattern. Trefoil cusped at the head, and kicking up at the foot, these shapes are of a similar form to the panelling in Adam's provisional design for the side aisle at Croome Church (discussed on p 168), derived ultimately from William Kent. However, in the Strawberry Hill chimney-piece design the Gothic finial is replaced by a line of five neoclassical husks. This pattern, like the neoclassical-Gothic moulding below the chimney piece's lintel, is not derived from the shrine, which suggests that Adam was specifically introducing these elements to Gothicise what is to all intents a neoclassical design.

Adam's conscious use of additional Gothic motifs – effectively 'Gothing-up' his neoclassical reinterpretation of the medieval shrine – reduces the potential imbalance between the chimney piece and the room's explicitly Gothic ceiling, which, as Walpole instructed in 1766, is based upon the rose window of Old St Paul's Cathedral in London.[74] As Adam's design indicates, this ceiling, though convincingly following the St Paul's tracery pattern, was originally intended to have been elaborately coloured in the manner of his

Fig 7.11
Four designs from a series of five by Robert Adam for the Round Room chimney piece, Strawberry Hill, 1766–7.

neoclassical ceiling designs, with accents of blue, pink, green and purple to the mostly yellow, brown and white plasterwork tracery. But as recorded by John Carter in his 1788 survey of Strawberry Hill, the ceiling design was executed with a plain white field and gilt blind tracery (Fig 7.12).[75]

Adam's other notable, though unrealised, contribution to the Round Room at Strawberry Hill was a proposal for the bed, mentioned in Walpole's letter quoted on p 171.[76] This design, dated 1767, survives at the Lewis Walpole Library in Connecticut and is an idiosyncratic

Fig 7.12
John Carter, The Round Room Looking into the State Apartment, *1788.*

Fig 7.13
Robert Adam's design for the bed for the Round Room, Strawberry Hill, 1767.

concoction of neoclassical and Gothic motifs (Fig 7.13). Scholars have claimed variously that this design was for a settle and bed, and observed that it would never fit in the Round Room; however, its projected scale is approximately 1ft (0.3m) less than the room's height.[77] But unlike the Round Room's chimney piece, which is primarily neoclassical, the headboard design is undoubtedly Gothic, though peppered throughout with Classical motifs. The three fan-vault pendants are, for example, bracketed out on Classical acanthus-based corbels, with Roman-style finials that match the flaming Roman candelabra included by Wren on the interior of the Sheldonian Theatre, Oxford (1664–9), and by William Kent on his screen in Westminster Hall (1739). The rose-window panelling on the headboard is based upon the Old St Paul's model used in the Round Room ceiling, but Adam has reconfigured the tracery's structural pattern towards the circumference. This 1766 neoclassical-Gothic design clearly concentrates on surface pattern, and it is an important step in Adam's reconfiguration of the Gothic style. Although unexecuted, it is the template upon which his later remodelling of Alnwick Castle, discussed below, was based.

Apex of neoclassical-Gothic: Adam at Alnwick Castle

Robert Adam's largest and most significant Gothic work was undertaken from 1769 at Alnwick Castle, Northumberland, for Hugh Smithson Percy (d 1786) and his wife, Elizabeth Seymour Percy, 1st Duke and Duchess of Northumberland, for whom he had just completed the neoclassical state rooms at Syon House, Middlesex.[78] Though Alnwick was but one of the Percys' properties, it was the family's original seat. Following the death of Elizabeth's brother in 1744, and later her father, Algernon Seymour, 7th Duke of Somerset, in 1750, Elizabeth and, by extension, Hugh Percy, inherited the Earldom of Northumberland, and in 1766 George III elevated them to the dukedom, thereby re-establishing the Percys' otherwise long and distinguished place in the British peerage. Their exaltation was marked in architecture, of which the 18th-century refurbishment and Gothicisation of Alnwick was an important and conspicuous statement, as were the neoclassical modifications to Syon and Northumberland House in London. The choice of Gothic was not accidental: like Walpole, whose Gothic villa at Strawberry Hill was a conscious expression or recreation of the historic and chivalric heritage of his lineage, for the Northumberlands an overtly Gothic Alnwick seamlessly professed their ancient and noble past. For example, the display of heraldry within the castle increased dramatically after the Percys' elevation to the dukedom, and in 1769, 110 painted coats of arms were sent up to Alnwick to enrich the great staircase's cornice.[79]

Adam took over as architect at the castle in 1769, replacing James Paine (d 1789) as he had done at Kedleston, and was responsible thereafter for fitting up its Gothic interiors. In the spring of 1759 John Adam, Robert's eldest brother, spent the night at Alnwick on his way to

London, and his diary records Paine's Gothic work thus far:

> The dining room & drawing room ... are all extremely noble & elegant in the Gothick taste but the drawing room please me most, at least it struck me with that idea. The ornaments of both these rooms on the walls & ceilings are done in very good Gothic stile, of stucco. My Lady's Bed Chamber, Dressing Room &c. are very suitably finished. These all in the old building.[80]

Although John Adam's assessment of Alnwick is positive, later reflections upon Robert Adam's and James Paine's involvement at the castle are mixed. In an undated entry in his collection of descriptions of noblemen's residences, the agriculturalist Arthur Young described the castle as 'new built, but not yet finished. The apartments are fitted up in the gothic taste, and ornamented in a light & elegant stile.'[81] The connoisseur and patron George Bowles (d 1817), of Wanstead Grove, praised the interiors, including those by Adam, such as the library and chapel:

> The Staircase after mounting about 10 steps breaks off to the right & left & is very pretty, at top of the Stairs you enter the Saloon – a very elegant room filled up in the Gothick taste 2 towers mocking circular windows: the Drawing room very handsome Gobelin Chairs the frames only cost 25G apiece a large French glass £1000 seedy & the silvering damaged not pleasing & hung over the chimney very high: the next dining parlour a large good room & then a Breakfast room, the library very pretty & the Chappell at the end of it the roof after Kings Chappell at Cambridge Window painted Glass from York Cathedral & the whole variously coloured rather glaring; a state Bed in the middle of a room curtain tent fashion on other blue & white Satin furniture with 2 dressing rooms very elegant. All these are on one floor & are very curiously furnished in the Gothick style heavily ornamented.[82]

Bowles correctly identified Adam's source for the chapel's window tracery – the nave's west window, also known as the 'Heart of Yorkshire' window, at York Minster – and Adam's other ornamental Gothic additions are picked out.[83] In 1807 Thomas Smith of Bruce Castle, Tottenham, however, was thoroughly critical of this work, especially Adam's affected style of Gothic in the chapel:

> To Alnwick ... the castle presents itself very magnificently and covers a great space of ground ... I was disappointed in the number & style of the rooms most of them ornamented with plaster work in the gothic taste the drawing room is handsome, The Library comfortable – the dining room I thought shabby especially the chairs – the chapel I think is perfectly ridiculous covered with gilt, filigree & more like a music room – a monument to the late Duchess which did not please me.[84]

These vivid descriptions and assessments of Alnwick, and a view of the saloon from 1823 depicting the dense, grid-like (though not strictly Perpendicular) blind tracery on the walls, matching Adam's design from *c* 1769, record his contribution to the castle.[85] Though Adam was responsible for the decoration of several rooms at Alnwick, including the saloon, drawing room, chapel and library, today the only significant trace of his work on the estate is his Gothic eyecatcher and observation tower called Brizlee Tower, designed in late 1777 and begun in 1778. The rest of his work was swept away from 1854 by Anthony Salvin under the orders of Admiral Algernon Percy (1792–1865), 4th Duke of Northumberland, who felt that the 18th-century Gothic interventions had removed much of the seat's 'original beauty and variety of effect', and that Alnwick lacked 'the domestic comfort and modern conveniences requisite in the residence of a nobleman of his Grace's rank'. Salvin preferred to 'devise Medieval decorations to a plan consistent with modern requirements', though in this instance his work was to be in an Italian Renaissance style instead.[86]

Adam's involvement at Alnwick Castle has been investigated thoroughly elsewhere, as has the duchess's interest in Gothic forms.[87] This study therefore concentrates on and evaluates Adam's surviving designs for the house and Brizlee Tower within the broader context of his Gothic work, demonstrating its continued development. The most overtly Gothic of Adam's designs for Alnwick – though still idiosyncratic – is a preliminary proposal for the end walls of the library; it is dominated by overbearing ornamental patterns of cusped reticulations and pendants that suggest blind windows (Fig 7.14).[88] Over-scaled, and without mullions or tracery, these blind windows of sorts are separated by attenuated quasi-buttresses: Adam has effectively applied external architectural forms familiar from medieval tomb canopies and

Fig 7.14
Adam's design for the end wall of the library at Alnwick, c 1769, unexecuted.

chantry chapels to Alnwick's interiors. Such a stylised Gothic scheme creates an uneasy distinction between Gothic ornament and plain wall space; it is applied also to the chimney piece, which repeats the room's four-centred arch-heads and buttressing.

Fig 7.15
Adam's design for a chimney piece for the saloon at Alnwick.

Unlike this exclusively Gothic, though stylised, design for the library, Adam's 1769 proposal for the saloon's chimney piece offers a mix of both Gothic and Classical sources.[89] This chimney-piece proposal freely references medieval forms, including a frieze of 'ten elegant Gothic brackets' and two quatrefoil roundels 'richly adorned with the crest [crescent] encircled by the Garter'.[90] However, the frieze's brackets, inspired by fan vaults, are interspersed with anthemion, albeit much in the manner of medieval window reticulations (Fig 7.15). The chimney piece's Gothic columns also include Classical ornament in the borders of scrolls; though having the appearance of crockets, these are essentially Vitruvian in form. Adam's choice of Classical motifs is carefully calculated to harmonise effectively with medieval forms. Much like the reuse of the Croome Church throne at Alnwick, this chimney-piece design reappeared almost verbatim in an Adam office drawing for Sir William Pulteney.[91] A contemporary Adam design for the Large Room at Pulteney's Shrewsbury Castle is less balanced in its mix of the Classical and Gothic: there the chimney piece is entirely Classical, and the columns have more than a passing resemblance

Fig 7.16 (above)
Robert Adam, carpet design for the chapel at Alnwick Castle, 'in two different ways', 1780.

to those used by Horace Walpole for the Farnese tables in the gallery at Strawberry Hill, supplied by William Vile in 1765 and based on the Islip table-monument at Westminster Abbey.[92]

The chimney piece in the Alnwick saloon illustrates Adam's increasing use of Classical forms in his Gothic designs. This tendency is even more explicit in a detailed proposal for the castle's library chimney piece, where the only Gothic forms are the Tudor arch, quatrefoil and cusping: the remaining ornament is otherwise firmly Classical. Some of Adam's plans for the chapel and Circular Room at Alnwick developed this stylised mode of neoclassical-Gothic still further, incorporating anthemion, oval paterae and scrolled acanthus. In a carpet design of December 1780, the dagger-like Gothic decoration of Adam's rose window motif is gradually transformed so that the linear panels become curvilinear mouchettes modelled upon the anthemion's C-shape curve (Fig 7.16). Perhaps the most stylised of these Classical-Gothic motifs are the squashed, lozenge-like rose windows in the Circular Room's ceiling, which are connected by quatrefoils containing putti, and surrounded by anthemion and neoclassical filament ornament (Fig 7.17).[93]

Fig 7.17 (left)
A detail from the ceiling design for the Circular Room, 1770.

This series of Adam designs for Alnwick elaborates upon his earlier, neoclassical-Gothic bed design of 1767 for Horace Walpole (*see* Fig 7.13). But whereas Walpole's bed remained largely Gothic in style, though interspersed with neoclassical features such as corbels, Adam's most extreme proposals for Alnwick retain only a modest and at best loose reference to Gothic motifs, such as the quatrefoil, cinquefoil and cusped panelling. Instead, these traditional Gothic forms are used as a structural framework within which to place neoclassical motifs, thereby creating a hybrid style. Even though the chapel's window is overtly modelled upon the 'Heart of Yorkshire' window at York Minster, upon closer inspection its apparently Gothic design is, in fact, neoclassical.[94] Thomas Smith's criticism of the chapel is understandable and valid: it is 'perfectly ridiculous covered with gilt, filigree & more like a music room'. Compared to the work in this style by his contemporaries, such as Essex or Wyatt, or Sir Roger Newdigate, 5th baronet, at Arbury Hall, Warwickshire, Adam's brand of Gothic rewrites medieval design according to his own, 18th-century neoclassical principles.[95] Had they survived, however, Adam's interiors at Alnwick would have offered a remarkably sustained and coherent glimpse of one of the most original and idiosyncratic exponents of late Georgian Gothic architecture. Nowhere else, nor under any other architect-designer, was the Gothic so systematically enmeshed within fashionable neoclassical design principles.

Interestingly, this overt form of neoclassical-Gothic seems restricted to Alnwick's interiors; Adam's most notable and only surviving contribution to the estate, Brizlee Tower (Fig 7.18), returns to his angular, gable-style Gothic as seen

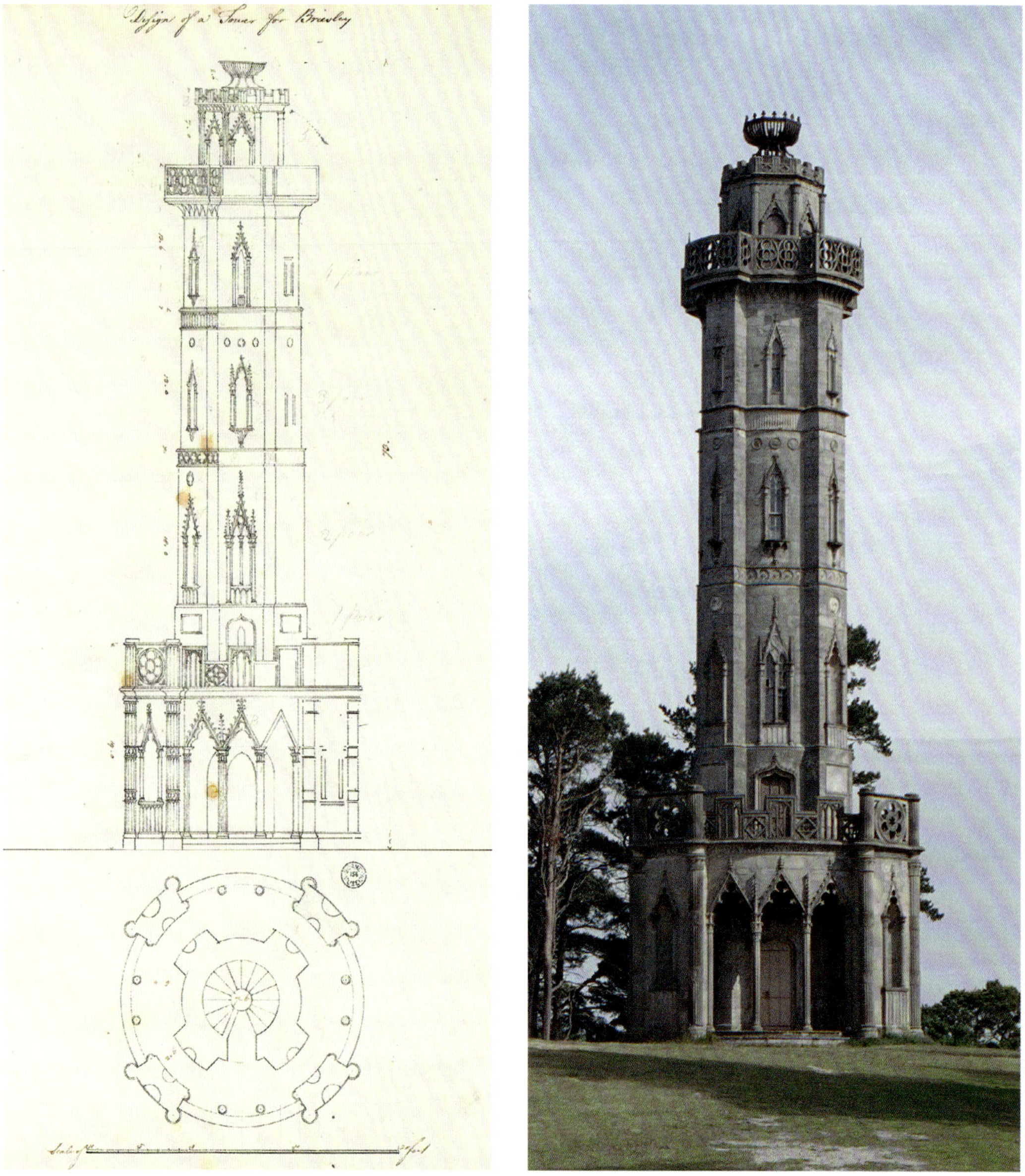

Fig 7.18
Adam's design for the tower at Brizlee, 1777 (right), and the tower today (far right).

at Croome and Strawberry Hill. Occasionally it incorporates Classical forms, such as the medallions and fluted engaged columns, but the fluting has angular termini (corresponding with Adam's Gothic panelling in Alnwick's saloon and a design for a ceiling in 'Huln Abbey'), and the overriding tenor of the ornament is Gothic.[96] Importantly, the openings on the tower's first floor – directly behind and above the crenulated entrance parapet – have ogee arch-heads and hence align much more closely with the mid-18th-century fashion for ogee-arch Gothic, as favoured by, among others, Batty Langley and William Kent.[97] The tower itself is a concoction of Gothic openings and castellations: its structure, however, makes no concessions to medieval forms and instead references Adam's early, pre-grand-tour Gothic architectural designs. It does, on the other hand, engage with his post-grand-tour Gothic architecture in the advancement and recession of planes (notably the four projecting buttress-like bays).

Coda

Adam's engagement with medieval architecture and its vocabulary of ornament extends beyond his surviving architectural works. Having sketched designs for imaginative Gothic structures in the years before his grand tour and his immersion in the glories of Rome, he was happy to return to the style occasionally once back from the Continent and established as a leading neoclassicist. His extant Gothic buildings and schemes of interior decoration are not numerous, but they represent an important, highly idiosyncratic and until now relatively unexplored interpretation and use of Gothic forms. Although not universally approved of in the Georgian and Victorian periods, Adam's Gothic designs reveal his overriding ability to collect and reuse such motifs in explicitly decorative schemes, in a manner akin to his neoclassical works.

Though the early Gothic designs of his youth appear tentative, Adam's application of the style and its motifs in later country-house commissions across an 18-year period suggest a highly inventive attitude towards the Gothic aesthetic. He did not follow contemporary architects like James Essex, or even James Wyatt, who increasingly applied antiquarian explorations of medieval architecture to the work of their practices. Instead, Adam was more interested in developing a new form of Gothic, one that was based essentially on his contemporary and highly successful style of neoclassical architecture and interior decoration, but 'Gothed-up'. Adam, consequently, although a minor exponent of Gothic design in 18th-century Britain, was an important and imaginative one, whose work reveals the potential the style offered to Georgian architects. The works and drawings considered here could never be framed as Adam's most significant contributions to 18th-century design: nevertheless they reveal a remarkably inventive evolution, a subtle and aesthetic reinvention of Gothic forms harnessing the work of others and his own architectural genius. Adam may be on the periphery of Georgian Gothic, especially when compared with James Wyatt, but he emerges without doubt as one of its most imaginative exponents.

Notes

1 Harris 2001, 4–5, 18–39; Harris 1987, 6–60.

2 For Osterley, *see* Harris 2001, 156–79. For Witham, *see* SM, Adam vol 54/7/219–20; vol 43/25–32; Wilson-North and Porter 1997; and Collinson, J 1791 *The History and Antiquities of the County of Somerset, Collected from Authentick Records, and an Actual Survey Made by the Late Mr. Edmund Rack*, vol II, 235.

3 Sir John Soane's Museum has Adam designs for in excess of 200 country houses: the figure of 89 known country-estate commissions is derived from Colvin 2008, 50–4. For the Hulne Priory debate, *see* Fleming 1958a and Rowan 1998. The Alnwick drawing-room chimney-piece design is in SM, Adam vol 22/53.

4 For Yester Chapel, *see* Cameron 1993 and Dunbar 1972. For Shrewsbury Castle, *see* SM, Adam vol 50/83–4, and the related designs for Alnwick (SM, Adam vol 27/37; vol 39/14, 16). For interpretations of the Shrewsbury Castle designs, *see* Williams 2016.

5 Harris 2001, 319–33. For Adam's castle architecture, *see* Fleming 1968a and 1968b; Tait 1981; Rowan 1985, 14–19, 94–151, and 1974, 679–94; Astley 2000.

6 Rowan 1985, 17.

7 Lewis 1937–83, vol IX (1941), 149.

8 Rowan 1985, 18; Fleming 1968a.

9 Aspects of Adam's relationship with the broader Gothic Revival are addressed briefly in Lindfield 2016, 154–6, 174–7.

10 For Esher Place, *see* White, R 2013 'William Kent and the Gothic Revival', in Weber, S (ed) *William Kent: Designing Georgian Britain*. New Haven and London: Yale University Press, 253–7. For York Minster, *see* Friedman, T 1995 'The transformation of York Minster, 1726–42'. *Architectural History* **38**,

81–3. Kent's Gothic designs for a chimney piece at Esher, and his work at Westminster Hall, York Minster and Gloucester Cathedral were published in *Some Designs of Mr. Inigo Jones and Mr. Wm. Kent.* London: Vardy, 1744, plates 36, 48–9 and 51. For a full analysis of this early phase of the Gothic Revival, see Lindfield 2016, 7–80. For the Halfpennys and Chippendale, *see* Halfpenny, W and Halfpenny, J 1752 *Rural Architecture in the Gothick Taste Being Twenty New Designs for Temples, Garden-Seats, Summer-Houses, Lodges, Terminies, Piers, &c.* London; Halfpenny, W and Halfpenny, J 1752 *Chinese and Gothic Architecture Properly Ornamented.* London; and Chippendale, T 1754, 1755 and 1762 *The Gentleman and Cabinet-Maker's Director.* London. *See also* Lindfield 2016, 81–130, and Lindfield, P N 2015 'New light on Chippendale at Hestercombe House'. *Burlington Magazine* **157**, 452–6.

11 Cocke, T H 1975 'James Essex, cathedral restorer'. *Architectural History* **18**, 12–22, 97–102; Cocke, T H 1984 *The Ingenious Mr Essex, Architect, 1722–1784: An Exhibition to Mark the Bicentenary of His Death.* Cambridge: Fitzwilliam Museum; Stewart, D R 1950 'James Essex'. *Architectural Review* **108**, 317–21.

12 For example, *see* SM Adam vol 12/109.

13 *See* Harris 2001; Harris 1987, 50–1, 59–61. For the fullest recent assessment of neoclassical-Gothic, *see* Lindfield 2016, 171–9.

14 Wren, C 1750 *Parentalia, or, Memoirs of the family of the Wrens: viz., of Matthew, Bishop of Ely, Christopher, Dean of Windsor, etc., but chiefly of Sir Christopher Wren in which is contained, besides his works, a great number of original papers and records.* London: Stephen Wren/Society of Antiquaries, 297. For the later view of Gothic, see also Cocke, T H 1987 'The wheel of fortune: The appreciation of Gothic since the Middle Ages', in Alexander, J and Binski, P (eds) *Age of Chivalry: Art in Plantagenet England, 1200–1400.* London: Royal Academy of Arts in association with Weidenfeld & Nicolson, 183–91.

15 Adam and Adam 1773–1822, vol I, no 1, preface.

16 Adam in *The Works* ridiculed formulaic design, arguing that 'among architects destitute of genius and incapable of venturing into the great line of their art, the attention paid to those rules and proportions is frequently minute and frivolous' (ibid); *see also* Robinson 2012, 37–41.

17 Adam and Adam 1773–1822, vol I, no 1, preface.

18 Robinson 2012, 14–28; Lewis 1937–83, vol XXIII (1967), 298.

19 *See* Robinson 2012, 48–50, 74, 219–20, 324, 345.

20 Frew, J 1979 'Richard Gough, James Wyatt, and late 18th-century preservation'. *Journal of the Society of Architectural Historians* **38**, 366–74; Robinson 2012, 223–31.

21 *See* Robinson 2012, 224, 227; Crook, J M 1995 *John Carter and the Mind of the Gothic Revival.* London: W S Maney with the Society of Antiquaries of London, 27.

22 Indeed, as early as 1785, when he had completed the external elevations of Lee Priory, Kent, a house that Walpole styled as 'a child of Strawberry prettier than the parent', Wyatt's significant knowledge of medieval forms was manifest in his architectural practice. Lewis 1937–83, vol XII (1944), 111. *See* Reeve and Lindfield 2015; also Lindfield, P N 2017 'Rediscovering Lee Priory's Lost Library Ante-Chamber'. *Georgian Group Journal* **25**, 207–12

23 Rowan 1998, 269.

24 *See* Lindfield 2016, 131–79.

25 Christie, J 1818, *A Catalogue of a Valuable Library ... the Property of the Late Robert Adam, Esq. F.R.S & F.A.S.* London, 141.

26 Sanderson 1992, 36; NRS, GD18/4764, GD18/4981/1.

27 RIBA Drawings Collection, SE16/2, fols 7, 8, 13, 15, 17, 19, 20, 22.

28 Rowan 1990, 12.

29 The Gothic folly drawing of 1749 in the collection at Blair Adam is illustrated in Fleming 1962 (plate 24) and Beard 1978 (plate 13).

30 SM, Adam vol 56/36. For the relationship between Adam's drawing and the Vertue engraving, *see* http://collections.soane.org/OBJECT705 [accessed 3 April 2018]. For Vertue's drawing, *see* Society of Antiquaries of London 1747 *Vetusta Monumenta, quæ, ad Rerum Britannicarum Memoriam Conservandam* **1**. London, plate LXI.

31 Despite Adam's overt concentration upon ornament, the gables and flying buttresses of the cross are stylised, indicating at least a partially imaginative, or inexperienced, assessment of its forms.

32 *See* Fleming 1962, 85.

33 The tower and church drawings of 1753, from the collection at Blair Adam, are reproduced in Fleming 1962 (plates 24–5), Stillman 1966 (plates 166–7) and Beard 1978 (plates 11–12). Other Gothic sketches of the period are reproduced in Fleming 1962 (fig 5, a Gothic ruin, *c* 1748; plate 28, a Gothic bridge, *c* 1750) and Tait 1971 (fig 4, a Gothic church, 1753; fig 5, a Gothic fortification, 1750s).

34 'His earliest architectural drawings', John Fleming writes, 'date from a few years later [after 1746] but many of them are in the same vein, being fantasies in an extravagant and whimsical Gothic *à la* Batty Langley, worked out with some sensitivity of line and precocious appreciation for the sophisticated charms of the Rococo'. *See* Fleming 1962, 81–2. For an account of Langley's plates, *see* Rowan 1975; Lindfield 2014; also Mowl, T and Earnshaw, B 1999 *An Insular Rococo: Architecture, Politics and Society in Ireland and England, 1710–1770.* London: Reaktion.

35 Halfpenny and Halfpenny, *Rural Architecture*, plate 11.

36 SM, Adam vol 56/36.

37 *See* Rowan 1975; Harris 1977; Lindfield 2014.

38 National Library of Scotland, Yester papers, 4862, 98/2.

39 RIBA Drawings Collection, SE16/2, fol 15.

40 National Galleries Scotland, D 448, https://www.nationalgalleries.org/art-and-artists/8234/rocky-landscape-castle-waterfall-and-bridge-riders-and-wagon [accessed 30 April 2018]. *See* Tait 1971, 164–8. These capriccios resonate with Culzean Castle's position perched on the cliffs above the Firth of Clyde. *See also* Harris 2001, 317–33.

41 NRS, GD18/4907.

42 SM, Adam vol 54/4/1, 54/4/15; compare with RIBA Drawings Collection, SE16/2, fol 19.

43 For examples of these grand tour capriccios, *see* SM, Adam vol 54/4/2–11; vol 55/16–24.
44 Colvin 1998, 28–32; Harris 2001, 41–53.
45 V&A Museum, W.76:1, 2, A/1 to 10, B-1975; SM, Adam vol 50/10; *see also* Beard 1953.
46 *See* Colvin 1998.
47 For Gothic follies, *see* for example Stewart, D 1996 'Political ruins: Gothic sham and the '45'. *Journal of the Society of Architectural Historians* **55**, 400–11; Adshead, D 1998 'The design and building of the Gothic folly at Wimpole, Cambridgeshire'. *Burlington Magazine* **140**, 76–84.
48 SM, Adam vol 25/209 (pulpit); vol 50/15–21 (gable-Gothic side aisle, Kent-Gothic side aisle, entrance door, gable-Gothic side aisle and vault, vaulting, plan and office copy after Adam's design for the priest's chair); vol 52/167–8 (Gothic railings for church). There are also designs by Adam of *c* 1760 for Gothic tracery stained-glass windows, now in the V&A Museum, nos 3436.4–5; *see also* Rowan 1988, 19, 50–1.
49 SM, Adam vol 50/16 (the Kent-Gothic side aisle) is based on William Kent's Gothic work and almost certainly predates the gable-Gothic side aisle and vault drawings in SM, Adam vol 50/15, 18.
50 SM, Adam vol 50/16 (Kent-Gothic side aisle).
51 *Some Designs of Mr. Inigo Jones and Mr. Wm. Kent,* plates 36, 48–9, 51.
52 *See* Colvin, H 2009 'Henry Flitcroft, William Kent and Shobdon Church, Herefordshire', *in* Jones, D and McKinstry, S (eds) *Essays in Scots and English Architectural History: A Festschrift in Honour of John Frew.* Donington, Lincolnshire: Shaun Tyas, 1–8; Lindfield 2016, 71–8.
53 *See* Lindfield, P N 2014 'The Countess of Pomfret's Gothic Revival furniture'. *Georgian Group Journal* **22**, 77–87.
54 *See* ibid, 77–94; Lindfield 2016, 154–5.
55 His arrangement of Gothic panelling quotes, though perhaps unintentionally, medieval sub-window blind arcading seen, for example, in chapter houses.
56 SM, Adam vol 50/15.
57 SM, Adam vol 2/7; vol 19/62–4.
58 The Gothic gable, for example, was not incorporated into any of Langley's designs in *Ancient Architecture* (1741–2), or by Chippendale in his *Director* (1754, 1755, 1762 edns). For the gable's role in English Decorated Gothic architecture, *see* Binski, P 2014 *Gothic Wonder: Art, Artifice and the Decorated Style, 1290–1350.* New Haven: Yale University Press, 81–279.
59 For the breakfast room at Kedleston, *see* Harris 1987, 52–4.
60 Colvin 1998, 31; Sutherill, M 1999 'John Hobcroft and James Essex at Audley End House'. *Georgian Group Journal* **9**, 17–25; Saffron Walden, Audley End House, Picture Store Scrapbook, fol 72.
61 SM, Adam vol 25/209 (Gothic pulpit), vol 50/15 (gable-Gothic side aisle) and vol 50/18 (gable-Gothic side aisle and vault).
62 Similar decorative flourishes were proposed for the family pew ceiling at the chapel at Audley End, Essex, designed by John Hobcroft (or Hobcraft). *See* Sutherill, 'John Hobcroft and James Essex', 19–23.
63 *See* McCarthy 1987, 63–91; Guillery, P and Snodin, M 1995 'Strawberry Hill: Building and site'. *Architectural History* **38**, 102–28; Snodin, M 2009 *Horace Walpole's Strawberry Hill.* New Haven and London: Yale University Press, 15–73.
64 Following Adam's involvement, Walpole used other professional architects – James Essex and James Wyatt – to work on Strawberry Hill's Gothic gates and offices. *See* Snodin, *Horace Walpole's Strawberry Hill,* 80–1, 349.
65 *See* Wilton-Ely 2011 for a brief assessment.
66 It has also been suggested that the first volume of the Society of Antiquaries of London's *Vetusta Monumenta* (1747) could have been the source for the Westminster Abbey ornament, as it contains a detailed plate by George Vertue of Edward the Confessor's shrine. *See* Wilton-Ely 2001, 149–51.
67 Lewis 1937–83, vol XLI (1980), 39.
68 Ibid, vol XXXV (1973), 406–7.
69 Walpole, H 1784 *A Description of the Villa of Mr. Horace Walpole ... With an Inventory of the Furniture, Pictures, Curiosities, &c. Strawberry Hill,* 53.
70 SM, Adam vol 18/59, 61; vol 22/227–9.
71 SM, Adam vol 59/227–8. Payne, M and Rodwell, W 2017 'Edward the Confessor's shrine in Westminster Abbey: Its date of construction reconsidered'. *Antiquaries Journal* **97**, 187–204.
72 Walpole, *A Description of the Villa,* 33.
73 Wainwright, C 1985 'Only the true black blood'. *Furniture History* **21**, 250–7.
74 Ibid; Lewis 1937–83, vol XLI (1980), 39.
75 SM, Adam vol 11/234; Lewis Walpole Library, Yale University, Farmington, CT, 33 30 Copy 11 Folio, fol 153.
76 Lewis 1937–83, vol XLI (1980), 39. The Round Room was originally designed to be a state bedroom, and a design survives by John Chute for a low, couch-type bed under a canopy based upon that of Sir Simon Burleigh's tomb at Old St Paul's, London (Lewis Walpole Library, 49 2523, fol 121.8).
77 Lewis Walpole Library, 49.3678. The design is inscribed with 'H Walpole Esq' on the verso. There is also a previously unknown preparatory design for the bed: SM, Adam vol 27/41. *See also* Wilton-Ely 2011, 7, and 2001, 152–3. The Round Room would accommodate the bed's height of 9ft 6in (2.9m) as specified in Adam's design. *See* Snodin, *Horace Walpole's Strawberry Hill,* 48, 57.
78 *See* Harris 2001, 65–83.
79 For Walpole, *see* Lindfield, P N 2016 '"Hung round with the helmets, breast-plates, and swords of our ancestors": Allusions to chivalry in eighteenth-century Gothicism?', *in* Stevenson, K and Gribling, B (eds) *Chivalry and the Vision of the Medieval Past.* Woodbridge: Boydell, 77–82. For Alnwick, *see* Alnwick Castle, MS U/I/44; *also* Harris 2001, 87.
80 Quoted in Harris 2001, 85.
81 Lewis Walpole Library, LWL MS vol 52, fol 197.
82 Ibid, vol 47.
83 For the west nave window, finished 1338–9, *see* Binski, *Gothic Wonder,* 172.
84 Lewis Walpole Library, LWL MS vol 189. For the duchess's monumental legacy, *see* Aymonino 2010a.

85 SM, Adam vol 39/16. The saloon was also described as 'the gayest and most elegant style of Gothic architecture' (quoted in Harris 2001, 87).
86 Donaldson, T L 1856 'Some description of Alnwick Castle, Northumberland, the seat of His Grace the Duke of Northumberland, K.G.', *Papers Read at the Royal Institute of British Architects* (1856–67), 14–16.
87 Harris 2001, 84–93; Mayer 2011, 34–9.
88 SM, Adam vol 24/214.
89 SM, Adam vol 22/50.
90 Alnwick, Alnwick Castle, Alnwick MS (P Wadell). *See also* Harris 2001, 87.
91 SM, Adam vol 50/85.
92 Pulteney: SM, Adam vol 50/84; Alnwick: SM, Adam vol 39/16. For the tables at Strawberry Hill, *see* Toynbee, P 1927 *Strawberry Hill Accounts*. Oxford: Clarendon, 10, 124. For Shrewsbury Castle, *see* Williams 2016, 65–80.
93 SM, Adam vol 22/54; vol 11/30.
94 SM, Adam vol 39/18.
95 For Arbury Hall, *see* Tyack, G 1994 *Warwickshire Country Houses*. Chichester: Phillimore, 9–15.
96 SM, Adam vol 19/156–9; vol 21/29; vol 5/74.
97 *See* Lindfield 2016, 16–19, 60–67, and White, 'William Kent', 247–69.

8

The ingenious Mr Adam

David King

Robert and James Adam made it clear in their preface to the first volume of *The Works in Architecture* that they regarded architecture as an art. For example, they noted that they had 'met ... with the imitation of other artists', and wrote that 'in this work, we aim not only at affording entertainment to the connoisseur, but wish also to convey some instruction to the artist'.[1] Their focus on the artistry of their profession reflected Robert's enthusiasm for landscapes, and it is undoubtedly the artistic appeal of their work that has inspired so many people to study it and write about its numerous strands.

However, great architects cannot be solely artists. Another important quality is the ability to design buildings that meet practical requirements, and this often requires ingenuity in devising plans. *The Works* says little about this aspect of their architecture, except for noting in volume I that the plans for altering Syon House in Isleworth, Middlesex, had to overcome 'some inequality in the levels of the old floors' and 'some limitations from the situation of the old walls', and explaining in volume II that the plans for 20 St James's Square and Derby House were constrained by 'the smallness of the scites [*sic*] upon which most houses in London are built'.[2]

These examples of the Adams' ingenuity in planning are well known, doubtless because the brothers themselves drew attention to them. Otherwise, their planning ingenuity seems to be largely overshadowed in the literature by discussions of their artistry. A major reason for this may be that anyone studying an Adam façade or interior has this artistry in full view. In contrast, their skilful planning is rarely immediately apparent from their drawings, and can often be appreciated only when one also has some knowledge of their clients' requirements and of any particular difficulties which those requirements created. Yet clients' specifications rarely survive in written form, and the difficulties in meeting them are seldom noted on the plans, so usually some effort and imagination are needed to determine when issues arose that required ingenious solutions.

I was fortunate that in writing both *The Complete Works of Robert and James Adam* and *Unbuilt Adam,* I frequently looked (on microfilm) through all the brothers' drawings held at Sir John Soane's Museum. In so doing, a number of designs gradually struck me as highly ingenious, and this essay looks closely at five of these. First, though, some flavour of the disparate issues which the brothers' partnership faced is given by looking briefly at six other ingenious designs, beginning with three that were executed, followed by three that were not. Throughout, the ingenious Mr Adam concerned can be taken to be the principal partner, Robert.

Three ingenious executed designs

The first of these to be considered is Robert Adam's 1768 design for the Adelphi housing scheme in London, much of which has been demolished.[3] The smartest houses there were built in a row, with their front windows facing the River Thames across a new street that is now called Adelphi Terrace (*see* Fig 0.7). Behind these stood two more rows of houses, lining a new parallel street called John Adam Street. One aim of the Adams' pioneering scheme was to provide the houses of Adelphi Terrace with elevated views, and this was achieved by building the entire development over a network of vaulted warehouses, with the basements of the houses set immediately above the vaults. However, this device created a potential problem, because if the street levels of John Adam Street and Adelphi Terrace were also placed immediately above the vaults, then the occupants of the houses there would have to climb many steps to reach their front doors above the basements. To avoid this, each street was raised to the level of the front doors. John Adam Street was raised chiefly by making the vaults below it exceptionally tall, but Adelphi Terrace was raised more ingeniously by placing on top of the riverside vaults a row of river-facing 'cottages', lit by Diocletian windows over the warehouse entrances, and then putting the street above the cottages. Raising the level of the two streets in these ways had the further advantage of bringing them into line with the busy and important thoroughfare of the Strand, which bounded the Adam development to its north and provided the main access to it, so allowing residents to reach their homes without the inconvenience of steep slopes. The Adelphi Terrace occupants benefited from one further feature, which was that the two rows of houses lining John Adam Street behind them deadened the sounds from the Strand, to create a tranquillity that is very rare in central London. In *Little Dorrit*, as Arthur Clennam follows Tattycoram and Blandois from the Strand to Adelphi Terrace, Charles Dickens describes the 'sudden pause in that place to the roar of the great thoroughfare', by comparing the effect to 'putting cotton in the ears, or having the head thickly muffled'.[4]

Another scheme to consider is the plan of around 1770 for Mellerstain in Berwickshire (*see* Fig 0.21).[5] It seems that the client here wanted the house to appear much bigger than it is. Adam achieved this by placing all the main rooms on the principal floor in a single row along the south side of a corridor; this corridor has one window at each end and more along its north side. The result is a building whose south front has an impressive 17 bays. However, there was clearly a risk that the house would look exceedingly thin at each end, and Adam ingeniously masked this thinness by adding a small service room and a back staircase at each end of the house, on the north side of the corridor, thereby creating respectably long five-bay end façades.

The third example is the final series of plans that Robert Adam devised for Culzean Castle, Ayrshire, in 1787.[6] Adam's first-phase designs, prepared in 1777, concerned remodelling and extending an existing tower house to create the southern portion of the present castle. To the north of this there stood assorted earlier buildings, and Adam's second-phase designs, prepared in 1785, concerned their replacement. The main feature of this phase was the northern part of the castle itself; this is the same length as the southern part and lies due north of it. However, the plans show these two building phases as mostly separated by a courtyard, with only a modest link at their western ends. Adam's third-phase designs, of 1787, concern filling in this courtyard, chiefly with a spectacular double staircase in an elliptical well, which is surrounded by colonnades on both the first and second floors.[7] This elaborate staircase is fitted in between the north and south parts of the castle so ingeniously that a visitor might easily assume that it was designed first, with the rest of the castle placed carefully around it; but in fact almost everything aside from this staircase was already built or planned when Adam devised it.

Three ingenious unbuilt designs

I now turn to three unexecuted designs. The first is Adam's astonishing scheme of 1774 for enlarging Barnbougle Castle at Dalmeny, near Edinburgh (*see* Fig 10.5).[8] The existing castle was roughly L-shaped with four main storeys, although a taller square turret projected upwards at its apex. Seemingly the client did not want to have any new work more than two storeys high, so for the new scheme to secure symmetry, the existing work had to be placed centrally. Adam ingeniously achieved this by designing the extended castle in the form of a right-angled isosceles triangle, with the old castle – or, more precisely, its tall corner turret – at the right-angled vertex of this triangle. Anyone who viewed the hypotenuse front of the Adam castle would see the old part and its tall corner looming up behind, to form a somewhat distant climax.

The second is his 1777 design for Bewley Castle at Kiltarlity in Inverness-shire (Fig 8.1).[9] This scheme envisaged a V-shaped castle, with the hall set in the angle between the arms, an idea which Adam was to use again in unbuilt designs for a house in Harwich, Essex, and for Barnton Castle near Edinburgh.[10] At Harwich and Barnton, Adam set the two arms at right angles, or 90 degrees, but at Bewley he wanted a wider angle, perhaps to make the arms seem more welcoming for visitors arriving at the front door, and he proposed an angle of 108 degrees. Perhaps he suggested 108 degrees because it is the angle between the arms of a regular pentagon, or because it can be constructed using only a compass and straight edge, as demonstrated by Euclid,[11] and was therefore an angle that Classical architects could have used. It was certainly used in the Renaissance, most notably at the pentagonal Villa Farnese in Caprarola, which Adam knew from his grand tour.[12]

One planning issue in all Adam's V-shaped castles was how best to link the rear room at the apex of the V to its neighbours. In each case these were the three principal rooms – namely a saloon flanked by a drawing room and an eating room. The neatest solution when the arms are at right angles is to make the saloon a regular octagon, as proposed for Barnton, so that two of its sides align perfectly with the flanking rooms and there can be a door directly from one to the other. However, Bewley's arms were not at right angles, so instead Adam ingeniously made the saloon circular and placed a shallow apse in the adjacent end wall of each flanking room, as shown in Fig 8.1. This means that the route from the saloon to each neighbouring room must be by a short passage with a slight bend, but this bend would doubtless appear to anyone in the passage to be necessary solely because of the curved room walls, and not because of the

Fig 8.1
Plan of the principal floor of Bewley Castle, Kiltarlity, Inverness-shire, dated 1777.

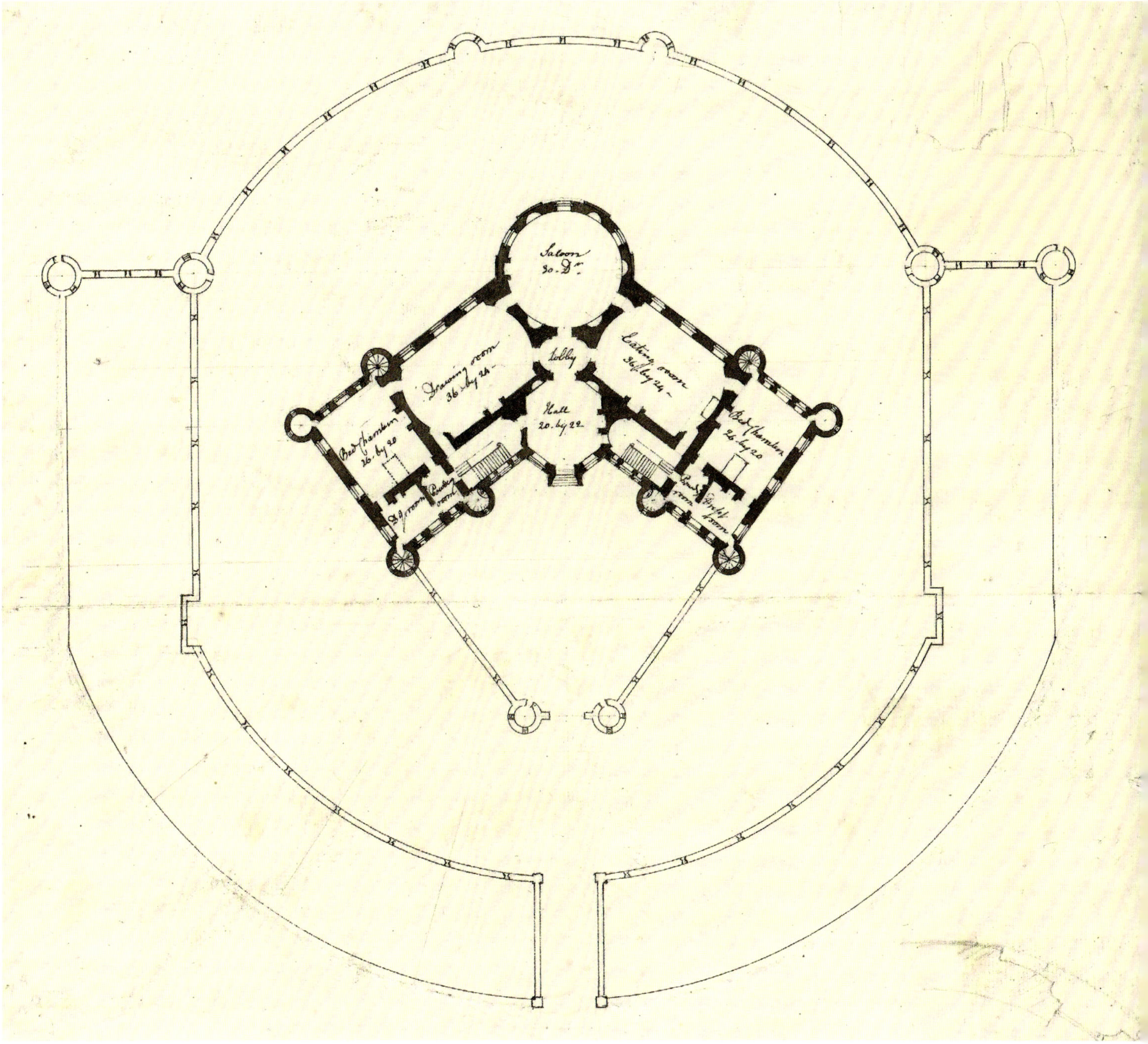

unusual angle between the arms. Another intriguing idea was to compose the central hall at the front of the house, in the re-entrant angle of the V, as an irregular octagon. Its shape was derived by making two of its walls align perfectly with the adjacent walls of the rooms flanking the saloon, while the door from the hall into the saloon directly faced the front door; the result was four angles of 126 degrees and four of 144 degrees.

The third scheme considered here is the 1791 design for Lasswade church, Midlothian.[13] This recalls many Byzantine churches by having the shape of a Greek cross and a central dome. Any Byzantine architect who looked at the outline of Adam's floor plan would envisage a square central space, with its sides 26ft (7.9m) long, and would expect each side to have a wide arch leading into an arm that was 22ft (6.7m) deep. The area of the square central space would be 676 sq ft (62.8 sq m), and the architect would envisage vaulting this space with a dome whose diameter was at most 26ft (7.9m). Now the diagonal of this central space is a little over 36ft (11m), and Adam's design actually incorporates a dome with a diameter of 36ft. He achieves this by having the dome spill over some distance into each arm. On the plan, the arch from the central space into each arm is seen to be curved, to follow the curved outline of the base of the dome above. The result is that Adam ingeniously contrived a central space with an area of 1,018 sq ft (94.6 sq m), about 50 per cent larger than the 676 sq ft (62.8 sq m) which a Byzantine architect would have created. Unlike many Byzantine churches, though, Adam's central space is not lit by windows in a drum below the dome but by large windows at the ends of three of the arms.

Case studies

Woodburn House

I now look more closely at five further projects. The first is an unbuilt design for Woodburn House at Dalkeith, Midlothian, where the client was James Ker (or Kerr) of Morriston and Kersfield, Berwickshire.[14] The drawings for this are dated 18 February 1792, only a fortnight before Robert Adam's death on 3 March, although most of them also bear the date 22 March 1792, by which time James Adam had taken over the practice. The earlier date suggests that the design was Robert's.[15]

The design brief was seemingly for a modest three-bay house whose principal floor was to have three main rooms: a large drawing room and a medium-sized eating room at the rear, and a small study at the front, along with the hall and staircase. The key design problem was how to accommodate three main rooms of unequal size behind symmetrical front and rear façades. Adam had already faced similar problems at Sunnyside House, near Edinburgh, and also in his designs for a villa for a Mr Wilson. At Sunnyside he initially proposed a rectangular house, but the executed design of 1786 was T-shaped, while the unbuilt design for Mr Wilson was quasi-triangular; perhaps these unusual shapes helped Adam resolve the undoubted planning difficulties.[16]

The most obvious solution at Woodburn might have been to allocate two rear windows to the drawing room and one to the eating room, as suggested in Fig 8.2a. These are the two public rooms, used for entertaining, so it would be nice to place their rear windows symmetrically when seen from inside, as shown here, but this means that the central rear window would be well off the central axis of the house, so that the rear façade would then be asymmetrical. Adam had the better idea of giving the drawing room and eating room equal widths but different lengths, as indicated in Fig 8.2b. Here, the rear windows inside each room are again placed symmetrically, while the rear façade is also symmetrical; admittedly this rear requires only two bays, but as Adam preferred odd numbers, I have added to my version a third bay with a blind central window.

Although the plan in Fig 8.2b meets the brief, the rear façade would not be very attractive because its three windows would be relatively close to each other and relatively far from the ends of the house. Adam ingeniously fixed this problem by slightly advancing the central part of this façade, as shown in Fig 8.2c, so that the eye relates the rear windows only to this advanced part, thus disguising the distance between the end windows and the ends of the house. But what now happens inside? One possibility would be to set the rear window of each rear room in a recess, and to have a thick wall behind the blind window, as suggested in Fig 8.2d.

However, Adam devised a far more ingenious plan (Fig 8.3). He gave both the drawing room and the eating room apsidal ends, and placed a small circular anteroom between them. This allowed a door to the garden to be located where a blind window was suggested in Fig 8.2d. The drawing room gains a second apse at the opposite end, and there is further interest with an apse in the hall facing the front door. It is doubtful if

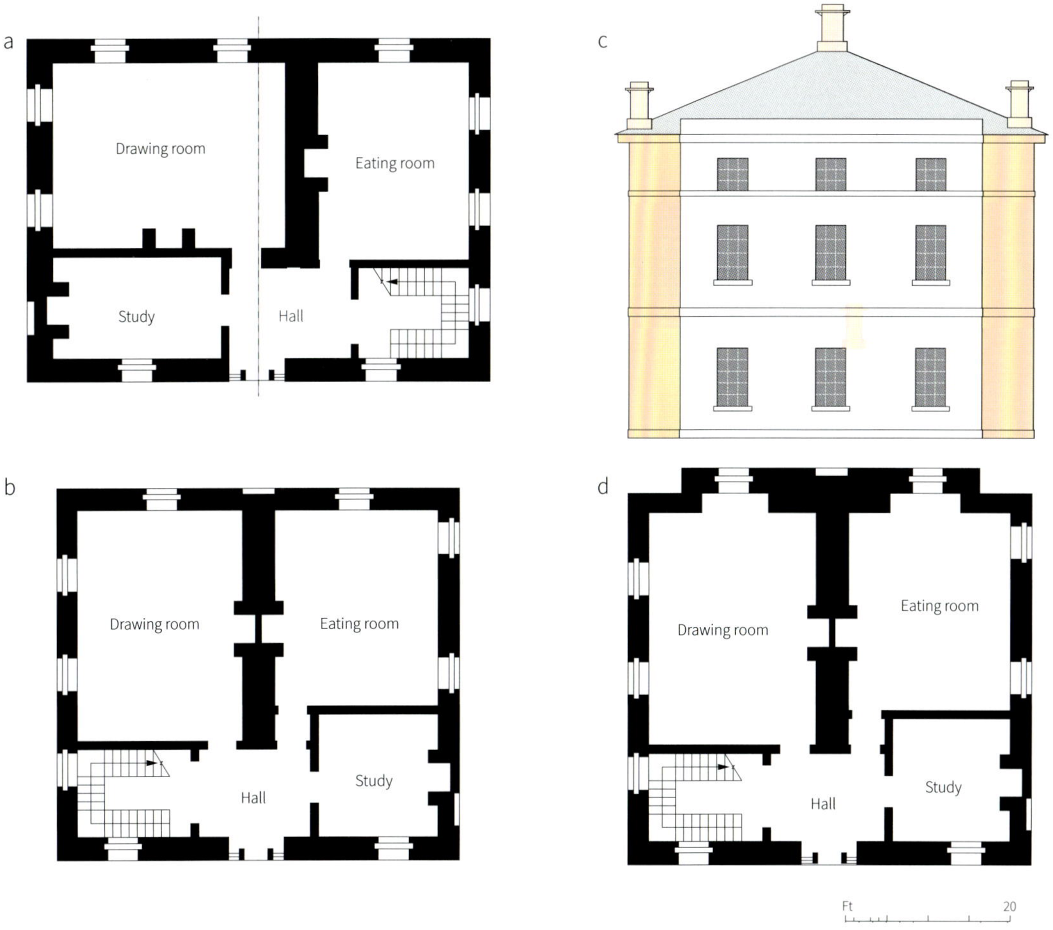

Fig 8.2
Woodburn House, Midlothian: the author's attempts to solve the problems faced by Robert Adam's design brief.

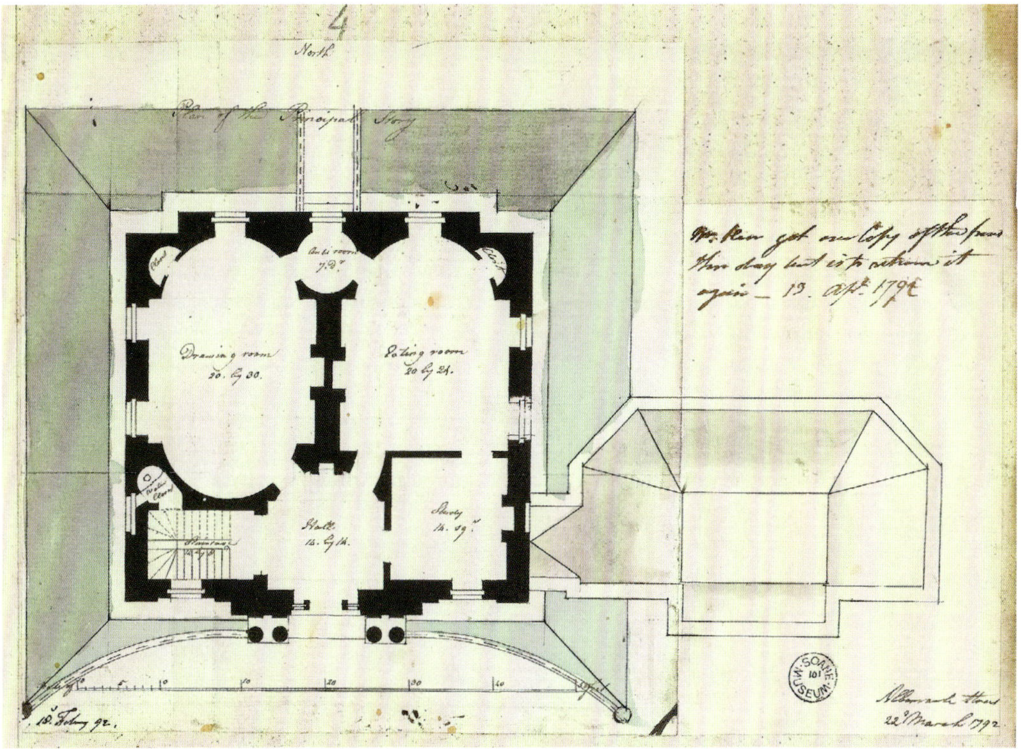

Fig 8.3
Woodburn House: Robert Adam's plan for the principal floor, February 1792.

anyone looking at the perfectly proportioned but plain façade in Fig 8.2c could conceive how lively a plan lay behind.

Another feature of this house is that it is essentially square, with three bays to each front. There is no recourse here to a T-shape, or any other complex form, to ease the planning problems. Interestingly, the end windows at the front are placed acceptably close to the ends of that façade, even though this leaves the study window well off-centre inside, but at least this was a private room. In the event, this house was not built, despite seeming modest for a man said to be 'very rich'; perhaps he abandoned the design because he decided to concentrate on his extensive property in Berwickshire, or because he tended to live mostly in London; he died in 1794.

Two designs for Airthrey

The discussion of Woodburn House outlined some of the problems inherent in designing a home with three main rooms of unequal size behind attractive symmetrical façades. But what about designing a larger country house with more rooms? If the architect is given some freedom to decide what rooms to include, along with their relative positions and sizes, then creating a plan might be relatively straightforward. But if the architect is denied these freedoms, then major problems might arise.

Robert Adam's many clients no doubt allowed him varying degrees of freedom, but it is rarely possible to estimate from his plans how much freedom he had in any individual case. Certainly his freedom cannot be estimated from his plans in the case of clients for whom he produced only one design. Nor can it be estimated with clients for whom he produced two or more designs for houses of differing size, except to say that these clients' briefs must have changed. Interestingly, though, some of these clients clearly had some constant requirements, even unusual ones: for example, both the built and unbuilt designs for Hatch House (now Mersham-le-Hatch) in Kent have a great first-floor dressing room; all the unbuilt designs for Lowther Hall (or Castle) in Westmorland have a large gallery; and both the unbuilt lodges designed for unknown sites for the Earl of Findlater lack drawing rooms.[17]

There were other clients for whom Adam produced two or more designs for houses of a similar size and shape. In these cases, it is hard to know if he initially faced a tight brief which the client later slightly adjusted, or a loose brief where his first plan met with substantial though incomplete approval. Examples include the unbuilt designs for two houses in Kent, Brasted Place and Eastwell Park, and for one in Sussex, Saint Hill House, along with the built and unbuilt designs for two Ayrshire houses, Caldwell House and Dalquharran Castle. In another case, Rosebank in Lanarkshire, the brief clearly changed considerably, because Adam produced two unbuilt designs with very different plans.[18]

More interesting are the four cases where Adam designed two large houses of similar size, and with closely related room arrangements, but where the two houses had very different shapes. In these cases, it seems clear that the clients gave him detailed, exacting briefs, while his designs show that he could handle these in two different ways. In three of these four cases, both designs were for Classical houses and neither design was built.[19] Thus for Wyke House in Isleworth, Middlesex, Adam designed an octagonal house and a rectangular house with broadly similar room arrangements and room sizes, especially for the drawing and dining rooms. For Great Saxham House (or Hall), Suffolk, he designed a three-block house and a D-shaped house, with very similar room arrangements, and with identically sized drawing and dining rooms, although the D-shaped design has a much smaller library. And for The Hill (or Hill House), at Putney Heath, near London, he designed an almost circular house and a squarish house, again with very similar room arrangements and with drawing and dining rooms of the same size, although the squarish design has a much smaller 'library or garden room'. An interesting inference from these three cases is that Adam's clients often seemed to give him target sizes for their drawing and dining rooms.

The fourth case is Airthrey, near Stirling. This stands apart from the others, partly because the shapes of the alternative plans were so different, partly because one design was Classical and the other in the castle style, and partly because one design was actually built.[20] The Airthrey estate was purchased in 1759 by Robert Haldane, who had made a fortune as a sea captain in the East India Company. He was the youngest of three sons of John Haldane, Laird of Gleneagles, and his second wife. Robert never married, and by 1767, when he died, his two full brothers had already predeceased him, along with all the issue of one of them. The other brother had a surviving son, James, and Robert left Airthrey to him. But James died the following year, leaving Airthrey to his oldest son, another Robert, then aged only four. This Robert and his younger brother, another James, were brought up by their mother and her family, seemingly away from Airthrey.

This Robert Haldane went on to serve in the Royal Navy in 1780–3. In 1786 he married Katherine Cochrane Oswald and settled in the existing house on the Airthrey estate. Katherine's father was George Oswald of Scotstoun and Auchincruive: George had received his second title in 1784, when he inherited the Auchincruive estate in Ayrshire from his uncle, Richard Oswald, who had employed Adam there in 1766 to decorate some rooms, and again in 1778 to build a garden tea-house in the form of a castle-style folly.[21] It was possibly this connection that led Haldane to approach Adam in 1790 when, following the creation of a loch in the Airthrey grounds and the birth of a daughter, Margaret, he decided to build a new house.

Adam's first design was for a Classical house with wings (Fig 8.4, top), and with the principal floor plan shown in Fig 8.4, bottom. Haldane must have specified the rooms on this floor in some detail, because this was the only commission where Adam's principal floor included a nursery suite. It is most likely that Haldane also specified the order of these rooms, with the

Fig 8.4
Airthrey House, Stirlingshire: unexecuted elevation (top) and principal floor plan (bottom) by Robert Adam for a classically planned house.

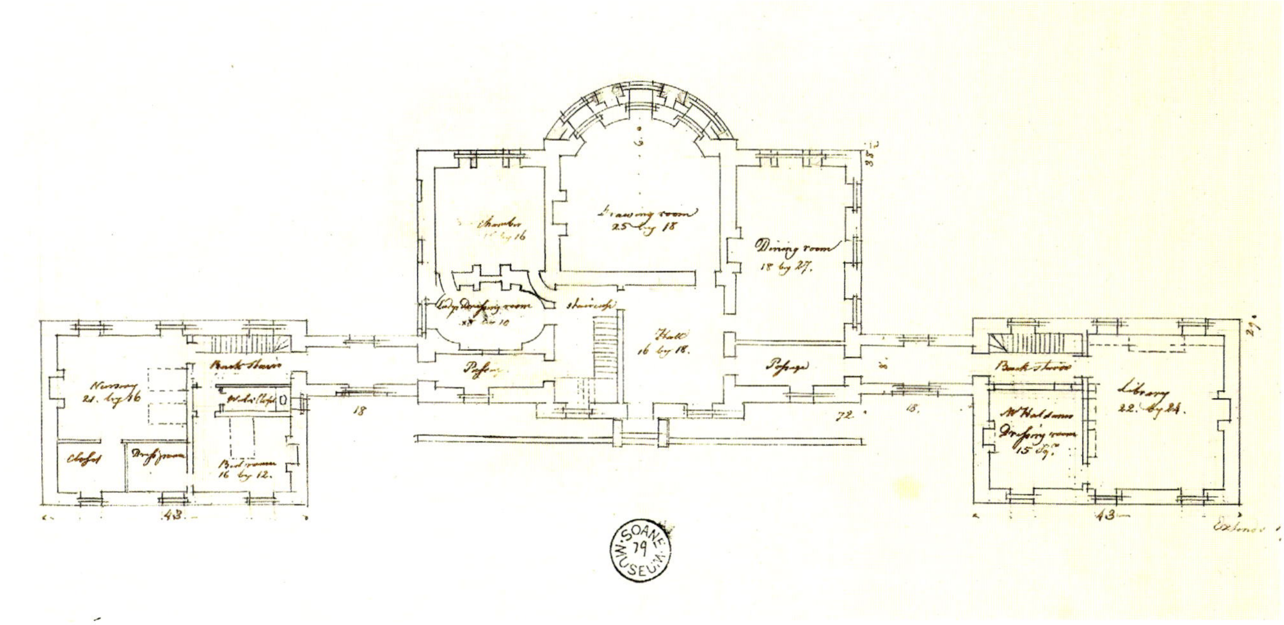

three-bed nursery suite in the left pavilion, followed in turn by the lady's dressing room, the principal bedchamber, the drawing room and the dining room, all in the main block, then Mr Haldane's dressing room, and – as far as possible from the nursery – his library, in the right pavilion. Moreover, the client probably specified target sizes of 530 sq ft (49.2 sq m) for the drawing room and 490 sq ft (45.5 sq m) for the dining room, and Adam's plan gets each room within 2 per cent of its target.

Haldane rejected this design, possibly because its potential cost was too high. As the design was for a Classical house, it was probably intended to be built in ashlar, and money could be saved by instead using rubble masonry, but Adam probably suggested that rubble would be more appropriate for a castle-style house. Money could also be saved by replacing the winged layout with a single block, to avoid the need for passages to the pavilions. Perhaps, too, Haldane agreed that some of the private rooms could be a little smaller, and that the three beds of the nursery suite could be placed in a single room. The revised D-shaped plan is shown in Fig 8.5. With considerable ingenuity, this contains the same rooms in the same order, and with both the drawing room and eating room again within 2 per cent of their target sizes, and yet all of these rooms are placed in a house of a very different shape and style, and with complex but symmetrical façades.[22]

The rear of Airthrey is shown in Fig 8.6. Like most of Adam's proposed curved façades, this one is interrupted by a central feature, here a bowed

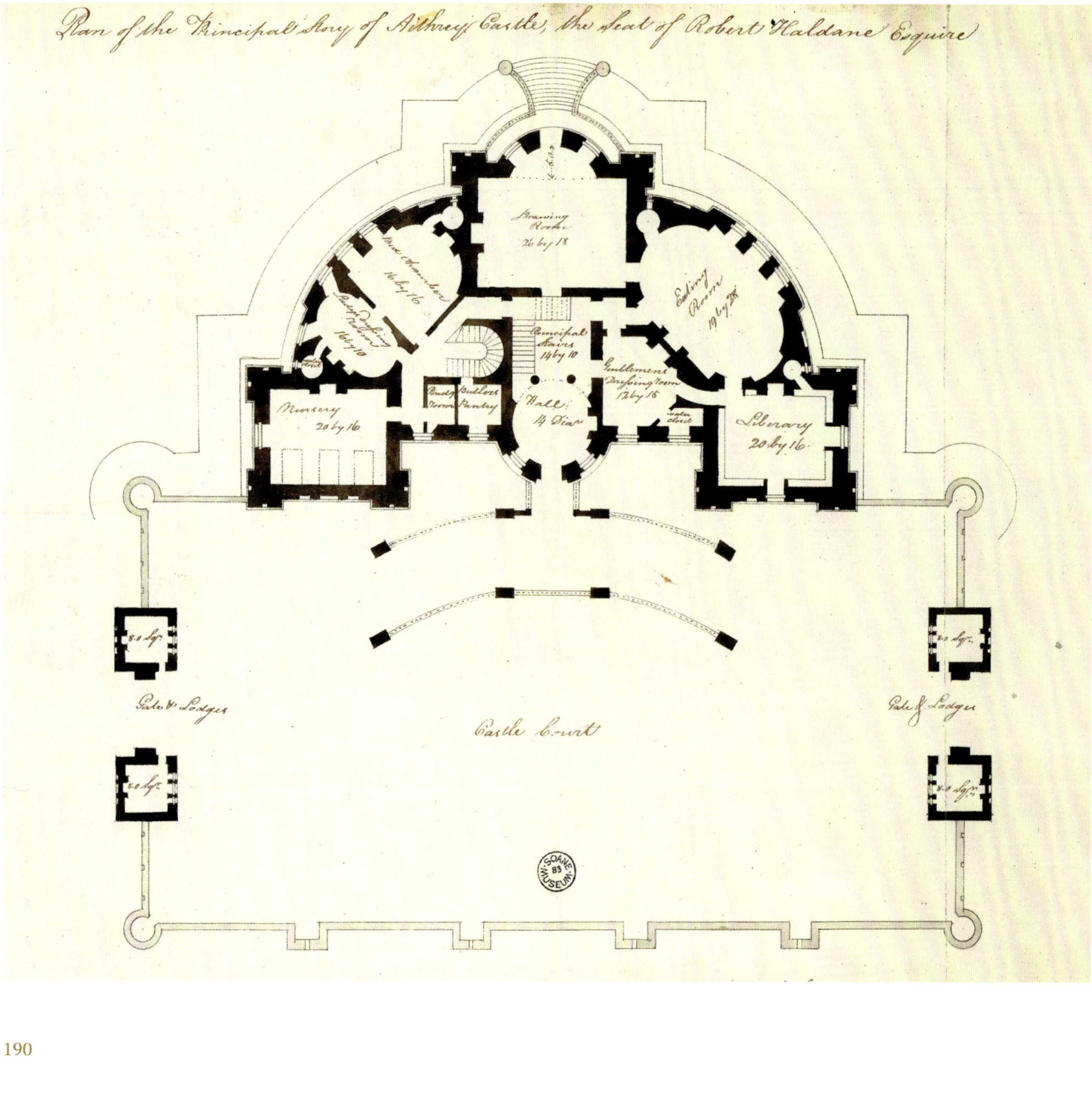

Fig 8.5
The revised principal floor plan for Airthrey Castle, Stirlingshire.

Fig 8.6
Airthrey Castle, south front, in the early 19th century, drawn by J P Neale (Neale 1818–23).

projection set between narrow square turrets. In almost every other case, however, this feature interrupts a single continuous curve, whereas at Airthrey it lies between two separate quadrants; the only other notable designs where the central feature is set between two separate curves are the sides – but not the front or rear – of the almost circular scheme for Hill House, and two preliminary schemes for the Edinburgh Bridewell, where the two curves are very shallow.[23] The Airthrey plan is so carefully constructed that Alistair Rowan suggests it was produced at leisure rather than in a 'bout of hectic activity'.[24] It would be fascinating to know whether it was during his leisurely meditations that Adam concluded that a D-shape was the most economical for including the rooms required, or whether he chose that shape because it was best for the site, as it offered panoramic views from its curved façade.

In this plan the bedchamber, Mr Haldane's dressing room and the library are all a little smaller than in the Classical plan. In contrast, the lady's dressing room, which was in any case relatively small, is of a similar size. Interestingly, the client's own copy of Adam's plan shows some late changes.[25] Chiefly, the proposed forecourt was abandoned. Also, the nursery was moved upstairs, to the room above the nursery in Fig 8.5. That room then became the main bedchamber, while the bedchamber in Fig 8.5 became Mr Haldane's dressing room, and his intended dressing room was reconfigured as an anteroom. These changes meant that Mr Haldane's dressing room also ended up close in size to its Classical counterpart, while the main bedchamber ended up a little larger.

The castle was built to this revised plan. However, the client dispensed with Adam's services when it came to supervising the builders, using the money thus saved to have a facing of fine drove masonry instead of rubble.[26] But the client lived there only until 1798, when he joined his brother James in Protestant evangelism and sold the estate to Robert Abercrombie, the uncle of James's wife. Robert Haldane later wrote, 'I gave up a place and a situation which continually presented objects calculated to excite and gratify the lust of the eye and the pride of life. Instead of being engaged in such poor matters, my time is now more at my command; and I find my power of usefully applying property very considerably increased.'[27] Despite his deep commitment to evangelism, though, might he have thought twice about selling Airthrey if he had had a son? As it was, Margaret was always the sole child in the upstairs nursery. The castle was much extended in the late 19th century, and is now part of Stirling University.

Office court for Alva House

The next project to be discussed is a splendid unexecuted 1789 design for a castle-style office court for Alva House near Alva, Clackmannanshire. The caption to Robert Adam's elevation itemises what were presumably the requirements of the client, John Johnstone, as follows:

> Stables for all the Coach and riding Horses, with Coach Houses for 4 Carriages, Brewhouse, Bakehouse, Laundry & Wash house, Cow house, Dairy, Hen house, Stewards House, Pig house, Slaughter house, Cart shed, Smith & Carpenter's shops, with all the Yards necessary to the different Offices, and Privys for the Servants of each, with a Dung Pit in the Center of this Court covered over with a Pidgeon House and ... rooms for lodging the Servants near the Offices which they superintend & to which they belong.[28]

Although this is a huge array of offices, the client was known to be 'very rich, from India', where he had apparently accumulated £300,000.[29] So here Adam's ingenuity was not exercised in including all the offices in an economical design but in using them to create a fine building. With so many offices, the design clearly needed a court, and assuming that the pigeon house was in a tower, then it would be natural to put this tower in the centre of the court above the dung pit. But what shape should Adam give to the court?

Of course, Adam asked himself this question with all his stable and farm court designs, and to appreciate fully his remarkable proposal for Alva, it is helpful first to consider briefly what he did elsewhere. Given his interest in houses with interesting shapes, one might expect a variety of geometrical forms in these courts; and given his penchant for symmetry, one might also expect any facing pair of ranges inside each court to match. In practice, though, the need for wide openings for coach houses and cart sheds, and the need for other offices to vary in shape and size, frequently constrained his hand.

Indeed, Adam's three unexecuted designs for a castle-style stable court at Kirkdale House, Kirkcudbrightshire, and his unbuilt Classical scheme for stables and other offices at Bush House, Midlothian, largely abandon any attempt at regularity.[30] The former all have a D-shaped court whose straight side is symmetrical but whose curved side contains a picturesque assemblage of disparate buildings and walls. The latter has a square court with a wall along one side and three different blocks along the others, and only one of these blocks is itself symmetrical.

Elsewhere, Adam managed more regularity, but rarely as much as he might have liked. For example, three other unexecuted Classical court designs envisage facing ranges along two sides of a square or rectangular court, with walls on the other two sides.[31] In the stables for Bromsberrow Place, Gloucestershire, and Panmure House, Angus, the ranges differ greatly, and only partly because of some wide openings. In contrast, the ranges in a sketch plan for farm offices for a Mr David Kennedy do match, but they are asymmetrical. Related to these is a more regular unexecuted Classical scheme for stables at Congalton, East Lothian; this has buildings on three sides and a wall on the fourth, and the facing ranges differ only because of some wide coach-house doors.[32]

There are four unbuilt Classical designs where the entrance range differs from the facing range only to incorporate some wide openings, while the side ranges match each other. These are the stables for Kimbolton Castle in Huntingdonshire, Castle Ashby in Northamptonshire and Mamhead in Devon, along with the farm offices for Kedleston Hall in Derbyshire; at Kimbolton, though, the matching side ranges are each asymmetrical.[33] Related to this group is the 1777 castle-style farm built for Culzean Castle. This has a square court with four very similar ranges: one facing pair match, while the other pair differ only because of some wide openings. This scheme is also interesting geometrically, because these four ranges are in fact the 'tops' of four T-shaped blocks, a device which necessitates entrances in the corners of the court rather than in the middle of one or more sides.[34]

Three other designs comprise more than one court.[35] Adam's plan for the 1788 castle-style stable and farm offices built at Castle Upton, County Antrim, shows two adjacent squarish courts, each with the entrance range differing from the facing range, and with small differences in the side ranges, and few of these differences were necessitated by openings. The design for an unbuilt Classical farm at Wonersh House, Surrey, has a squarish central court flanked by semicircular courts; in the central court the entrance and facing ranges differ because of some wide openings, while the side ranges match. The third design is shown only in a very accomplished plan for a castle-style farm at Findlater Castle, which is probably another name for Cullen House, Banffshire; here, a squarish central court is flanked by thin rectangular courts. This plan shows the squarish court with almost identical entrance and facing ranges, along with fully identical side ranges, and all this despite numerous wide openings.

Two other designs have some interesting geometry. One is a seemingly preliminary sketch for the Classical farm built around 1790 at Kirkdale House which shows an octagonal court with alternating larger and smaller blocks; the sketch suggests that the four larger blocks would match each other, and that the four smaller blocks would also match.[36] The other example is the unbuilt castle-style design for stables at Findlater Castle, which has a circular court 160ft (48.8m) in diameter.[37] This court is surrounded by a low range of buildings that is divided into four matching quadrants by four matching entrance arches; the design also includes four matching pavilions, one in the centre of each quadrant. But while Adam's plan here is very satisfying, his elevation shows that the low buildings around the circle would be almost overly dominated by a dramatic circular riding room in the centre of the court.

For the Alva House offices, Adam proposed a scheme with far more complex geometry than any other. His plan (Fig 8.7, right) shows a hexa-decagonal court, with its 16 straight sides containing 8 trapezium-shaped blocks alternating with 8 rectangular blocks.[38] Moreover, seven of the trapeziums match exactly, with three-bay arcades; the eighth, at the bottom of the plan, is the entrance block, with a wide entrance arch set between a scullery and a bakehouse. The eight two-bay rectangles also match very closely: three have wide arched openings, while five have windows and doors in wide relieving arches. The arches all around the edge of the court are echoed by an arcade around the central dung pit.

The topmost trapezium contains the carpenter's and smith's shops, with a passage between them leading to a yard with troughs for washing water and drinking water. Moving clockwise, the other trapeziums contain a cow house, a stable and a stable with 'bail stalls', all three with yards behind; the entrance block; a complex hen or poultry house, with a poultry yard behind; a dairy with a drying yard behind; and a pig house and slaughterhouse with a pig yard behind. As for the rectangles, again moving clockwise from the top, these comprise a coach house, two stables, a cart shed, a brewhouse, a steward's house, a wash house and laundry, and another coach house. Each rectangular block contains a staircase: the brewhouse stairs lead to the upper parts of a tall room, while the other stairs lead chiefly to servants' rooms, and also to some haylofts above the stables.

This scheme might have been tolerably easy to devise if Adam had been allowed to adjust the sizes and details of the various offices, but the number of carriages and the other requirements were doubtless carefully specified. For example, one stable is inscribed as having bail stalls, which are indicated with dashed lines. Bail stalls are divided by wooden bars suspended on straps or chains, instead of the more usual timber partitions, and were very rare in 18th-century Britain, except in military stables; the only other Adam stable court design where he was asked to include them was the one for Castle Ashby.[39] Adam's attention to detail is best illustrated by the fittings in the brewhouse and by the composition of the poultry house; the latter includes a closet, two hatch rooms, a fowl house with nest boxes for hens, and a 'poultry inside house', possibly for night-time use by the ducks for which the water pool in the poultry yard was presumably intended. Other touches include harness rooms behind the coach houses and the cart shed, boiling rooms by the slaughterhouse and some of the stables, two ovens in the bakehouse, an office for the steward, a scullery and a dairymaid's room next to the dairy, a linen room next to the wash house, and a water pool for the pigs to wallow in.

Adam's elevation for these offices (Fig 8.7, bottom) is as impressive as the plan. Considerable vertical movement is created by the upper storey and ball-topped pyramidal roofs of the eight rectangular blocks, whose outer walls have three bays, and by the central pigeon house, which takes the form of a round crenellated tower; the central portion of the pigeon house supports an octagonal timber steeple with arched openings at its base for the pigeons and a stepped roof above. It is a striking testament to Adam's ingenuity that he could fit this wide array of precisely detailed offices into a complex and rigid geometrical shape, and simultaneously create a wholly unified and atmospheric ensemble.

Adam doubtless hoped that this court would be built. Aside from John Johnstone's wealth, his family were well known to Adam. One of his older brothers, James, was 4th baronet of Westerhall,

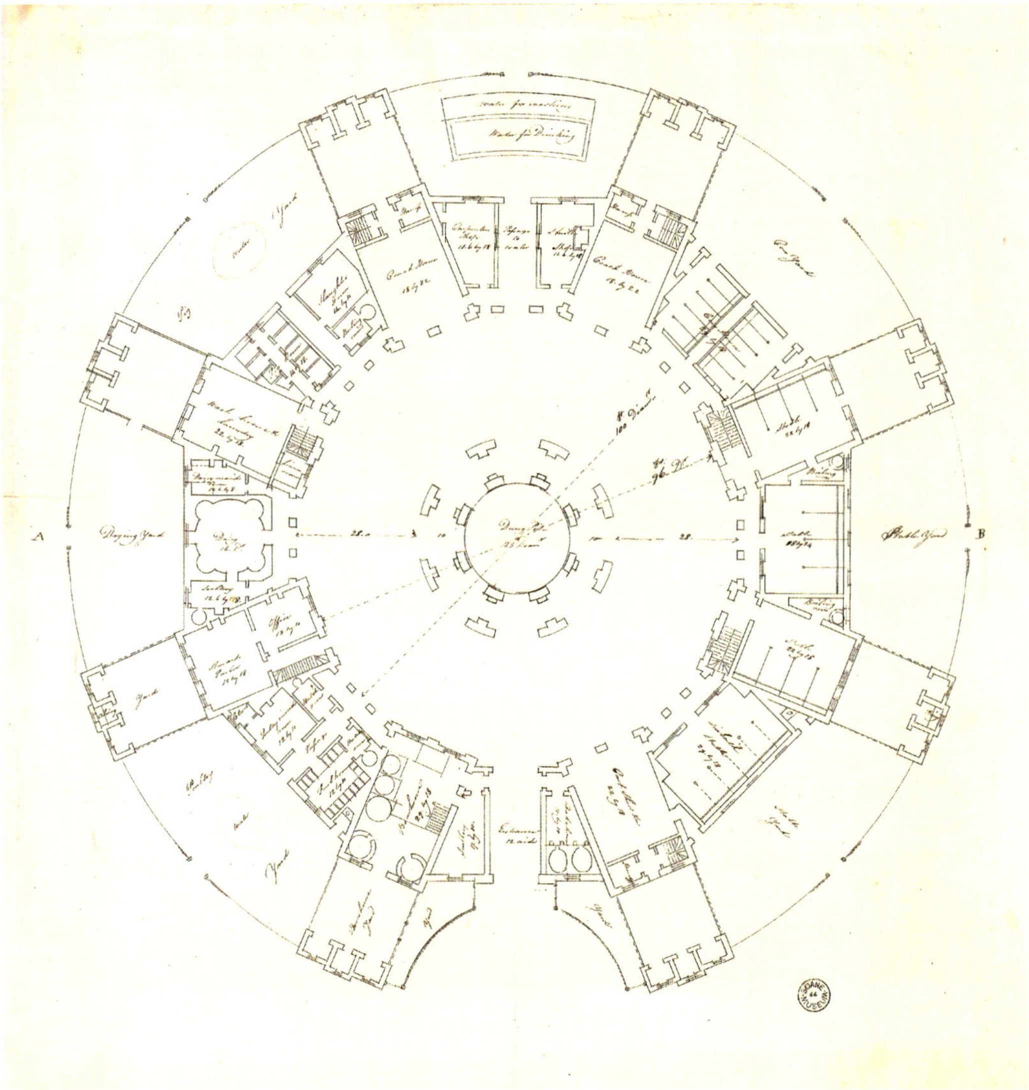

Fig 8.7
Alva House, Clackmannanshire: plan for the office court (right) and elevation (bottom).

Dumfriesshire, and in 1790 he was to commission designs from Adam for a mausoleum which was built nearby at Westerkirk. On James Johnstone's death in 1794 another brother, William, became 5th baronet; in 1767 he had changed his surname to Pulteney when his wife Frances (née Pulteney) inherited the Pulteney estates in Bath, and it was he who commissioned Adam's fine executed design, prepared around 1768, for Bath's Pulteney Bridge. Moreover, John Johnstone in 1789 commissioned designs from Adam not only for these offices, but also for remodelling the house itself, for a new entrance and for a mausoleum in the nearby graveyard at Alva.

In the event, the only Adam design which John Johnstone carried out was the mausoleum. One reason why he abandoned the office court may have been that by 1790 he had also bought an estate in Dumfriesshire. Another reason might be that there seems to be no flat area in the grounds of Alva House big enough for them. John died in 1795, and between 1805 and 1809 his son James built some more modest Classical-style offices, to designs by William Stirling.[40]

A cottage for Mr Dalzell

I now turn to one of Adam's designs for thatched cottages, in this case a cottage 'for Mr Dalzell'.[41] Robert Adam's drawings include an appealing front elevation (Fig 8.8, top) but give no date or location. It has been suggested that the client was Andrew Dalzell, who in 1786 married Ann Drysdale, the daughter of Robert's sister Mary and her husband, the Very Reverend Dr John Drysdale.[42] This may be so, although if that Andrew Dalzell were the client, then the drawings might perhaps have been inscribed 'for Professor Dalzell', because he was professor of Greek at Edinburgh University from 1772.

This cottage has two small office wings, shown also on the associated plan (Fig 8.8, centre), but the discussion here will focus on the central block. Like Woodburn House, but on a smaller scale, the principal floor of this block was to contain three main rooms, but here two were to be of equal size, which greatly facilitates the creation of symmetrical façades. These two rooms are placed at the rear and given apses on their adjacent walls. The smaller third room is also at the rear, with the form of an irregular pentagon, and is placed partly between the apses and partly in a central rear projection. The hall at the front is flanked by passages, one also containing a staircase, which lead to low links to the pavilions.

In Adam's elevation (*see* Fig 8.8, top), the central block has three main features at wall level: a semicircular pantiled log porch, a slight gabled advance for the hall and the room above, and a gabled projection on each side, where the large rear rooms stick out a little behind the links to the pavilions. But the most striking feature of this elevation is the pyramidal roof, which Adam intended to cover most of the central block.

Unfortunately for Adam, the idea that this plan will support a well-defined pyramidal roof is illusory. The reasons can be explained with the help of Fig 8.9a, which is the roof plan that would result from Adam's floor plan. This roof plan shows that the front face of the putative pyramid would indeed be clearly defined, as it is in the elevation behind the gable for the room above the hall (*see* Fig 8.8, top). But the roof plan also shows that if the cottage were viewed from either side, then the gabled roof of that side's projection would obscure much of that side's face of the putative pyramid. At the rear, the side projections would be aligned precisely with the rear of the pyramid, so that all along the rear there would be a continuous roof; this would effectively remove any outline of the rear face of the pyramid, except for an awkward un-Adamesque triangle which would stick up in the middle, where its top would rise above the ridges of the side projection roofs.

The key problem which Adam faced here was this awkward projecting triangle in the centre of the rear roof, and this was a problem which he himself created with his desire for a pyramidal roof. That he found this problem troublesome is shown by the fact that he considered five different schemes for removing the triangle. The elevation in Fig 8.8 (top) shows in faint pencil the first of these, which is to make the side projection roofs steeper, to give their ridges the same height as the pyramid. This means that the gables at the ends of the side projections would be steeper than before, and perhaps that is why the pencil changes in Fig 8.8 (top) hint also at steepening the gables in the pavilion roofs to match, and then giving each pavilion a dormer window. The resulting roof plan is shown in Fig 8.9b. The front face of the putative pyramid is here shown unchanged from Fig 8.9a, with its highest point in the same position. At the rear, the entire continuous roof must actually be steeper than in Fig 8.9a to ensure that its height throughout is the same as that of the highest point on the pyramid. So there must now be a horizontal ridge between the centre point of that steeper continuous roof, where the chimney stack is, and

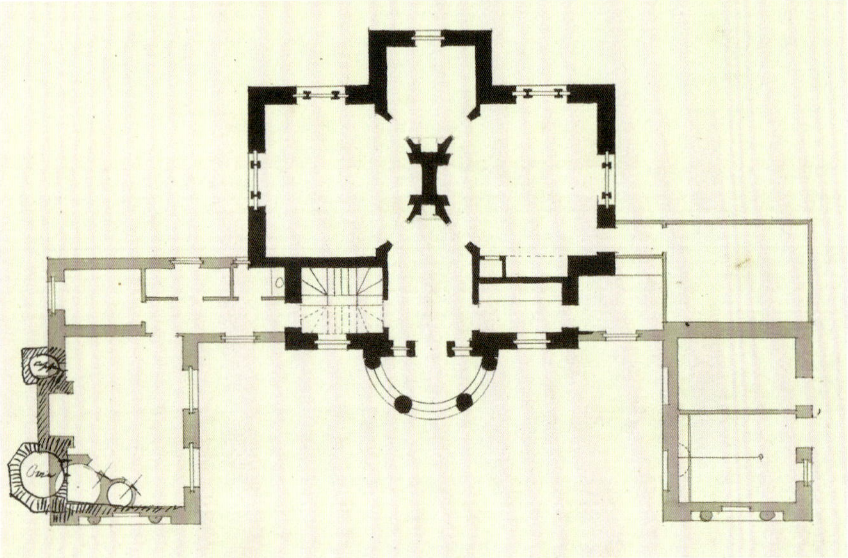

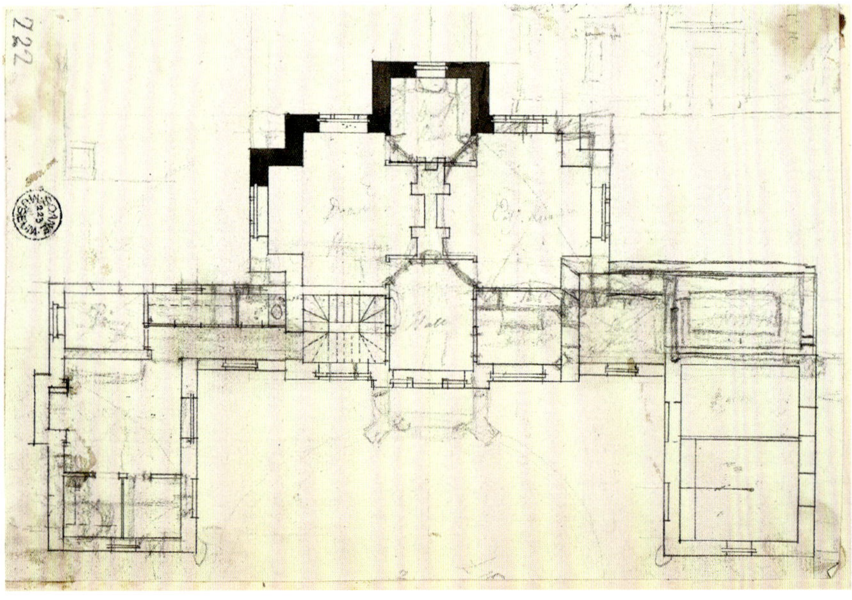

the top of the pyramid. Fig 8.9b shows that there would now be no hint of the pyramid at the rear, because the triangle has gone, nor indeed on either side, because in each case the roofline would be topped by the ridge. Only on the front would any vestige of the pyramid remain. A further issue here is that the higher roofs on the side projections would threaten to make the whole cottage appear roof-heavy.

Another Adam drawing has pencilled indications of the second, third and fourth schemes for removing the triangle; each of these also removes the front face of the pyramidal roof. The second scheme, which is indicated in Fig 8.9c, raises the side projection roofs as in the first, and Adam's pencilled thoughts show him again envisaging steeper roofs with dormers on the pavilions. However, this scheme differs from the first by removing the slight advance for the hall and the room above it, and instead placing a large gable across the whole front of the main part of the building. Once again the cottage seems roof-heavy.

The third scheme is indicated in Fig 8.9d. This is identical to the second (*see* Fig 8.9c), except that it removes the roof-heavy effect by making the entire roof flatter and lower. Adam's sketch for this scheme shows the roof still thatched, even though its pitch is now very low for thatch. It also suggests reducing the wall height a little. The sketch omits the pavilions, but perhaps their roofs could now be left as they were in the original design.

The fourth scheme is indicated in Fig 8.9e. This retains the roof design from the third (Fig 8.9d), but raises the height of the walls to create a taller and more elegant cottage. It seems that Adam was minded to make this handsome building Classical: his sketch replaces the log porch with a Classical one, and it also omits the thatch, perhaps because with this low pitch he would prefer slates, as suggested in Fig 8.9e.

The fifth scheme for removing the awkward triangle, which is implied by a rather crudely drawn ink plan (*see* Fig 8.8, bottom), has the further merit of achieving a pyramidal roof with four well-defined faces. The ingenious idea here is for each large rear room to have a bit chopped off each of its two corners on the side of the cottage. This means that if the cottage were viewed from behind, then each side projection would now be slightly recessed, so there would be no continuous roof at the rear. This in turn means that both the front and rear elevations could have roof profiles that match the appealing one for the front shown in Fig 8.8 (top), so that both the front and rear faces of the pyramid would be clearly defined. Moreover, the associated roof plan (Fig 8.9f) shows that its side faces would also be clearly defined, because the rear of the side projection roofs would no longer be aligned with the rear of the pyramid, which is what happened with the original roof (*see* Fig 8.9a).

But while this corner-chopping is an ingenious solution to the roof triangle problem, it might threaten to create irregular shapes for the two main rooms. However, the plan in Fig 8.8 (bottom) shows Adam's ingenuity in also fixing this problem by making these rooms cross-shaped and, in the process, making the small third room square. Notice, though, that this plan also has some pencilled additions showing the differences from the plan in Fig 8.8 (centre); it is hard to be certain whether these additions were made simply to show how the plan in this ingenious fifth scheme differs from the plan used in the other four, or whether this was possibly Adam's original plan, which he here considers changing in pencil for some unknown reason,

Fig 8.8 (opposite) Elevation (top) and alternative plans of the principal floor (centre and bottom) for a cottage for a Mr Dalzell.

Fig 8.9 Author's musings on the possible roof structures suggested by Robert Adam's designs for Mr Dalzell's cottage.

only to find later that he could not then achieve a well-defined pyramidal roof.

The plan in Fig 8.8 (bottom) has a further intriguing feature, as it shows that Adam considered having a rectangular porch with diagonal buttresses instead of a semicircular log porch. It seems, then, that he considered making this building in his thatched style (*see* Figs 8.8, top and centre, 8.9a–d), or Classical (*see* Fig 8.9e), or in either a Gothic or castle style (*see* Figs 8.8, bottom, 8.9f).

Alterations for the Edinburgh Assembly Rooms

The final ingenious project concerns Edinburgh's Assembly Rooms, which still stand, though somewhat altered, in George Street. In 1782 the committee that commissioned these rooms had raised enough funds from public subscription to appoint John Henderson as the architect. The building was begun in 1783 and opened in 1787, although it was not then complete. The interior was much admired, but the street front was regarded as heavy and inelegant, even an eyesore.[43]

Sir John Soane's Museum has some undated drawings by Robert Adam showing unexecuted ideas for 'altering' the Assembly Rooms.[44] It is necessary to conjecture from these drawings when he was consulted and what problems he was asked to tackle. One problem may have been the dull front. As executed, it had two floors and seven bays, with little interest aside from a rusticated ground floor, Tuscan pilasters between the first-floor windows, and paired first-floor Tuscan pilasters at either end. The paired Ionic pilasters shown on Adam's elevation (Fig 8.10) indicate the length of this executed front. Adam's chief ideas for improvement were replacing the two bays at each end with one tripartite bay, and adding an Ionic portico, some statues at roof level and a central dome.

However, Adam's elevation extends the building at either end a little beyond the paired pilasters, towards the neighbouring houses, and his drawings suggest that he was also consulted about the plans. Henderson's intention was to have an entrance hall three bays across, set between a pair of two-bay shops; the hall was to lead to a central lobby with access on each side to a double staircase. Both double staircases would lead to a saloon above the lobby. All along the front of the building on the first floor would be a huge ballroom. At the back, the ground floor would have a kitchen flanked by other offices, while the first floor had a tea room flanked by two smaller card rooms; the tea room would also be used for dancing during card assemblies.

Adam's drawings show that the committee considered replacing the proposed hall and the two shops with a long supper room, and as it would be inconvenient for people who wished to go upstairs to pass through this supper room, the committee also considered having an entrance vestibule in the middle of each side of the building, with a double staircase leading up from each vestibule. To protect these entrances from the elements, Adam proposed having covered passages with arcaded external walls along both

Fig 8.10 (below) Elevation of the north front of the Assembly Rooms in Edinburgh. Unexecuted Adam office design, 1780s

Fig 8.11 (opposite) Assembly Rooms, Edinburgh, as intended by Robert Adam, 1780s: plan of the principal floor (top) and sectional drawing looking east through the ballroom and tea room (bottom).

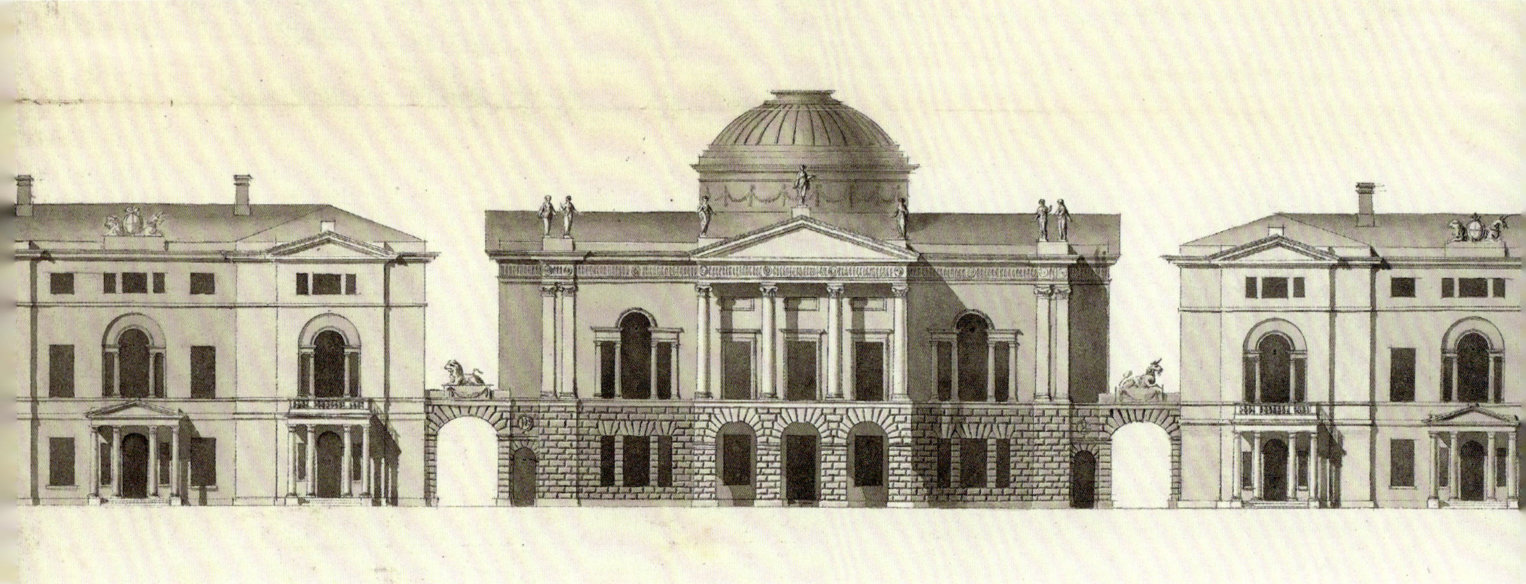

sides of the building, and also along the rear; the rear passage would need a curved middle section to take it around a bowed projection which Adam proposed introducing for the kitchen and the tea room above. Pedestrians and sedan chairs would be able to access the side passages through the narrow arches which his elevation shows at each end of the building, while the wider arches between this building and its neighbours would provide access for coaches.

Adam's first-floor plan for the Assembly Rooms (Fig 8.11, top) shows at the bottom the paired pilasters at each end of the front, and it also shows that he proposed pushing the side and rear walls on this floor out over the passages, except that there would be a balcony over the curved middle section of the passage at the rear. The grey wash here shows how few of Henderson's intended walls would survive.

These proposals cannot have been prepared once the building was complete, as surely no committee would contemplate shifting the first-floor walls of a new building. It is more probable that Adam was consulted during its construction. The most likely time was soon after February 1786, when John Henderson died and a new architect was needed, and seemingly the committee were then considering replacing the hall and shops with a supper room.

The most ingenious feature of Adam's planning here is his proposed use for the extra first-floor space that he would create, chiefly above the passages. At the front, he proposed making the ballroom cross-shaped, by extending its ends over the side passages, its front into the portico and its rear into part of Henderson's intended saloon, which would become a smaller octagonal lobby (*see* Fig 8.11, top). At the rear, he proposed making the tea room circular by pushing part of it into the intended saloon and part over the rear bowed projection; this splendid room would also have a fine balcony set behind a tall Ionic screen. In addition, the flanking card rooms would end up about 40 per cent larger by being extended over the side and rear passages. Incidentally, Adam also proposed extending the ballroom upwards into the dome shown in Fig 8.10, and in the tea room behind he proposed having a second lower dome, set below a shallow conical roof. Both domes are shown in Fig 8.11 (bottom), which is a section through the ballroom, lobby and tea room 'as proposed to be finished'; below these, the section shows the much lower supper room, a passage and a cellar to replace Henderson's lobby, and the kitchen.

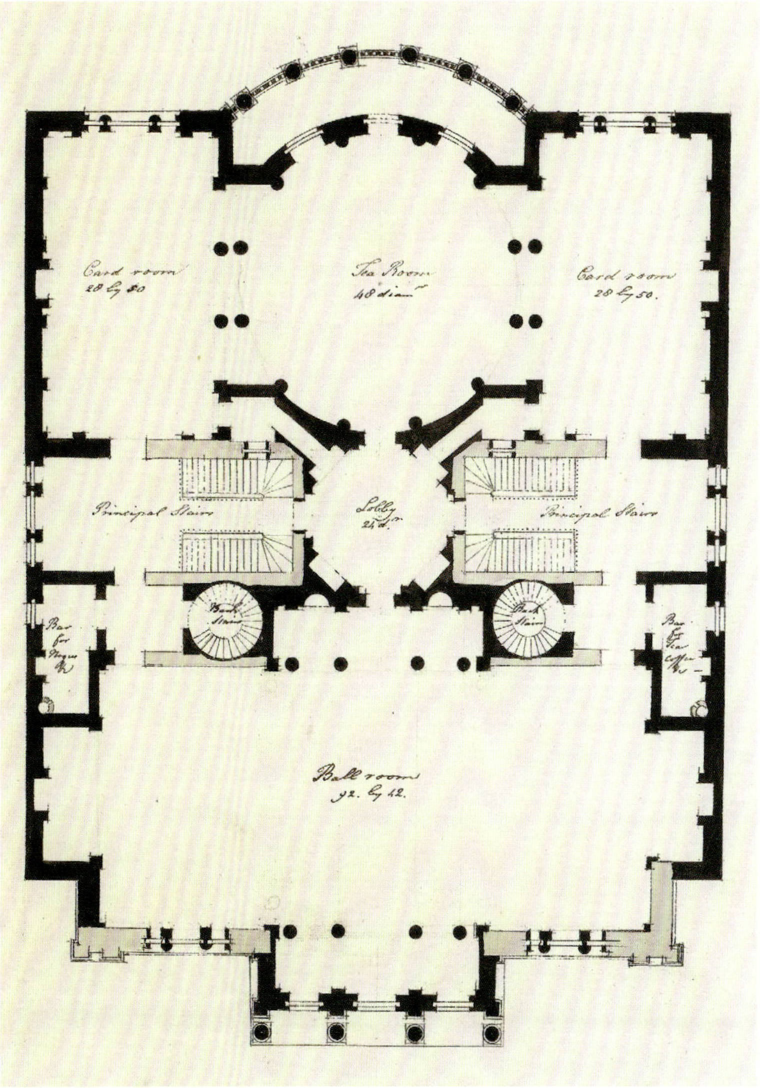

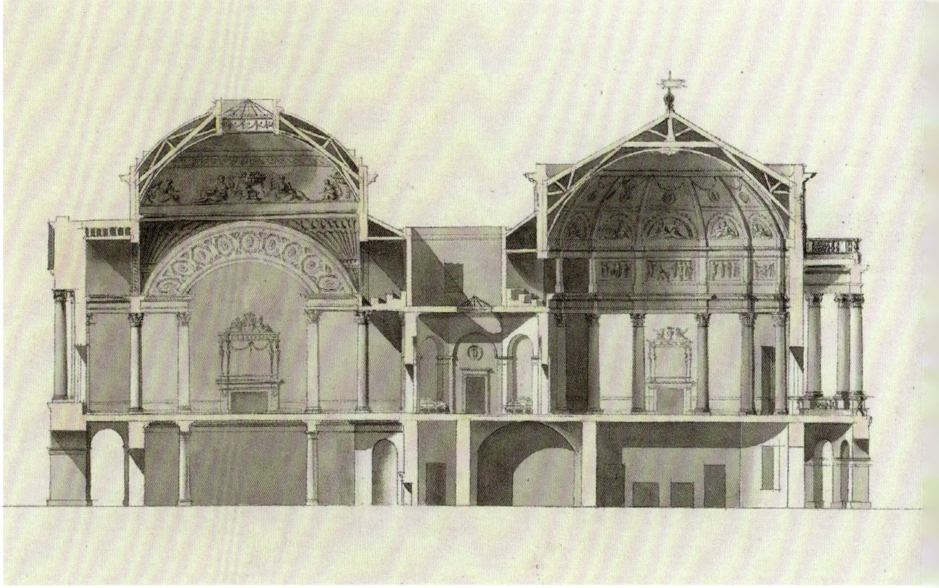

However, this scheme has one weak feature which results from using side entrances in a building that is close to its neighbours. Having one's sedan chair carried through a narrow archway into a narrow passage, or one's carriage hauled into a poky side alley, is not the way to arrive at an elegant ball. Of course, if the committee were set on having side entrances to facilitate a supper room, then Adam's passages were doubtless the best scheme that could be devised. But one can imagine the committee studying Adam's plans and concluding that if even the ingenious Mr Adam could not devise appealing side entrances, then it would be better to revert to the original idea of having an entrance vestibule between two shops and abandon the idea of having a supper room; and that is exactly what they did, with John Henderson's position as architect passing to his father, David. But while this was very understandable, Adam's admirers will regret that his proposals for the façade were ignored.[45]

Notes

1 Adam and Adam 1773–1822, vol I, no I, preface; vol I, no II, preface.

2 Ibid, vol I, no I, 'Plans of Sion House'; vol II, no I, 'Explanation of the plates'.

3 Details and illustrations are in Bolton 1922, vol II, 18–47; *Survey of London*, vol 18, 1937, 99–119; Rowan 1974, 661–7; and King 1991, 77–85.

4 Dickens, C 1855–7 *Little Dorrit*, book II, ch 9.

5 Details and illustrations are in Bolton 1922, vol II, 252–62, and King 1991, 160–3.

6 Details and illustrations are in Bolton 1922, vol II, 263–77; Rowan 1985, 128–33; King 1991, 165–8; and Harris 2001, 316–33.

7 King 1991, 165, gives phase two as completed before phase three began, but Harris 2001, 321, has since found that, although these phases were designed some years apart, the building work on both started in 1787.

8 Details and illustrations are in Rowan 1985, 98–101, and King 2001, 148–56.

9 Details and illustrations are in Rowan 1985, 118–21, and King 2001, 150–7.

10 Rowan 1974, 704, illustrates an elevation of the Harwich design. Details and illustrations of the V-shaped Barnton design are in King 2001, 156–9.

11 Euclid, *Elements*, book IV, proposition 11.

12 Adam used a wide array of angles between internal walls, but between major external walls he usually chose angles that could be drawn with a compass and straight edge alone. The most complex of these angles were used in a conservatory designed for Wyke House, Isleworth, Middlesex. This includes angles of 157.5 degrees (that is, 1.75 right angles) and angles of 162 degrees (that is, 1.5 of the 108-degree angles found in a regular pentagon); details are in King 2001, 210–11. But Adam sometimes chose angles which a Classical architect could draw only by including a 20-degree angle. Thus a house designed for a Mr Wilson has some angles of 140 degrees, which would be drawn as a 120-degree angle plus a 20-degree one; details are in Rowan 1985, 50–3, and King 2001, 101–2. Also, a design for Fullarton Castle, Ayrshire, has two angles of 125 degrees, that is, a 120-degree one plus a quarter of a 20-degree one; drawings are in SM, Adam vol 36/79–85. A Classical architect could draw a 20-degree angle only by trisecting a 60-degree angle; this is impossible with a compass and straight edge alone, but it can be done with other methods devised in ancient Greece. Adam was doubtless very familiar with the method of Nicomedes, which starts by using a straight edge with one distance marked on it to draw a curve known as a conchoid, for the sides of all Adam's Classical columns were conchoids that were drawn with a device devised by Nicomedes: *see* Adam and Adam 1773–1822, vol I, no II, preface; and King 1991, 3.

13 Details and illustrations are in Bolton 1922, vol II, 224, and King 2001, 18–19.

14 Information about the client can be found in Bolton 1922, vol II, index, 77, and the Coldstream & District Local History Society, *Little Swinton: Early History*, www.coldstreamhistorysociety.co.uk/index.php/projects/historic_document/little-swinton-early-history [accessed 20 September 2016].

15 Drawings are in SM, Adam vol 33/97–102. The earlier date seems to have been cut off from two of the drawings.

16 For the rectangular Sunnyside design, *see* King 2001, 106–9. For the T-shaped design, *see* Rowan 1985, 78–9, and King 1991, 131–6. King 2001, 137, notes an uninscribed rectangular design, which is illustrated in Rowan 1985, 70–71, and five uninscribed loosely T-shaped designs, two of which are illustrated ibid, 72–7, which might all be other preliminary ideas for Sunnyside; if so, they strengthen the view that Adam found the brief tricky. For the design for Mr Wilson, *see* note 12.

17 Details and illustrations of Hatch House as built are in King 1991, 113–15; drawings for the unbuilt design are in SM, Adam vol 41/96–101. For details and illustrations of the designs for Lowther, *see* Colvin, H, Crook, J M and Friedman, T 1980 *Architectural Drawings from Lowther Castle, Westmorland*. London: Society of Architectural Historians of Great Britain, and King 2001, 86–94, 131, 138–45 and 163. Drawings for one Findlater lodge are in SM, Adam vol 36/42–5, 47–50, and for the other, vol 36/55–9. Other houses for which Adam produced two or more designs of varying size include Kirkdale House, Kirkcudbrightshire; Llanrhaeadr Hall (or Llanadder House), Denbighshire; Luton Park (now Luton Hoo), Bedfordshire; and The Oaks,

Carshalton, Surrey; only at Kirkdale and Luton was one design built.
18 This paragraph omits two further examples, Balavil, Inverness-shire, and Balbardie House, West Lothian. In each case, one design was built. However, only one set of Adam plans survives for Balavil and none survives for Balbardie, so detailed comparisons of his two schemes are impossible. For Rosebank, *see* Rowan 1985, 88–91.
19 For details and illustrations of the octagonal Wyke House design, both Great Saxham designs and both Hill House designs, *see* King 2001, 20–3, 96–102 and 108–11. Drawings for the rectangular Wyke House are in SM, Adam vol 45/98–102. Adam also produced two similarly sized unbuilt designs for Barnton Castle, near Edinburgh, one D-shaped and the other V-shaped, but the D-shaped plans do not survive, so comparisons are impossible.
20 Drawings for both designs are in SM, Adam vol 48/75–87.
21 Despite retaining the 'of Auchincruive' part of his title, George Oswald seems on inheriting the estate to have passed it immediately to his eldest son, Richard. Details and illustrations of Adam's work at Auchincruive are in King 1991, 242–45 and 322.
22 For further details and illustrations, *see* ibid, 170–75, and Rowan 1985, 142–5, 153–4.
23 For Hill House, *see* note 19. Principal floor plans for the relevant Edinburgh Bridewell schemes are in SM, Adam vol 33/26–7.
24 Rowan 1985, 142.
25 Stirling University Library and Historic Environment Scotland have photographs of this plan.
26 Rowan 1974, 701.
27 Information from http://ukwells.org/revivalists/robert-haldane [accessed 20 September 2016].
28 SM, Adam vol 34/64.
29 Bolton 1922, vol II, index 77, and History of Parliament online, www.historyofparliamentonline.org/volume/1754-1790/member/johnstone-john-1734-95.
30 For Kirkdale, *see* King 2001, 240–42, 245. Drawings for Bush House are in SM, Adam vol 35/90, 92.
31 Drawings for Bromsberrow House are in SM, Adam vol 44/76; for Panmure House, vol 41/33–4; and for David Kennedy, vol 4/161.
32 Drawings in SM, Adam vol 45/84–5.
33 Drawings for Kimbolton are in SM, Adam vol 30/134–7. Adam's preference for symmetrical sides is evident in the unbuilt domestic office block which he also designed for Kimbolton, as its front and side ranges are identical to the stables, except that its side ranges are made symmetrical; drawings are in SM, Adam vol 30/138–41. Drawings for Castle Ashby are in SM, Adam vol 29/23–4; for Mamhead, vol 32/35–7; and for Kedleston, vol 40/28–31.
34 Details and illustrations of the Culzean farm are in King 1991, 330–2.
35 Details and illustrations of Castle Upton are in King 1991, 325–8, and of Wonersh are in King 2001, 203–4. The plan for Findlater farm is in SM, Adam vol 36/78.
36 *See* King 1991, 342–3. The sketch is in SM, Adam vol 4/172.
37 Details and illustrations are in King 2001, 241–2. Adam's unbuilt design for Classical stables at Kedleston Hall, shown only in two elevations in SM, Adam vol 40/27, may be a related design for a circular court divided into four quadrants by four pavilions, and with an arch in the middle of each quadrant.
38 Drawings are in SM, Adam vol 34/64–7. The internal angles in this court are 157.5 degrees, which, as explained in note 12, can be constructed with a compass and straight edge.
39 I am very grateful to Dr Frances Sands at Sir John Soane's Museum for her help here. Bail (or bale) stalls are discussed in Worsley, G 2004 *The British Stable*. New Haven and London: Yale University Press, 41–2.
40 Colvin 2008, 987. These offices survive, although they are omitted in Gifford and Walker 2002.
41 There are three inscribed drawings for this in SM, Adam vol 46/152–4. Two are reproduced in Fig 8.8 (top and centre), and the third is a first-floor plan; Figs 8.9a–b are based on these. SM, Adam vol 1/171, is uninscribed but has the three elevations on which Figs 8.9c–e are based. SM, Adam vol 1/223 is the uninscribed plan in Fig 8.8 (bottom) on which Fig 8.9f is based.
42 Bolton 1922, vol II, index 67.
43 This brief history is based on Grant, J 1882 *Cassell's Old and New Edinburgh: Its History, Its People, Its Places*. London: Cassell, Petter, Galpin, vol III, 148–50; Gifford *et al* 1984, 281–2; Cosh 2003, 182 and 263; and www.assemblyroomsedinburgh.co.uk/about-us/the-origins [accessed 20 September 2016].
44 Drawings are in SM, Adam vol 31/33–8.
45 In 1817–18 William Burn added a portico. In 1843 Burn and David Bryce rebuilt the rear part of the building as a concert hall, but Historic Environment Scotland has a plan of the original building, EDD/95/24, which shows that David Henderson included Adam's proposed rear bowed projection.

9

The Adam brothers and Portland Place: a reassessment

Colin Thom

Portland Place, in the Marylebone district of London, was by far the biggest of the Adam brothers' various street improvement schemes to reach fruition, but it is also the least understood. Thanks largely to the labours of Alistair Rowan, we know much about their two other great speculative developments in London, at the Adelphi (1768–72) and Fitzroy Square (1790–4), and there is a considerable body of knowledge regarding their other planned street improvements in Edinburgh and elsewhere. For Portland Place, however, there has been little appreciation of the Adams' role and intentions, despite the survival of original houses in significant numbers. Several fundamental questions, for example as to its authorship, chronology and full extent, have not, until now, been properly addressed.[1]

All four Adam brothers shared to varying degrees their father's instincts and abilities as a speculator. John, the eldest sibling and leader of the family practice after William Adam senior's death in 1748, was naturally the first to show such tendencies, as has been touched upon in the Introduction to this volume. William junior, the youngest – who trained as a banker and merchant from an early age – was by 1758, at only 20, already a senior partner in a forceful agents' business in London (*see* p 5). But when the Adam brothers' architectural design and construction practices in London were operating at full tilt in the mid- to late 1760s and early 1770s, it was clearly Robert and James who were the arch speculators.

The huge Adelphi development, as is well known, was a joint project undertaken by all four brothers through their builders' merchants and contracting firm William Adam & Co, founded in 1764–5, in which they were equal partners.[2] But it was a scheme evidently devised and driven ahead by the two architects in London. So much is clear from Willie's oft-quoted letter of February 1768, excitedly informing big brother Johnnie in Scotland that 'Bob and Jamie have secured Durham Yard for the Brotherhood ... which you will say is a bold stroke'. In a later letter between the same parties, written during the tense period of family wrangling and dispute in the 1780s, Willie confirmed his and Johnnie's peripheral involvement, reminding his older brother that Robert and James had 'consulted' them both about the Adelphi, which was then 'universally esteem'd a very promising Undertaking'.[3] Robert even took plans of the scheme to Scotland to show Johnnie; for though occasionally sneering and dismissive of what he regarded as his older brother's outdated 'Scotch' tastes, he valued John's knowledge and experience of large-scale building projects.[4]

Willie was using the example of the Adelphi to help explain to Johnnie his interpretation of the family's very different approach to Portland Place – something John seems to have either misunderstood or perhaps (more likely) been deliberately misinformed about – that it was an initiative which Robert and James had kept to themselves, at least to begin with. Audacious and ambitious though the Adelphi undoubtedly was, it now seems doubly so if we consider that it was begun *after* the two architect brothers had engaged with the 3rd Duke of Portland to develop a large swathe of his estate in Marylebone as Portland Place, and not before, as was previously thought.

Before the Adelphi? Robert and James Adam and the Portland Estate

> It was enter'd into a considerable time before the Adelphi occurr'd – It was made by Robert & James Adam in their own names, without consulting either You or me upon the Subject, and I declare I consider'd that transaction, as distinct as their Architecture business, from any thing in which I was engaged.[5]

So wrote William Adam to John Adam in 1787 of the brothers' first Marylebone building agreement. That agreement was made in October 1767 in James Adam's name only; it has not survived, but is referred to in a much later lease from the Duke of Portland for a house erected on part of the ground covered by it, in Duchess Street.[6]

It has become customary to credit James Adam with the design of Portland Place – wrongly in my estimation – and we will return to this subject shortly. But he emerges from surviving documents as the chief negotiator with the Portland Estate, certainly at first – corresponding in 1767 with the Duke's steward Thomas Isatt (d 1771) about the particulars of the agreement, and signing this and all the early building leases as the Adams' sole representative.[7] Even as late as February 1774, when the development was under way, it was Jamie whom the 7th Earl of Findlater, writing from Aix in the south of France, urged his mother to speak to in order to secure a piece of ground at Portland Place for a new house.[8]

This conforms to what we know of the brothers' architectural practice and personal relations. Robert had launched the Adam style in England almost single-handedly in the early 1760s, while Jamie was on his grand tour, and he was always the undisputed leader in terms of design – something Jamie was acutely aware of from the outset.[9] But Jamie was of great importance to Robert, as ally, *confidant*, sounding board and, not least, as an extra supervisor in the busy Adam drawing office. Additionally he was a clever and effective negotiator, as is shown by his acquisition of the great Albani collection of drawings in Rome for George III from under the noses of other antiquaries, and also his successful wooing of the Royal Society of Arts as early takers of a site at the Adelphi for their new headquarters.[10]

James Adam understood the potential offered by speculation, praising while still in Italy in 1761 his brother John's 'forsight' in buying ground in Scotland on which to build villas when the market improved, and bemoaning the scarcity of similar land 'about London'.[11] Perhaps there is some truth in their brother-in-law John Clerk of Eldin's suggestion that the Adams' involvement in speculative building and 'Ornamental business', which began in the mid-1760s after Jamie's return, was 'particularly moved by James'. Sometimes his ideas for new projects were novel or prescient but went unfulfilled; at other times, as we shall see, they were misguided and too hastily pursued, dragging the Brotherhood deeper into dangerous financial waters.[12]

But whichever brother was ultimately behind the Portland Place endeavour, it and its sister project at the Adelphi were essentially driven by Bob and Jamie's desire to make a lasting impression on the capital. Starved of the important royal and public commissions they craved, which so often went to their rival, the royal favourite William Chambers, they turned instead to private town-planning speculations as a means of expressing their architecture on a grand scale.

At Marylebone, the ground covered by the Adam scheme belonged to the estate of William Henry Cavendish Cavendish-Bentinck, 3rd Duke of Portland (d 1809), a future prime minister. Formerly known from its previous owners as the Cavendish–Harley estate, this well-situated property on the fringe of London's West End had come to the Duke via his father's marriage in 1734 to Lady Margaret Cavendish Harley (d 1785), the sole heiress to these lands, which she inherited in 1755. Development of the estate had begun in the mid-1710s with a grid of fashionable streets centred on Cavendish Square and had stuttered on in fits and starts during the 1720s and 30s. After something of a hiatus in the 1740s, building was on the rise again in the 1750s and 60s under the new owners.[13]

What is sometimes overlooked is that the Adam brothers' development covered much more than Portland Place, which was its focus and showpiece. Many other buildings in the adjoining streets were constructed at the same time (Fig 9.1). Mansfield Street, for instance, had ten big, Adam-designed houses facing each other in two short terraces, as well as (at its south end) a mansion designed by Robert Adam for his friend General Robert Clerk; and Chandos House, a speculative gentleman's town house, built and lavishly decorated by the brothers in the hope of luring an aristocratic big fish to their estate. For a time, too, in the early 1770s, there were plans for an Adam-designed town 'palace' for the Duke of Portland, looking down Mansfield Street from the north side of New Cavendish Street, but this never came to pass. All the adjoining frontages of the narrower cross streets to east and west – such as Devonshire, Weymouth, New Cavendish and Duchess streets – were included in the land taken by Robert and James, and were built up under them with smaller houses, though not to their designs. The same was also true of most of what is now Hallam Street (formerly Charlotte and Duke Streets, *see* Fig 9.1) and the west side of Great Portland Street (originally Portland

Fig 9.1
Portland Place c 1800, showing the full extent of the Adam development. The red dashed line indicates the division between the land taken by the brothers under the 1767 building agreement, to the south, and that to the north, agreed in 1776.

Road and Street). So this was a vast undertaking, of around 28 acres, with at least another 200 houses and 150 or so coach houses and stables in mews streets, in addition to the 68 fine residences of Portland Place.[14]

The timing of Jamie's negotiations with the Portland Estate in 1767 was crucial. Not only was it a boom time for London housebuilding, but the ground at Marylebone had only recently become available for development through an agreement by Act of Parliament earlier that year, which ended a long-running dispute between the Duke of Portland and his tenants, the Barons Foley. The dispute related to Foley House – a detached suburban villa built by Lord Thomas Foley in 1755 on what were then empty fields north-east of Cavendish Square (Fig 9.2, *see* Fig 9.1), on the site now occupied by the Langham Hotel – and more particularly to the still vacant land to its north. Foley had managed to secure from the 3rd Duke's parents an interest in this land, so as to protect his view towards Hampstead and allow him first refusal on any future development; but when the Duke inherited the estate he refused to honour this agreement. Foley's death in 1766 paved the way for a compromise under the 1767 Act, by which his heir and successor (another Thomas Foley, a cousin) gave up any claim to this ground, but in return was guaranteed that no future buildings would interfere with his view north for the full width of Foley House (*see* Fig 9.2). Hence the unusual size of Portland Place – at around 125ft thought to be the widest street in London – which contributed so much to the majesty of the Adams' design.[15]

Planning vision

Very few original plans or elevations survive for Portland Place, and those that do tend to date from the mid-1770s or later, making it difficult to determine Robert Adam's original vision at the time of the brothers' first building agreement with the Duke of Portland in 1767. Also, that early agreement covered a little over half of the ground eventually taken by the Adams (*see* Fig 9.1); a second agreement, of 1776, was required for the northern part of the development. It seems certain that from the outset the Adams' intention was to take the whole ground and build on an extensive scale. In February 1772, before any work had begun in Portland Place itself, Robert Adam discussed 'in great confidence' with Peter Burrell, Surveyor-general of Crown lands, the possibility of transferring to the Duke and adding to the development the piece of Crown land at the top of Portland Place, next to the Marylebone Road (now the site of John Nash's Park Crescent, *see* Fig 9.1). 'From my description of the new proposed streets in that quarter', Adam told the Duke, '[Mr Burrell]

Fig 9.2
Portland Place c 1840, looking south towards Foley House, watercolour by T H Shepherd.

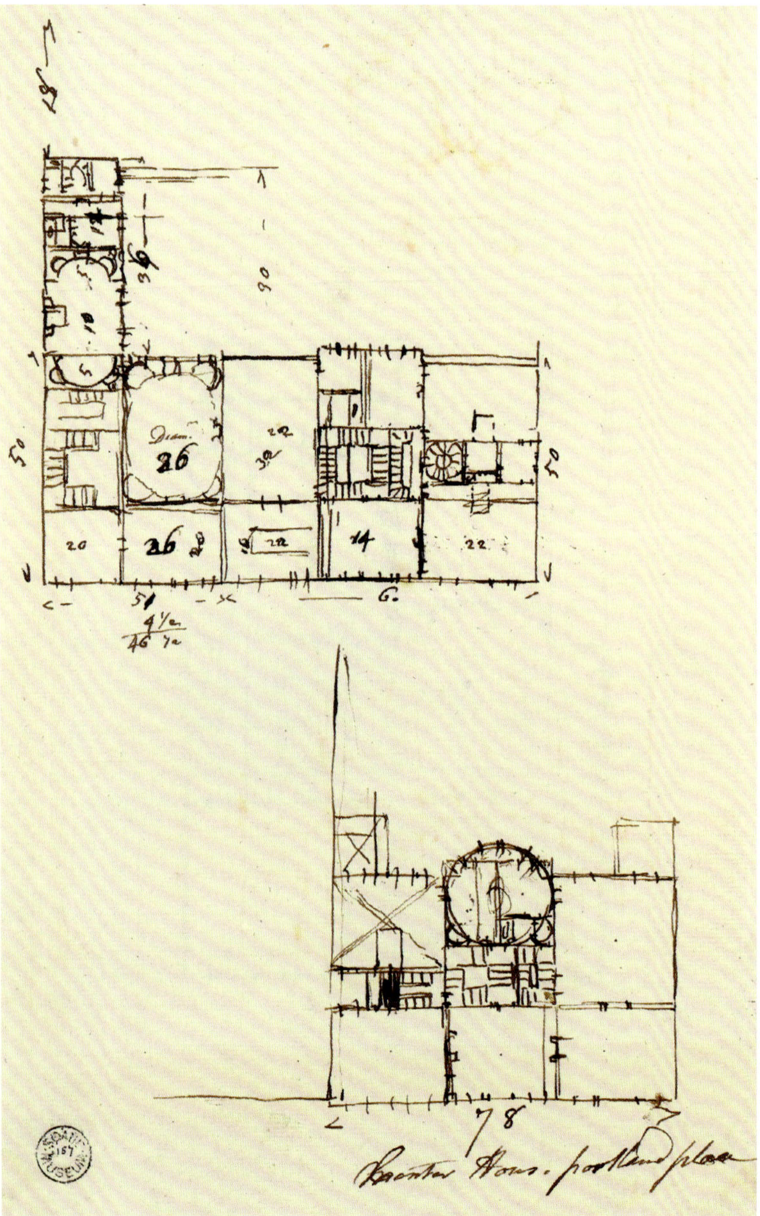

seemed very desireous of promoting every improvement there.'[16]

In the end the Crown retained its land, and the exact nature of Adam's 'description' is not known. It would seem that, from the beginning, his desire was to create a progressive, mixed social environment, similar in ethos to the Adelphi, but without the industrial–commercial element there and with more of the bigger sorts of houses aimed at the upper classes, offering elaborate interiors, as befitted a West End location. This was certainly what was under way on the Portland Estate by 1769 in the earliest phase of Adam construction in and around Mansfield Street. There, two mansions – Chandos House and the large, adjoining residence for General Robert Clerk – rubbed shoulders with two short rows of terraced houses and two mews streets. The same thinking permeates the earliest known Adam designs for Portland Place itself, of *c* 1772–4. Two very grand detached 'palaces' with gardens were planned for the east side, for the Earls of Findlater and Kerry (Fig 9.3), on plots of around 200ft frontage. A similarly ambitious town residence for the Duke of Portland was to stand on the west side, moved there from New Cavendish Street. And there are also sketch plans by Robert for imaginative terraced houses on the west side, on an impressive scale of around 50–60ft frontage (Fig 9.4), with a commanding 'Center House' 78ft wide – these were later worked up into finished drawings by an Adam office draughtsman.[17] For various reasons – most notably the after-effects of the 1772 banking crash, the poor liquidity of the clients involved, and the increasing political tension in America – none of these large houses got beyond the drawing board. So the concept of a diverse development, embracing detached residences spread amid terraced houses, was only achieved on a limited scale at Mansfield Street – though even in its revised form, the Portland Place development still offered groups of terraced houses and mews buildings of various sizes.

One aspect of the Adam brothers' planning for Portland Place that remained constant was its relative seclusion and privacy. Its rare width undoubtedly contributed to this, but far more important was the fact that it was not originally a through road. Our modern conception of the street is guided by Nash's later additions of Regent Street to its south and Park Crescent to its north, which made it part of one long, continuous (and very stylish) thoroughfare leading all the way to the Marylebone Road and beyond. But during the Adam period Portland Place was separated from the Crown lands to its north by gates and fences, and from the other streets to its south by Foley House and its grounds (*see* Figs 9.1, 9.2). Access was only possible via the side streets to the east and west, so residents could enjoy what was in essence a private enclave, with much of the character of an elongated square.

As already noted, the conventional scholarly assessment is that Portland Place in its final form was Jamie's design, Robert Adam having supposedly lost interest when his so-called *Strada di Palazzi* failed to materialise, preferring instead to focus on other, private commissions – though there is precious little evidence to support this interpretation. For a start, as well as the early designs discussed above, there are many later drawings for the development in Robert's hand (Fig 9.5), including several of 1783–9 for an unexecuted villa of nearly 100ft frontage for the Earl of Findlater (Fig 9.6), for vacant ground at the north end of the street; also, the intricate planning of some of the terraced houses erected points to his continued involvement.[18] Secondly, there were in fact very few other commissions to distract him: the brothers' own correspondence of the 1780s talks of a 'great falling off in the Architecture Business', to the extent that Robert complained they had been unable to pay their clerks and draughtsmen for several years. As a result, Jamie flung himself into new business speculations, and spent ever more time on a farm he had acquired at Shenley Lodge, Hertfordshire, where he lived, eventually publishing a two-volume study on the practice of agriculture – none of which suggests a man at the centre of a major London improvement.[19] In the winter of

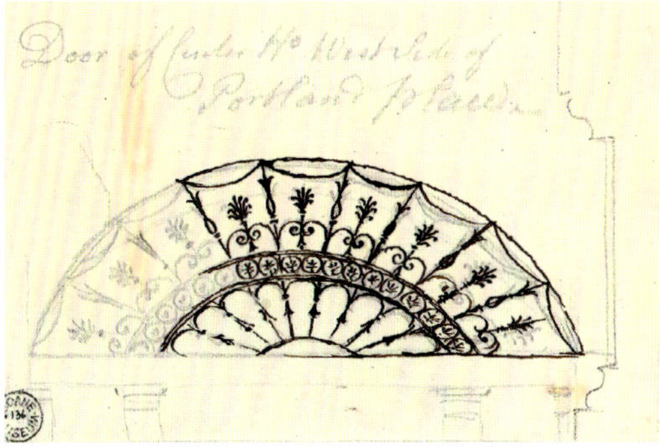

Fig 9.3 (opposite top) Robert Adam's unexecuted designs for Kerry House, c 1773, one of the three large detached mansions originally intended for Portland Place, of about 200ft (61m) frontage.

Fig 9.4 (opposite bottom) Early plans by Robert Adam for terraced houses on the west side of Portland Place, c 1773.

Fig 9.5 (above left) Fanlight design by Robert Adam for 37 Portland Place, largely as executed.

Fig 9.6 Findlater House, August 1783, a large villa on the west side of Portland Place, of 97ft (30m) frontage, for the Earl of Findlater and Seafield, designed by Robert Adam. Unexecuted.

1788–9 it was Robert, not James, who was still trying (unsuccessfully) to persuade the Earl of Findlater to proceed with his house, and was negotiating with the builder James Gibson, the owner of the adjoining plots, who had offered to begin the foundations. The tone of these letters suggests Robert was by then working largely alone, with little in the way of support, one of his clerks having just left him. What we know of the Adams' increasing focus on Scottish commissions at this time confirms this, as there Robert seems to have been labouring almost single-handedly, with the support of successive assistants and clerks of works in his Edinburgh office, and occasionally with his youngest brother Willie or John Clerk of Eldin in tow. Following Robert's death in 1792, Jamie had to be coaxed back and briefed comprehensively on the Adam projects then in hand, so prolonged had been his detachment from the 'architecture part of the business'.[20]

A 'very precarious Aspect': leasing, finance and slow progress

At Portland Place, as at the Adelphi, the Adams operated as architects, developers and builders simultaneously – a holistic, modern approach to development normally associated with later entrepreneurial contractors like Sir Thomas Cubitt.

House construction was undertaken via the speculative leasehold system common to London in the 18th and 19th centuries, whereby Robert and James, having agreed with the Duke of Portland for two large parcels of land, divided them into house plots which they then sublet to tradesmen and craftsmen. These sub-lessees built and fitted out the houses to the Adams' specifications and received a leasehold interest as payment in kind. Many were men who had worked for or with the Adam brothers before – for example the plasterer Joseph Rose junior, mason John Devall junior, and carpenter John Hobcroft – and were presumably keen to participate in (and hopefully profit from) such a large and progressive London development. Sometimes other individuals took leases of several plots and acted as a second rung of small-scale developers, under the Adams, and as intermediaries with the builders. There was no direct labour force at Portland Place, as there had been under William Adam & Company at the Adelphi, where apparently as many as two to three thousand men, the bulk of them brought down specially from Scotland, were thought to be involved in the construction process.[21]

At Marylebone, Robert and James reserved some of the best plots for themselves. These were the sites of: Chandos House; the corner houses at 15 and 22 Mansfield Street; 17 Portland Place (another corner house); and 37 and 46–48 Portland Place – the striking, stucco-fronted houses at the centre of the new street. These sites they then transferred to William Adam & Co, as, according to Willie, it was thought 'a very eligible Speculation to build some Houses there on our own Account, especially as it helped greatly to extend our Connections in the Brick & Timber business'.[22] This moving around of property and assets between one branch of the business and another is entirely typical of the Adams and inhibits a clear understanding of their financial position at any one time.

They had set to work on General Clerk's house and Chandos House by 1768–9, and these and the terraced houses of Mansfield Street were mostly complete by the time of the Adelphi Lottery sale in 1774 – in fact Chandos House and the two Mansfield Street properties were included in that sale.[23] In Portland Place, however, the business of building and selling houses was perilously slow, and hampered by continual delays. Work there seems not to have begun until 1775, after the completion of the Adelphi and the lottery sale, by which time the proposals for large detached mansions for the earls of Kerry and Findlater had been dropped. By 1776 leases had been agreed between the Adams and their builders for all the Portland Place houses as far north as Weymouth Street (*see* Fig 9.1), but only three houses had been finished and occupied by May 1778; sixteen more were then still under construction. The air of desolation in the street seems not to have troubled 'An Idler' in the *Morning Post*, who on strolling into Portland Place in July 1777 saw 'art, and industry ... well employed'. But another visitor at the end of the year noted: 'That sumptuous pile, called Portland Place, where the houses were expected to fetch six hundred per ann. each, is at present as destitute of inhabitants as the deserts of Arabia.'[24]

The brothers' own financial difficulties, as we shall see, were a factor in these delays, but so too were the unfavourable economic conditions at the time. Often thought of as lucky as architects, the Adams were unfortunate in their timing when it came to housing speculations. They took on both the Adelphi and Portland Place in 1767–8 at the height of a London building boom when all seemed set fair and prospects were good. Very soon that surge abated and things began to slide, taking a serious turn for the worse with the banking crisis of 1772. Despite a rally in the mid- to late 1770s, when building in Portland Place finally began, the economy then dipped even lower into a prolonged recession in the late 1770s and 80s with the escalation of the revolutionary war in America and the entry of France on the American side.[25] One immediate effect of these delays was the accumulation of a substantial debt to the Duke of Portland. Under the terms of their building agreements, the ground rents that Bob and Jamie were paying him were low for the first few years of their lease, when they would hope to see houses completed and sold, but rose steeply in ensuing years to a very high figure indeed – more than they were able to pay.[26]

As the work proceeded slowly from south to north, the brothers tended to mortgage their houses at optimistic, pre-crash valuations. With prices falling and credit harder to secure, this made it increasingly difficult for the Adams to dispose of their houses, many of which remained vacant or unfinished for years. The fine stuccoed and pedimented pair at the centre of the east side (now Nos 46–48, *see* Fig 9.8) stood empty and forlorn for nine years, and was considered by Willie 'a dead loss' – he later told his brother Johnnie in Scotland that the 'Marybone Grounds wore a very precarious Aspect a long time after'. With the prospect of further losses to William Adam & Co from its involvement at Marylebone, Robert and Jamie in 1777 transferred to the company's account almost £15,000 they had made from the sale of improved ground rents there.[27]

Although they reckoned themselves 'licked whole again' by the success of the Adelphi Lottery, the London Adam brothers' financial stability was still far from certain, founded on shifting sands, and shored up by sizeable unrepaid loans from big brother John in Scotland. More losses from a series of rash business speculations in the 1770s, undertaken while Portland Place was under construction, sapped their shaky resources and morale. These were driven forward by the London brothers in the mistaken expectation of vast profits that could right the sinking ship. For example, their disastrous involvement with the Swiss clergyman John Liardet, whose patent for a new but ultimately faulty brand of stucco they acquired in 1774, brought only litigation, more debt, further discredit and the bad feeling of fellow building tradesmen.[28] Less well known is the agreement made by all four brothers in 1771 with the London-based Scottish merchant Robert Campbell, as partners in a paving business known as Adams, Campbell & Co. This company held leases of several stone-quarries in Scotland and Guernsey, and won many contracts in London for paving new roads. But Campbell had dabbled in too many other ventures, including the purchase of land in St Lawrence to open up trade with North America, and when he died in 1776 left the Adams encumbered by his debts, calculated by Willie at over £15,000.[29]

Then, in typical Adam fashion, Jamie became embroiled in another speculative business venture at the end of 1779. This was a factory manufacturing saltpetre at old Sandford manor house, beside a creek in the river at Sand End in Fulham, recently acquired by Peter Theodore De Bruges, a French chemist, with other mostly French associates. De Bruges, who claimed to have discovered a new and improved method of saltpetre production, needed assistance with buildings on the site and so 'applied to Jamie'. No doubt enthused by the possibilities of involvement in the lucrative arms trade, Jamie acquired the premises. The Battersea & Sand End Company, as it became known under the Adams, never got off the ground, and predictably the affair simply plunged them deeper into debt.[30] The enormous scale of the works at Fulham and a related barrel works at Battersea delayed production and swallowed up yet more money. In the end De Bruges fled with the extra capital and the company was wound up at a loss of around £30,000. Though it was Bob and Jamie's initiative, they had dipped into company resources to help fund it and also encouraged others to acquire shares. Even John's prudent son, the lawyer and politician William Adam of Blair Adam, was persuaded to invest around £2,500. Willie tried to assuage Johnnie by arguing that its success would have allowed them to repay his loan, but, like much of the content of the family letters travelling south to north at the time, this may have been disingenuous.[31]

By 1780 the number of houses finished in Portland Place had risen only to eighteen, and it took another upturn in the building cycle in the mid- to late 1780s for the brothers to be able to complete it. Forty-two houses had been finished by 1790 but work on the last runs of terraces at Nos 66–90, on the east side, and 53–75, on the west side, had yet to begin. Most of these houses were not constructed until *c* 1791–5, and at least two of them were still incomplete in 1797 – by which time Robert Adam had been dead five years and James three.[32]

Fig 9.7
Adam office elevation for a terrace on the west side of Portland Place, at Nos 27–47, not exactly as built.

Fig 9.8
Nos 46–48 Portland Place.

Portland Place in the context of Adam street improvements

When work finally began on Portland Place itself in the early 1770s and the idea of detached houses stalled, the brothers fell back on the tried and tested formula of rows of palace-fronted terraces, for which a single office elevation survives (Fig 9.7). In general these terraces followed the customary Adam model for street improvements, each of the blocks having a symmetrical temple front, with extra emphasis added at the centre and ends. As suggested by the office drawing, the grandest central houses at Nos 37 and 46–48 were picked out with stucco, giant order pilasters and pediments – a refinement of the earlier 'Center House' concept (Fig 9.8). Only here was there any additional, applied decoration – in the form of stucco figurative roundels of Classical tableaux and rectangular panels of paired griffins and urns, occupying the space between the first and second-floor windows, with a further roundel in the pediment (*see* Fig 9.8). Characteristically, Adam and his assistants based these on antique Roman models. On the surviving houses at Nos 46–48, the second-floor roundels depict the sun god Sol driving his *quadriga* and the moon goddess Luna in her two-yoke chariot (or *biga*), as found on the short east and west ends of the 4th-century AD Arch of Constantine in Rome, but reversed – suggesting a printed source, such as Montfaucon's *L'Antiquité expliquée et représentée en figures* (Figs 9.9 and 9.10).[33] The use of paired griffins or sphinxes is common in Adam's work; a possible source here is the Roman frieze of ancient griffins from the Forum of Trajan, as reproduced by Piranesi in his *Vasi, candelabri, cippi* (published 1778, Fig 9.11), and for which a drawing by Nicolas-François-David Lhuillier survives in the Adam collection at the Soane Museum.[34]

Elsewhere in Portland Place the Adams relied on subtle variations for effect. There was considerable inherent diversity in any case, given the

Adam source material:

Fig 9.9 (left)
Roman relief sculpture of Sol in his quadriga, *from the Arch of Constantine.*

Fig 9.10
Bernard de Montfaucon's L'Antiquité expliquée et représentée en figures *(vol V, 1719); Sol in his* quadriga *(far left) alongside Luna in her* biga *(left).*

Fig 9.11 (left)
Piranesi's engraving of the griffin frieze from the Forum of Trajan, from his Vasi, candelabri, cippi *(1778).*

range in size of the terraces – which had to be varied in order to accommodate connections with the side streets, laid out to continue the routes of pre-existing roads (*see* Fig 9.1). At either end were shorter runs of three or five houses, treated mostly in brick but with stucco pilasters and a pediment marking the centre house (Fig 9.12), and (in the southern ranges) slightly projecting frontages to those at either end. As shown in Shepherd's view, of *c* 1840 (*see* Fig 9.2), the other, longer terraces at Nos 49–69 and 86–90, which also incorporated stucco pilasters and pediments on their central houses, had their end 'pavilion' houses stuccoed entirely – those on the east side with pilasters but no pediments, those on the west with pediments but no pilasters.

There is obvious continuity here with the architecture of the Adelphi, especially in the flat, almost two-dimensional surface treatment of the elevations, with only the slimmest and most delicate of projections and recessions at the centres

Fig 9.12
No 28 Portland Place. The entrance porch, ground-floor rustication and upper floor are all later alterations. The pediment and its cornice would formerly have been above the second-floor windows; compare with Fig 9.8.

and ends. There is none of the greater plasticity or 'movement' found in many later Adam street schemes, as at nearby Fitzroy Square, or Charlotte Square in Edinburgh, where the elevational forms are much more boldly modelled – a consequence, perhaps, of designing terrace fronts in stone rather than brick and stucco.

A similar application of neoclassical elements, such as the use of pedimented centres and ends within monumental terrace façades, was to resurface in Robert Adam's later designs of the 1780s and early 90s for town planning schemes in Edinburgh – most notably in his visionary plans for the South Bridge improvement of the later 1780s (*see* Fig 0.18).[35] What appears novel though at Portland Place was the rhythmic counterbalancing of stucco and brick within the regular façades, so clearly evident from Shepherd's view (*see* Fig 9.2). As they had done at the Adelphi, the Adams at Portland Place made a distinctive contribution to London's street architecture, and one that was picked up almost immediately only a short distance away in Bedford Square (built *c* 1775–80), on the Duke of Bedford's Bloomsbury Estate, where the four sides of the square repeat the Adam model of symmetrical brick Classical terraces with stuccoed, pedimented houses at the centre.[36]

Portland Place and Mansfield Street interiors: mass-production of the Adam style?

The Adelphi when completed could boast many fine interiors in the Adams' characteristic manner of the later 1760s and early 70s. These were largely concentrated in the grandest houses of the Royal Terrace, which enjoyed a riverfront location and were aimed at the better-class residents, though intricate ceiling designs could also be found in the earliest houses in Adam Street. Elsewhere, in the smaller blocks of Robert Street and John Street, decorative features such as ceilings and chimney pieces tended to be in simplified, or distilled versions of the Adam style, with the chimney pieces often of timber.[37] In contrast, at Mansfield Street and Portland Place – especially in the earlier houses at its southern end – Robert and Jamie seem to have upped their game, providing much more in the way of elaborate internal decorations, and often of an exceptionally high quality. This may seem surprising for such a large speculative street development, but it indicates their aspirations to attract the upper echelons of society.

Fig 9.13
A Portland Place ceiling painting: roundel of Bacchus and Ariadne, thought to be by Cipriani. From the Adam first-floor ceilings at 23 Portland Place (demolished), installed in the 1930s at the present No 23.

These were sizeable town residences, usually with three large reception rooms to each of the main floors, bed and servants' rooms above, kitchen and other facilities in the basements, and stables and coach houses in the mews streets behind. All the same they were not of a scale to accommodate the large decorative items that were such an important feature of Adam's earlier, aristocratic country and town-house commissions. For instance, the entrance halls usually adopt his characteristic restrained approach for such spaces, with simple plaster mouldings and a subdued palette, but there are none of the big military trophy panels as found at Osterley Park or Newby Hall, or the anteroom at Syon. Nor are there any of the large panels of figure reliefs or plasterwork grotesques, as found extensively at Bowood, Lansdowne House, Harewood House or Osterley. Similarly, expensive painted wall panels or *capricci* on a large scale, like those provided by Antonio Zucchi and others for the private commissions, would have been inappropriate; any figurative paintings at Marylebone tend to feature only in ceilings, and then generally on a small scale (Fig 9.13). Adam did devise a particular form of plasterwork decoration for the walls of the staircase halls – the fulcrum of the sequence of reception rooms – that can still

be seen in Mansfield Street and Portland Place today (Fig 9.14). The most common motifs are plaster figures of dancing Bacchantes, often with cymbals or tambourines, or other mythological characters, usually in oval or circular medallions – all surrounded by framing patterns of light and delicate plasterwork swags and festoons, formed of husks, anthemion medallions and other foliate ornaments.[38] Sometimes the decorative elements are set within large rectangular panels, at others they are simply plastered directly on to the walls. The effect is to emphasise the importance of the staircase to the success of these smaller town houses as venues for the high-society 'routs' and 'levees' that characterised the London season, and its central role in the strictly choreographed route that guests took through the house on such occasions. Two good examples are Mrs Broadhead's 'grand Masquerade' at No 27 in May 1792, and the Hon Mrs Wyndham's 'Gala' a few doors up at No 33 in April 1799. At the former, over 1,000 fashionable guests had descended upon the house by one in the morning, including the Prince of Wales and Mrs Fitzherbert, disguised as 'black dominos', Mr T Vaughan as a Surrey farmer, 'with a simple female rustic on his arm' (thought to have been one of his 'admired daughters'), and Mrs Este as an armed 'but rather overgrown' Cupid. Mrs Wyndham's post-Opera event attracted 'near five hundred of the Nobility', mostly 'dressed in white, their head dresses ostrich feathers and diamond sprigs'. After dancing reels, Strathspeys 'and other quick movements', guests paraded downstairs to enjoy supper, delivered in three sittings on tables laid out in the front parlour and adjoining library with delicacies transformed into unlikely tableaux: 'neat's tongues [the tongues of oxen or cows] in the resemblance of dolphins, swans and rein deers', and 'hams, decorated at the top with sailors'.[39]

Some of the plaster figures of Bacchantes in the stair halls look very similar from house to house, and were no doubt cast in moulds on site or in the workshop by Joseph Rose & Co, and applied to the walls along with other cast items, such as the decorative string courses and bands, as well as hand-modelled elements. Many of the sculptural figures and panels found throughout Mansfield Street and Portland Place were repeated in this way. This 'mass production' aspect to the Adams' work was to a degree disguised by their skill (and that of the Roses) in mixing and matching the various decorative elements in limitless combinations – as with the identical relief panels of pudgy dancing *putti* incorporated into very different drawing-room ceiling designs at both 19 and 46 Portland Place (Fig 9.15).

Of course, these and the other favourite building blocks of the Adams' interior style – the friezes, borders, string courses and plaques composed of repeated Classical motifs such as anthemia, husks or animal heads – lent themselves admirably to the mass-production methods then coming into vogue through manufacturers like Josiah Wedgwood and Matthew Boulton, and were no doubt partly designed with that capacity in mind.[40] More evidence for the repetition of plasterwork items such as friezes and overdoors comes from the younger Joseph Rose's sketchbook, now at Harewood House, in which he occasionally notes on the drawings comments such as 'Done in Severall Houses in Portland

Fig 9.14 (opposite) Staircase decoration at 33 Portland Place.

Fig 9.15 Details of first-floor ceilings at Nos 19 (below) and 46 (bottom) Portland Place, the latter designed by Joseph Rose senior.

Place'. So, too, with some of the Adam designs for chimney pieces, which, though generally intended for particular rooms in particular houses, occasionally are marked as 'Chimney Piece for the Halls in Portland Place', or something equally generic. Reuse of chimney-piece designs had also taken place at the Adelphi: research by Brigid von Preussen has shown that there Adam and his draughtsmen were recycling and readapting some of their back catalogue of chimney-piece designs, often writing references to the Adelphi houses on the older drawings.[41]

Repetition of larger-scale decorative elements – for example entire plasterwork ceilings – is something found only very rarely in Adam's *ouevre*, such was his facility for invention and variety. But there were several examples in Mansfield Street and Portland Place where the designs were extremely similar (Fig 9.16), and, in at least once instance, apparently identical.[42] As to the

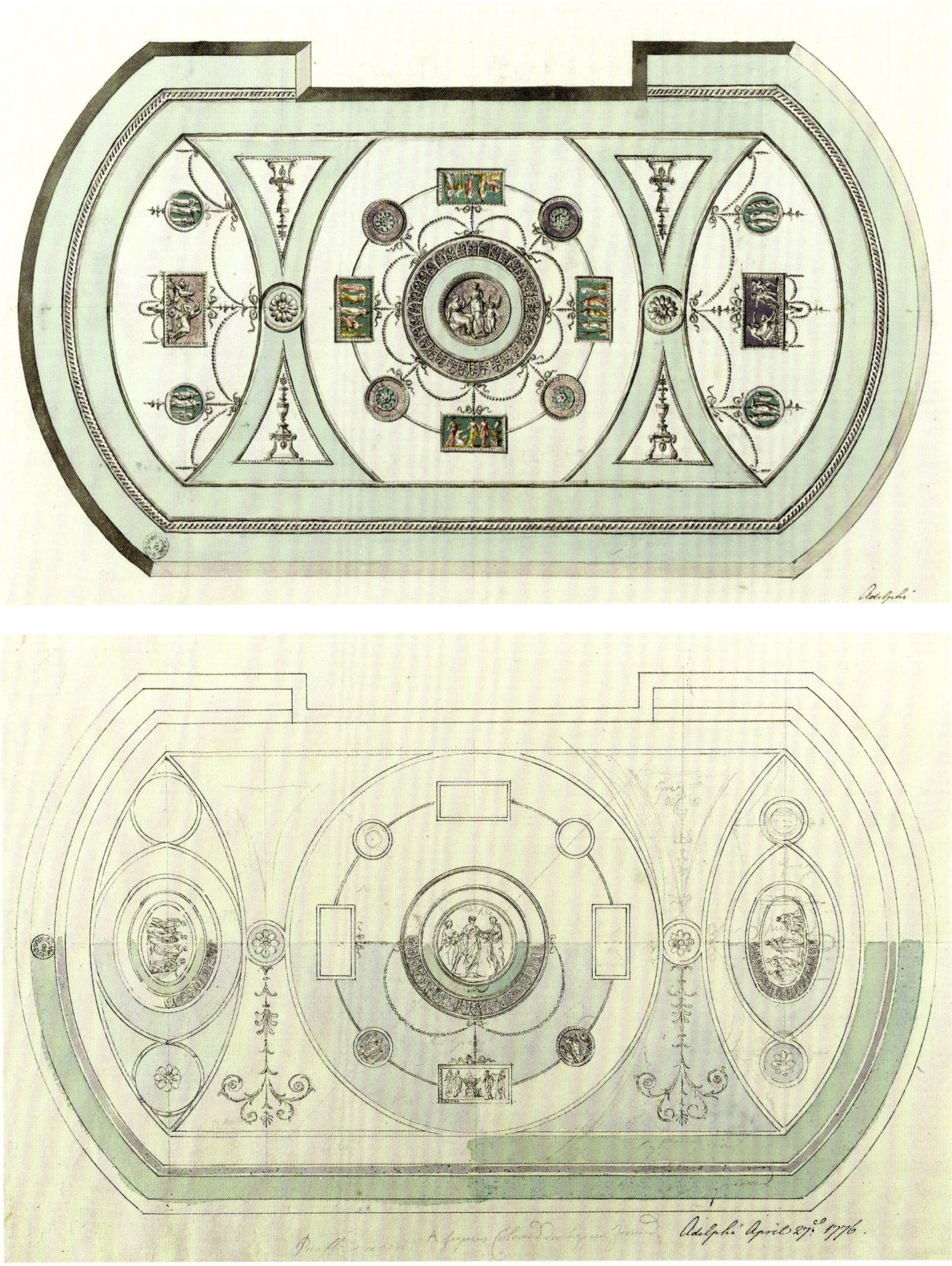

Fig 9.16
Adam Portland Place ceiling designs: for the rear drawing rooms at No 48 (top) and No 33 (bottom).

manner of the ceiling designs, Damie Stillman explained many years ago that by this date, in the mid- to late 1770s, they had become shallower in profile when compared to Robert Adam's earlier and heavier, more rococo-influenced work – more linear and geometric in form, and more refined. Patterns based on antique Roman and early Christian ceilings became increasingly common, as either seen and recorded by Adam on his grand tour, or otherwise readily available to his draughtsmen via antiquarian publications or the accessible Topham collection of drawings after the antique at Eton (*see also* pp 110–11).[43] A favourite Adam device was the central feature of a curved cross within a circle, derived from the vaulted ceilings of early Christian catacombs and Roman tombs.[44]

One notable aspect of the ceilings at Portland Place is the degree to which the Adam brothers' plasterers Joseph Rose, senior and junior, were involved in their design, as well as their installation. There are several drawings in the latter's sketchbook, labelled 'Mr Rose's Desine', and, in one instance (the rear drawing room at No 46), 'Old Mr Rose's Desine' (*see* Fig 9.15, bottom). In terms of their style, these ceilings differ little from those emanating from the Adam office, relying on a similar stock of antique ornaments, but overall they tend to lack the refined elegance and sense of proportion of the best Adam work. Historic photographs of interiors elsewhere in the street, including at its north end, suggest that the Roses' involvement in designing plasterwork for the Adams may have been more widespread at Portland Place than was previously thought – perhaps a consequence of their financial difficulties and the delays in progress.[45]

Surviving ceiling designs for Portland Place suggest Robert Adam's increasing use of varying shades of green to decorate the principal reception rooms of these houses. Many have a pale green as their main body colour, highlighted with accents in lighter and darker greens – examples of the 'green chiaroscuro' palette noted by Ian Bristow as common to Adam interiors of the 1770s (and see also Conor Lucey's analysis of Adam's designs for Headfort House in Chapter 6 of this book). In other ceilings a stronger contrast is provided by small areas of cream or grey-blue ground, and occasionally bursts of a dark purple for decorative borders and mouldings. Just occasionally a range of blues takes over and dominates entire ceilings, as in the Adams' own speculation at No 37, which became Viscount Stormont's town residence.[46]

Also worthy of note in these houses today is the number of original Adam chimney pieces and friezes to have survived (Fig 9.17) – though the latter are often obscured beneath several layers of paint. As with the ceilings, the variety found in the chimney-piece designs is remarkable, with as many as four or five carved marble pieces being provided for each house. Sometimes the degree of care and sophistication is surprising for speculative work – as with a delicately bowed marble piece designed to fit the curved walls of a rear room at 25 Portland Place (now demolished); sometimes more elaborate pieces were designed, such as those of *c* 1776 with delicate coloured and inlaid marble, intended for the front drawing rooms at Nos 34 and 43. As for the many friezes, each was individually suited to the overall decorative schemes of the main reception rooms in each house. By this date, Adam's frieze designs had become far more intricate and sophisticated than the simple running patterns of Greek keys or anthemions of his early houses, with elaborate repeating arrangements of urns, amphorae, cupids, paired sphinxes and griffins and many other Classical motifs – again arranged in endless variations and subtly executed in plaster (Fig 9.18, *see* Fig 9.17).

Fig 9.17
Adam chimney piece at 33 Portland Place. The frieze reflected in the mirror is the bottom of the five Adam frieze designs for the house illustrated in Fig 9.18.

Fig 9.18
Adam frieze designs for 33 Portland Place.

Conclusion

If we wish to understand what the Adam brothers were trying to achieve at Portland Place, and protect the surviving fabric, then we need to ensure that the extent of their work and influence there is recognised, and that Robert Adam is given due credit as its principal designer. The early years of the 20th century witnessed insensitive redevelopment in the street; and the destruction in the 1970s of the well-decorated pair of Adam houses occupied by the Chinese Embassy at Nos 49–51 to make way for a new building was in part fuelled by the attitude of Colin Penn, the new embassy's architect, who was happy to discard all the Adam fittings, despite opposition, because, he said, they were the work of a 'minor' architect (ie James Adam). Pressure for change remains strong in this affluent district of London, and today's historic buildings consultants, in their reports for developers, often like to stress (incorrectly) how little the Adams were responsible for in Portland Place – Robert in particular. Indeed, reports on buildings at the street's north end, for which there are no Adam drawings, stress that there is no evidence to connect them with Robert or James, placing any surviving fabric under increased threat. Again, this is incorrect, as family papers at Blair Adam include at least one agreement which states explicitly that builders in this part of the street must follow drawings supplied by the Adam practice – even down to the railings.[47]

Robert Adam was driven by a passionate desire to bring improvement to our towns and cities, often on a monumental scale, and he was willing to jeopardise his own and his family's financial security in pursuing that desire entrepreneurially – to speculate and take risks in order to make things happen. Indeed, like his father, he seems to have viewed these activities as an intrinsic part of his role as an architect. Though overlooked when it came to the big public commissions in London, the Adam brothers' impact on the capital through their speculative street schemes was considerable, transforming whole areas at the Adelphi, Portland Place and Fitzroy Square, which rank among their highest achievements. Portland Place today may be truncated and damaged, but its exceptional scenography and picturesque qualities were recognised by John Nash, who incorporated it into his Regent Street scheme of the 1810s–20s – the royal processional route from Carlton House to Regent's Park – giving the Adam brothers' work a deserved place at the heart of London's most outstanding piece of urban planning.

Appendix: Builders and craftsmen of the Portland Place estate

The list below notes all the known builders and craftsmen who were issued with leases in the Portland Place area between 1772 and 1792, either directly by the Adam brothers or by the Portland Estate via the Adams. The street names and dates for these leases are also given. For Portland Place itself and Mansfield Street the entries are italicised and the modern numbering given where known. (The Adams for a time retained the leases of several houses, eg Nos 17, 20 and 37 Portland Place, and Nos 15 and 22 Mansfield Street, so these do not appear here.)

Builders' and craftsmen's addresses at the time of the leases are Marylebone parish unless stated otherwise.

Architects

John Johnson (New Cavendish Street, 1775)
George Mercer (*32 Portland Place*, 1776)

Bricklayers

Joseph Babb (Great Portland Street, 1774)
John Brown (Duchess and Duke streets, 1778)
Edward Gray of St George, Hanover Square; and Queen Street, Grosvenor Square; then of Marylebone (*21 Portland Place*, 1775; Great Portland Street, 1775)
William Gray (Charlotte and Great Portland streets, 1775)
William Harley of Chigwell, Essex (*88 & 90 Portland Place*, 1791, with Richard Dole)
John Stone (Duke Street, 1775)
Simon Wakefield of St Pancras (Devonshire Street, 1781)
Jonas Wigley (*78 Portland Place*, 1791)
John Winstanley (*11 Mansfield Street,* 1772; Harley Street, 1772; Charlotte and Great Portland streets, 1775)

Builders

Hepburn & James Hastie (Great Portland Street, 1774–5; Charlotte Street, 1775; Cavendish Mews, 1776; *32 Portland Place*, 1776)
Thomas Hepstinstall (Devonshire Street, 1776)
William Phillips (*16 Mansfield Street,* 1773)
John White (Weymouth Street, 1779)

Carpenters, joiners and sash makers

George Buraston (Charlotte Street, 1776)
James Chapman (Devonshire Street, 1776)
Peter Day of St Giles in the Fields (Devonshire Street, 1791)

James Gibson (*37 & 39 Portland Place*, 1788; *?63 Portland Place*, 1790; Devonshire Street, 1791)
John Gibson of St James, Westminster (*38 Portland Place*, 1776)
Thomas & James Gibson (*41 Portland Place*, 1775; *49 Portland Place*, 1778)
John Grantham (*20 Mansfield Street,* 1772; Weymouth Street, 1779)
John Hall (*27 & 39 Portland Place*, 1775)
Lawrence Hall (Charlotte and Devonshire streets, 1776)
John Harrison of St James, Westminster (*80 Portland Place*, 1791)
James Hilson (*36 Portland Place*, 1776)
John Hobcroft, or Hobcraft (*18 Mansfield Street,* 1772)
Charles Jacks of St James, Westminster (Great Portland Street, 1774; Duke Street, 1779)
William Jaques of St George, Bloomsbury (Great Portland Street, 1774)
Samuel Loftus (Devonshire Street, 1781, 1784; Weymouth Street, 1789)
Robert McKenzie (Duke Street, 1775; Charlotte Street, 1776)
Nathaniel Moore of St George, Hanover Square (Great Portland Street, 1774)
Robert Otter (Charlotte and Great Portland streets, 1775)
James & John Reid (Weymouth Street, 1789)
William Stokes of St James, Westminster (Devonshire Street, 1776; Charlotte Street, 1778)
James Swinton of East Greenwich (*30 Portland Place*, *c* 1776)
Daniel Wassney (Weymouth Mews, 1788)
Joseph Webb (Charlotte Street, 1782)
John White (Weymouth Mews, 1777)
William Wood (Great Portland Street, 1774)
Robert Woodward of Kentish Town and Old Compton Street (Devonshire Street, 1789, 1791, 1793)
William Woolcott (*40 Portland Place,* 1776; *71 & 75 Portland Place*, 1790)

Carvers

Thomas Fusedale of St Giles in the Fields (Devonshire Street, 1777)
Solomon Hudson (*26 Portland Place*, 1775)
Thomas Nicholl (*5 Mansfield Street,* 1771; Great Portland Street, 1774; *22 Portland Place*, 1774; *15, 21 & 28 Portland Place*, 1775; Chapel Row, 1775)

Glaziers

William Cobbett of St George, Hanover Square (*9 Mansfield Street,* 1771; *25 Portland Place*, 1774)
Thomas Halliday (Great Portland Street, 1775)
William Thompson (Great Portland Street, 1774; Duke and Duchess streets, 1775; *24 Portland Place*, 1775; Devonshire Street, 1791)
David Williams (Charlotte Street, 1776, 1778; *42 Portland Place*, 1776; *64 Portland Place*, 1779; *76 Portland Place*, 1790; Weymouth Street, 1776, 1778, 1779, 1789; Williams's Mews 1779)

Ironmongers and smiths

William Bent of St Martin's Lane (*76 Portland Place*, 1790)
Daniel Cooper of St George, Hanover Square (*33 Portland Place*, 1775)
James Ricketts (Great Portland Street, 1774)
James Smith of St George, Hanover Square (Devonshire Street, 1790)

Masons

David Alston (*82 Portland Place,* 1791)
John Devall the Younger (*13 Mansfield Street,* 1772; *23 Portland Place*, 1774)
John Hamilton (Duke Street, 1775; Cavendish Mews, 1776; Charlotte Street, 1776)
Patrick Henderson (Weymouth Street, 1789)
Robert Lee of St James, Westminster (Great Portland Street, 1774)
James MacDouall (*44 Portland Place*, 1776)
David Piper (Devonshire Street, 1776)
Peter Matthias Vangelder (Devonshire Street, 1790)

Painters and paper-stainers

William Grant of St Ann, Soho (Duke Street, 1777)
Daniel Mawley of St Giles in the Fields (Chapel Row, 1775)
Stephen Moore (Devonshire Street, 1791)
William Thompson (Duchess Street, 1786)

Plasterers

John Davis (Duke Street, 1776)
Richard Dole of Tottenham Court Road (*88 & 90 Portland Place*, 1791, with William Harley)
Richard Hughes of St Pancras in the East (Charlotte and Great Portland streets, 1774; Devonshire Street, 1776)
Thomas Lee (Weymouth Street, 1779)
Anthony Maderni (Charlotte and Great Portland streets, 1775)
Thomas Morgan (Devonshire Street, 1781)
David Piper (Charlotte Street, 1776)
Joseph Rose (*7 Mansfield Street,* 1772; Duke Street, 1773; *19 Portland Place*, 1775; *34 Portland Place*, 1777; *74 Portland Place*, 1790)
Thomas Turner of St Ann, Soho (Charlotte Street, 1779; Devonshire Street, 1790, 1791)
John Utterton (Great Portland Street, 1774)

Plumbers

Abraham Dakin (Duke Street and Little Queen Anne Street, 1773; *31 Portland Place*, 1775; Weymouth Mews, 1778)
Richard Scurrah (Charlotte Street / Great Portland Street, 1775)

Slaters

John Westcott of St Andrew, Holborn (*35 Portland Place*, 1775)

Others

John Allday of St Ann, Soho, mahogany merchant (*?65 Portland Place*, 1790; *67–69 Portland Place*, 1791; Devonshire Mews, 1791)
John Fraser, cabinetmaker (Weymouth Street, 1789)
William Newsom, victualler (Devonshire Mews East, 1786)
Thomas Wyse (or Wise), scavenger (Great Portland Street, 1774)

Notes

1 For the Adelphi *see* Bolton 1922, vol I, 18–44, and King 1991, 77–82; but more especially Rowan 2007, and Rowan 1974; for Fitzroy Square *see* Rowan 2007, 32–5, and also King 1991, 97–9. Useful material on the Adam brothers' work at Portland Place is given in Bolton 1922, vol I, 118–20; King 1991, 86–96; and especially Rowan 2007, 25–7, and Rowan 1974; but the most comprehensive and accurate account was published only recently in *Survey of London*, vols 51–52, *South-East Marylebone*, 2017, 488–542, which this chapter is intended to complement.

2 *See* Rowan 2007, 52; Rowan 1974.

3 BAM, 1454/4/5/1, William A to John A, 20 Feb 1768; 1454/4/1/6, Wm A to John A, 12 April–17 May 1787.

4 Rowan 2007, 13. Examples of the Adam practice's ability to take on big engineering challenges include the Royal Exchange (now the City Chambers) on Edinburgh's High Street (designed by Johnnie with Robert's help and built in *c* 1753–61), and Fort George, constructed overlooking the Moray Firth at Ardersier, near Inverness (of 1748–69); for these *see* Fleming 1962, 99, 100–6, 339; also King 1991, 31, 32–4; and Sanderson 1992, 23–9.

5 BAM, 1454/4/1/6, William A to John A, April–May 1787.

6 London Metropolitan Archives, Middlesex Deeds Registry (hereafter MDR) 1790/10/282.

7 University of Nottingham, MSS and Special Collections, Pl C44/21/6; early leases signed by James Adam alone include MDR 1769/4/108–9, 1770/1/381, 1770/2/40, 1772/6/378.

8 NRS, GD248/56794, fols 118r, 120r.

9 In a letter of Feb 1761, written from Florence to his sister Nell when Robert's practice was taking off, Jamie admitted candidly: 'You may assure Bob I shall also pardon him for superior merit, I am much less ambitious than Caesar, I am contented to hold a second place'. NRS, GD18/4884.

10 For James Adam as an art dealer *see* Chapters 2 and 3; *also* Fleming 1958b, Fleming 1959. For his negotiations with the RSA *see* Bolton 1922, vol II, 40–3; *and* Allan 1974, 6–14.

11 NRS, GD18/4888, GD18/4911.

12 NRS, GD18/4981/6.

13 *See Survey of London*, vols 51–52, 2017, 7–19.

14 Marylebone parish entries in MDR 1769–94 *passim, sub* Adam and/or Portland.

15 For the Foley House dispute *see Survey of London*, vols 51–52, 13–14, 556–8; *and also* Sheppard, F H W 1958 *Local Government in St. Marylebone 1688–1835*. London: Athlone Press, 112–17.

16 BAM, 1454/4/238; University of Nottingham, MSS and Special Collections, PwF 35.

17 SM, Adam vol 4/157, vol 32/17–20.

18 For the continuing attribution to Jamie *see, for example,* Gordon, J 1851 *A General Biographical Dictionary*, vol I, *sub* Adam (Robert); Bolton 1922, vol II, 102–11 (who coined the term *Strada di Palazzi*); Lees-Milne 1947, 128–9; Summerson 1962, 164–5 (where he refers to Jamie as 'for once, the avowed designer'); Saunders, A 1988 *The Art and Architecture of London: An Illustrated Guide*. London: Phaidon, 183; and also all Historic England list descriptions for Portland Place. Portland Place drawings in Robert Adam's hand include SM, vol 4/157, vol 6/136, vol 21/67, vol 23/244 and vol 13/108v (info kindly provided by Dr Frances Sands).

19 The 'great falling off' is referred to in BAM, 1454/2/Bundle 567, also in BAM, Box 3: General Correspondence, 1784–5. For Jamie's agricultural publication *see* Adam 1789; also, in a bond of 1791 he gave his address as Shenley Lodge (BAM, 1454/2/Bundle 495).

20 NRS, GD248/3395/1; GD248/570/2/44; GD248/590/1/7; GD248/588/15/42; for the late Scottish period *see* Sanderson 1992, 108–29 and Sanderson 1982. Jamie's reluctant later return to the business is referred to in Sanderson 1992, 128–9, 132 n131, and cited in NRS, GD18/4961/40, GD18/4973A.

21 *See* Rowan 1974, 666; also Bolton 1922, vol I, 116.

22 BAM, Series 4, 1454/4/1/6.

23 'Adelphi Lottery, the property of Messrs. John, Robert, James, and William Adam; to begin drawing on Thursday, the 3d of March next…' (copy in BL, General Reference Collection RB.23.b.1535).

24 *Morning Post & Daily Advertiser*, 15 July 1777; 15, 17 Dec 1777.

25 For fluctuations in the London building cycle at this time *see* Sheppard, S, Belcher, V and Cottrell, P 1979 'The Middlesex and Yorkshire deeds registries and the study of building fluctuations'. *London Journal* **5** (2), 176–217.

26 The second agreement with the Duke, of 1776, specified ground rent of £57 15*s* for the first year – less than the Adams expected to receive themselves in yearly rents for just two Portland Place houses (usually set at £30–35 apiece). But after five years this figure had escalated to £857 2*s* – and this for only the northern half of the development. (It is likely that their original agreement of 1767 for the southern half was similar.) *See* BAM, Series 4, 1454/4/238, 1454/4/7/6; MDR *passim.*

27 BAM, Series 4, 1454/4/1/4, 1454/4/1/6, 1454/4/21/15; Series 1, Box 3: General Correspondence, 1784–5. 'Improved ground rents' are the increased rents charged to a subtenant by a head lessee, who pays a smaller ground rent to the landholder. The increase is often calculated so as to help repay the costs incurred by the head lessee in erecting a building on the site.

28 BAM, Series 4, 1454/4/21/15. The Liardet affair has been covered in detail by others, notably Frank Kelsall. *See* Kelsall 1984; *also* Rowan 2007, 28–9. Material relating to the court cases is in TNA, C12/921/11, C12/1346/22.

29 BAM, Series 4, 1454/4/21/16; TNA, PROB31/656, PROB11/1025/449–489.

30 BAM, Series 4, 1454/4/1/2: Rowan 2007, 29–30

31 BAM, Series 4, 1454/4/6/3, 1454/4/16/17, 63/Box 2 (General Correspondence 1780–3), 1454/2/Bundle 567 ('Answers to Observations of the balance sheet', Dec 1784); Rowan 2007, 30.

32 This chronology is based on Marylebone ratebooks in City of Westminster Archives and Portland Estate deeds in the MDR; *see also Survey of London*, vols 51–52, 492–3.

33 Two plaster roundels of the same subjects (also reversed, painted to resemble terracotta, and titled *Morning* and *Evening)* were among the Adam antiquities and casts sold at Christie's in 1818 along with a third roundel from the Arch of Constantine showing Hadrian and others departing for a hunt (of terracotta, not reversed). All three formed one lot and were purchased at the sale by Sir John Soane for £3 and are now on display at Sir John Soane's Museum (SM M888, M893, M898). *See* Bolton 1922, vol II, Appendix C, 333; also SM, notes by Stephen Astley on 'Antiquities and Casts from the collections of Robert and James Adam in Sir John Soane's Museum'. Similar roundels of *Sol* and *Luna* appear on the south front of Kedleston Hall, though there they are larger and carved in stone, to complement the architecture.

34 Adam 1773–1822, Vol III (1822), Plate 11; Piranesi, G B 1778 *Vasi, candelabri, cippi, sarcofagi, tripodi, lucerne, ed ornamenti antichi disegnati ed incisi dal Cav. Gio. Batt. Piranesi*, vol 2, Plate 79; SM, Adam vol 26/50.

35 Adam's South Bridge designs are reproduced and discussed in detail in Fraser 1989, Appendix 2; but *see also* Rowan 1997, and Sanderson 1992, 102–7.

36 The east side of Mecklenburgh Square (by Joseph Kay, *c* 1810–20) also uses the brick-and-stucco model, though there it was applied to architecture derived from Robert Adam's designs for Fitzroy Square. For Bedford Square *see* Byrne, A 1990 *Bedford Square: An Architectural Study*. London: Athlone Press; also *Survey of London*, vol 5, *St Giles-in-The-Fields, Pt II*, 1914, 150–84.

37 For the Adelphi interiors *see* Rowan 2007, 50–7; also King 1991, 80–5; and *Survey of London*, vol 18, 1937, Plates 74–84, 101–6.

38 The nearest parallels to this type of decoration are the plasterwork panels in the staircase wells designed by Adam for Wormleybury, Hertfordshire, or, to a lesser extent, Roxburghe House, Hanover Square; or perhaps the 'drops' and other plaster wall decorations in Lady Home's dining room or back parlour at Home House, 20 Portman Square, all of which are roughly contemporary with the Portland Place staircases. For Wormleybury and Roxburghe House *see* King 1991, 250–1, 292–4 and Plates 362, 364, 415. For the Home House 'drops' *see* Harris 2001, 309–11 and Plate 458.

39 *Morning Herald,* 14 May 1792; *Morning Post and Gazetteer*, 25 April 1799.

40 For a guide to the various elements of Adam's decorative language *see* Gerson 1981.

41 Rose sketchbook, Harewood House; SM, Adam vol 24/69, 70, 71; information kindly provide by Brigid von Preussen of Christ Church College, Oxford University, but *see also* Sands 2016, 20.

42 For instance, compare Adam's ceiling designs for the front drawing room at 23 Portland Place with those for the rear drawing rooms at Nos 19 and 47 (SM, Adam vol 13/107, 115, 124); also for the rear drawing rooms at Nos 48 and 33 (SM, Adam vol 13/137, 139), and the front drawing rooms at Nos 34 and 31 (SM, Adam vol 13/108, 128). For more on repetition in these houses *see* King 1991, 94–5 (he noted identical drawing-room ceilings front and back at 11 Mansfield Street). On Adam and repetition generally *see* Stillman 1966, 27.

43 Stillman 1966, 17, 18, 19. On Adam's ceiling designs *see also* Harris 2001, *passim*, and King 1991, 19–24. For the Topham drawings at Eton *see* Aymonino *et al* 2013.

44 *See* for example the ceiling design for the rear drawing room at 20 Mansfield Street, and those for the front drawing rooms at 21 and 37 Portland Place, and 15 Mansfield Street (SM, Adam vol 13/64, 72, 111, 131). Well-known Roman sources were the Tomb of the Nasonii (2nd–3rd C); the Catacombs of Priscilla (3rd C); and the Catacombs of SS Pietro and Marcellino (4th C).

45 Rose sketchbook, Harewood House, courtesy of the Earl of Harewood: Beard 1967 and 1992; LMA, GLC photo collection for Portland Place, *passim*.

46 For Adam's use of colour *see* Bristow 1996 (esp 110–12 for the 'green chiaroscuro'); Bristow 1986; Fowler and Cornforth 1978; *also* Harris 2001, *passim*, but esp 11–13.

47 Colin Penn's views on the building are quoted in *Country Life,* 12 March 1981, 636. Report on 75 Portland Place by The Architectural History Practice Ltd, September 2001 – a recent report on the same building by Donald Insall Associates, of May 2015, still gives James Adam as its architect. The Adam building agreement in question is in BAM, 1454/4/7/6.

10

Temporal sublime: Robert Adam's castle style and geology in the Scottish Enlightenment

Marrikka Trotter

Prelude: 'dead life'

The Scotland of Robert Adam's day witnessed the emergence of modern temporality. At the scale of daily experience, the germinal Industrial Revolution brought with it the persistent chronological regulation of quotidian existence that E P Thompson has termed 'clock time'.[1] At the scale of national consciousness, the machinery of history began to be assembled by figures like Adam's friend David Hume, who claimed in the wake of his *History of England* (1754–61) that this was 'the Historical age and this the historical nation'.[2] Yet only when these two scales are paired with the rise of geological awareness – the sense of the earth's timescale – do the particular contours of temporality as it began to be apprehended in the second half of the 18th century come into focus. Britain's great era of mineral exploitation had commenced. Locating, extracting and transporting coal, lead and iron ore exposed an uncanny underworld beneath Britain's familiar agricultural surfaces. As more mines were opened, caves explored and quarries established, the more the newly synchronised tick-tock of daily existence and the freshly systematised historical past paled beside the yawn of time exposed by mattocks and mapped by commerce.

This new and threatening temporal scale comes into sharp relief in the work of the geologist James Hutton. A prominent member of the Scottish Enlightenment, Hutton was an agricultural theorist and deistic philosopher who viewed the planet as a productive base for human life. In his schema, the earth was locked in an immense cycle of erosion and lithification where old land decayed at the same rate as new land formed under the sea. There was no such thing as primary rock, existing since an originary event; there were simply different densities of soil – some recently deposited from existing rocks, others hardening into new stone under the waves. The earth of today was composed of the recycled detritus of an endless succession of former worlds, and ceding its own material, granule by granule, to a world yet to come.[3]

Hutton famously concluded the published version of his ideas, *Theory of the Earth, or, An Investigation of the Laws Observable in the Composition, Dissolution, and Restoration of Land upon the Globe* (1788), with the unsettling remark: 'We find no vestige of a beginning, – no prospect of an end.'[4] As Stephen Jay Gould has observed, this view of geological time is radically ahistorical. Time passes, but the record of its passing is being continually effaced, and while evidence of great duration can be identified, it cannot be interpreted as part of a directional progression.[5] In Hutton's work the narrative arc that had been so central to the Western view of the world since the rise of Christianity was lost – and with it, any assumption of an intuitive human affinity with the landscapes that ultimately supported life.[6] Time and matter looped in upon themselves. What had happened before was happening again, very slowly, and would happen once more in a future too remote to imagine.

The incommensurability Hutton's theory proposed between lived experience and the vast and indifferent rhythms of the planet exemplifies the Enlightenment's contradictory attitude towards the earth. In his study of geology's impact on British poetry, Noah Heringman puts the situation succinctly: on the one hand, the Industrial Revolution was based on an instrumental view of the mineral landscape as a static reserve of economic resources. On the other, its meticulous investigation of this reserve revealed abundant evidence of large-scale activity and change, forcing romantic culture to confront the earth 'in its otherness; its non-human aspect'.[7] For example, Hutton attributed fundamentally lifelike qualities of motion, circulation and self-repair to inorganic matter. In his *Theory of the Earth*, he insisted that the world is not simply a machine but also an 'organized body' with a 'reproductive operation, by which a ruined constitution may again be repaired, and duration of stability thus procured'.[8] The transfer of organic rhythms to a mineral kingdom so old that it might well have existed forever imbued this previously inert substrate with eerie, even frightening, agency.

Thinkers had begun to probe such ramifications in the philosophical field of aesthetics, shaping a new linguistic facility required for such work along the way. As Denis R Dean has shown, early 18th-century authors struggled to articulate their reactions to the contours of the natural world. For instance, Thomas Gray and Horace Walpole seemed incapable of describing the Alps during their grand tour of 1739 in anything but the most generic of terms: mountain scenes were 'poetical', 'romantic' and 'astonishing'.[9] Contrast such inarticulacy with the terminological precision achieved by John Playfair, Hutton's biographer, as he recounted the experience of comprehending Hutton's interpretation of the rock formation at Siccar Point, Berwickshire, in 1788:

> We felt ourselves necessarily carried back to the time when the schistus on which we stood was as yet at the bottom of the sea, and when the sandstone before us was only beginning to be deposited, in the shape of sand or mud, from the waters of a superincumbent ocean. An epocha still more remote presented itself, when even the most ancient of these rocks, instead of standing upright in vertical beds, lay in horizontal planes at the bottom of the sea, and was not yet disturbed by that immeasurable force which has burst asunder the solid pavement of the globe. Revolutions still more remote appeared in the distance of this extraordinary perspective. The mind seemed to grow giddy by looking so far into the abyss of time.[10]

Playfair's dramatic depiction represents one end of a sophisticated gradation of sensual responses that had been devised over the intervening half-century: if the landscape is beautiful when cultivated for pleasure, and picturesque when worked for profit, then when it reveals what Hutton called 'the exquisite mechanism and active powers of things', it is sublime.[11]

Playfair's account of Siccar Point followed a particularly Scottish notion of this effect. Although geological features had been included in the repertoire of sublime objects since this aesthetic category had been formulated – Longinus, the ancient Greek author whose treatise formed the basis for 18th-century theories, included the volcano of Mount Etna in his catalogue[12] – Scottish philosophers placed a unique emphasis on temporality. In his *Treatise of Human Nature* (1738) Hume asked, 'Why a very great distance encreases our esteem and admiration for an object: Why such a distance in time encreases it more than that in space: And a distance in past time more than that in future.'[13] He concluded that, while the imagination perceived space as a smooth continuum, time was envisioned as a 'broken up jumble'. The greater difficulty in imagining a remote past than a remote place thus produced a more complete sense of disorientation, followed in turn by the attainment of 'a more vigorous and sublime disposition'.[14] Here the sublime is couched in terms of a contest between a tableau that fires the imagination and a core of reason that derives strength from resisting its onslaught. Similarly, Hutton's text emphasised his theory's appeal to reason while acknowledging his material's threat to imaginative sensibility: 'The hideous mountains and precipitous rocks, which are so apt to inspire horror and discontentment in minds which look at sensible objects only for immediate pleasure, afford matter of the most instructive speculation to the philosopher, who studies the wisdom of nature through the medium of things.'[15] The effort inherent in examining the true power and extent of the mineral world was worth the mental edification that would ultimately result.

Yet there is also the sublime menace of time itself, couched as a ruinating counterforce that erases not just mankind's physical productions but also mankind's mental constructs. The Scottish associationist philosopher Lord Kames, a contemporary of both Hume and Hutton, and also an influential friend and correspondent of Robert Adam's, speculated that the most basic reduction of a sublime encounter could be produced by 'a huge mass of rubbish, the ruins perhaps of some extensive building, or a large heap of stones, such as are collected together for keeping in memory a battle or other remarkable event'.[16] This ambiguous index of opposite processes is sublime because it reveals the indifference and unmeaning of Hutton's 'medium of things'. The past reaches up to touch the present in muteness. Ruins perpetually enact forgetfulness; cairns forecast failures of memory to come. The stones will outlive their arrangements, will again become part of the larger effacing upheaval of the earth, will return to the darkness and dust from whence they have come only temporarily, reluctantly, into the landscapes of the living.

The historian John Dalrymple, an associate of Hume, Hutton and Kames, and also a friend and patron to Adam, wrote in an influential 1774 treatise on landscape: 'The chief natural defect of a highland situation is that, being generally ill-inhabited, it has too much the appearance of dead life: that appearance, added to the vastness of the objects, creates a kind of despair in the mind, which considers itself as nothing amidst that stupendous and solitary scene it beholds.'[17] Coupled with Hutton's extension of an eerily biological 'reproductive operation' to the entire surface of the globe, this sense of 'dead life' becomes part of a deeper 18th-century intuition of a negative force, hostile to life itself. For Dalrymple, the ever-present threat of the barren Scottish landscape required extensive defensive bulwarks. His solution was architectural. As he put it, 'though the little finishings of art on the face of the ground would in such a situation be lost, yet the great efforts of art would please, because that very art is a sign of cultivation'. The 'great efforts' he specifically had in mind were castles (Fig 10.1):

Fig 10.1
'In a highland situation, the principal house should be in the form of a castle.' Robert Adam, River Landscape with a Castle, Lake and Sailing Boat, *1782, pen, ink, watercolour. The mountains in the background of this composition are recognisable as the 'Three Sisters' of Glencoe.*

> The slenderness of an Ionic or Corinthian pillar, placed at the side of a vast mountain, would create a ridiculous comparison; and therefore in a highland situation, the principal house should be in the form of a castle. The elegance and fineness of execution belonging to the Grecian architecture, would be here totally misplaced. If in that castle, added to the greatness and solid appearance of the main building, there should shoot up in the middle a Gothic tower, pierced and of hardy execution, a sentiment similar to the sentiment of terror, added to that of grandeur, would still more correspond to the natural genius of the place.[18]

In this, Dalrymple echoed a roughly contemporaneous remark by Lord Kames that a Gothic ruin 'exhibits the triumph of time over strength; a melancholy, but not unpleasant thought: a Grecian ruin suggests rather the triumph of barbarity over taste; a gloomy and discouraging thought'.[19] But embedded within Dalrymple's preference for hardiness over slenderness was a sense that architecture possessed a sublimity of its own – one capable of rising up against the inorganic hostility of an ancient and alien landscape. This illuminates one of the underlying attractions of the sublime in the context of the enlightened estrangement between culture and the natural world. From Longinus to Burke, the sublime had permitted a sense of parity, however fraught, between the works of man and the works of nature. However doomed such efforts were ultimately acknowledged to be, there was yet something in the artifice of human achievements – and architecture in particular – that could resist the overpowering force of nature even as it gave way before it.

Dalrymple inherited his own castle, the 16th-century Oxenfoord tower house in Midlothian, Scotland, in 1779, and promptly hired Robert Adam to update and expand the property (Fig 10.2). His choice of architect was hardly a simple matter of patronage by association, however. Adam was friends with Hume and Kames and closely associated with Hutton via his brother-in-law, fellow architectural enthusiast and sketching companion John Clerk of Eldin, a polymath who produced the illustrations for Hutton's *Theory of the Earth* and was so involved in that project's conceptualisation that Playfair felt it was impossible to tell where the work of one man stopped and the other began.[20] These connections tied him to the intellectual circle most responsible for framing a late 18th-century geological awareness of the landscape. Meanwhile, Adam had embarked upon the second great episode of his career, in which castles, sublimity and time emerged as central concerns.

Fig 10.2
Robert Adam, View across Park to Oxenfoord Castle, 1780s, pen, ink, watercolour.

'Majesty, gravity, and force': Adam's castle style

By the time he designed his major castle-style projects in the 1770s and 1780s, it seemed to Adam's early biographers that his best years were behind him. Perhaps Adam felt this way himself: an uneasy worry that the grand commissions and lightly worn accolades were fading, together with the originality he prized above all else.[21] Certainly there is an insistent quality to the well-known preface he authored with his brother, James, for publication in the first volume of their *Works in Architecture* (1773):

> We have not trod in the path of others, nor derived aid from their labours. In the works which we have had the honour to execute, we have not only met with the approbation of our employers, but even with the imitation of other artists, to such a degree, as in some measure to have brought about, in this country a kind of revolution in the whole system of this useful and elegant art.[22]

As the above passage obliquely suggests, Adam had held his own alongside his 'more architectonic' rival, William Chambers, only to see his 'picturesque' style poached and (in the eyes of prominent connoisseurs) improved by up-and-coming architects like James Wyatt and Henry Holland.[23] Other events augured ill. Clients in England tightened their purse strings and delayed or decided against ambitious building projects as the revolt in the American colonies ramified. Robert and his brothers had only superficially recovered from the financial fiasco that surrounded their speculative London development at the Adelphi, which had sorely damaged their reputations and strained the bonds of what had, in happier times, been characterised as 'a wonderfully loving family'.[24] The Adam flair for ambitious ventures that had served the family so well earlier in the 18th century finally rendered them vulnerable to unscrupulous absconders and specious investments. As the architectural commissions dwindled, James lost interest in the profession, retired to his farm at Shenley, Hertfordshire, and took up agricultural theorising.[25] Robert found himself with more time on his hands and more care on his shoulders. His health suffered.[26]

Yet through all of this, or perhaps because of it, the chief scion of Georgian architecture remained prolific in what were to be the final decades of his life, producing designs that Alistair Rowan has characterised as 'amongst the most original creations of eighteenth-century European architecture'.[27] As Adam's London-based practice dried up, he increasingly turned to his native Scotland, picking up residential commissions from friends like Dalrymple, and working on ambitious urban and public building projects in Edinburgh and Glasgow. Though many of the city improvement schemes relied upon Adam's traditional neoclassical vocabulary, the country residences were designed in a completely novel idiom. Together with hundreds of uncommissioned paintings of invented landscapes, these late, so-called 'castle style' works became the main focus of Adam's increasingly solitary design practice until his death in 1792.[28]

The castle-style buildings are notoriously difficult to cognise, both in relation to Adam's delicate approach to neoclassicism (known today simply as the 'Adam style') and in the wider context of late 18th-century British architecture. Arthur T Bolton, author of the first authoritative survey of the Adams' works, suggested that these residences belonged to the whimsical 'Gothick' exemplified by Horace Walpole's Strawberry Hill and the work of Richard Bentley and Sanderson Miller. He lamented them as 'atrocious attempts' and argued that 'without the requisite knowledge of the detail and character of Gothic architecture, Adam was unable to embody his romantic visions in serious architectural form'.[29] Yet, as Rowan has noted, the Adam castles intentionally lack the basic characteristics common to both authentic Gothic architecture and 18th-century 'Gothick' confections, such as pointed arches, traceried windows, pinnacles and flying buttresses.[30] Part of the issue here is the greater fluidity of Enlightenment nomenclature: as Dalrymple's description of 'a Gothic tower, pierced and of hardy execution' makes evident, 'Gothic' could be associated with castles as much as cathedrals. Nonetheless, Rowan has posited a more sophisticated interpretation of the castle style as one also inspired by Diocletian's Palace at Split, in present-day Croatia (which Adam had surveyed for publication early in his career), as well as by other vestiges of ancient Rome, including the remains of military fortifications in southern Italy.[31] Certainly Adam's castles share features, not just with Diocletian's Palace, but also with Diocletian's Baths in Rome – a structure that the Adam brothers also studied and admired.[32] These include the semicircular tripartite windows that bear the Roman emperor's name, strong horizontal banding, round windows and relieving arches (*see* Figs 10.3, 10.4 and 10.5 below). But such elements also feature prominently in Palladian and neo-Palladian architecture, and Adam used Palladian windows in several castle façades. Moreover, all of these elements are present in Adam's neoclassical designs.

The influences unique to the Adam castle style often have a more local source: the pepper pots, crow-stepped roofs, crosslet loops and bartizans that appear in many designs are typical of 16th-century Scottish tower houses. As William Kay has suggested, their presence alongside Roman and Palladian features seems to conjure an alternative British past in which the Romans had stayed in these distant and troublesome colonies rather than withdrawing at the end of the fourth century AD.[33] Likewise, John Dalrymple once remarked that Adam's late 18th-century update of his 16th-century castle 'really made it much older than it was'.[34] By invoking a remote past without remaining tethered to the ornamental indices of any specific period or place, the castles contrive to belong either to no particular history, or to one that never existed – a notional, elemental and universal period of architecture from which all styles ultimately derived.

The most striking difference between Adam's neoclassical architecture and his castle-style buildings is the comparative blankness of the façades of the latter. At Seton Castle, for example (Fig 10.3), ornament is stripped down to shapes and lines – the curves of arches, the stark rectangles, crosses and circles of windows, the double stripes of string courses, and the black bars of machicolations. The machicolations are in fact the only 'fortification' elements that span all of Adam's castle designs; even crenellations give way in several productions to this one minimum signifier of the castle typology. For all their defensive vocabulary, the castles also share a drawing-like neatness that serves to emphasise their artifice over their mass. Dalquharran Castle gives the impression of being wrapped by a single, paper-like plane that has been embossed, punched, creased and curved to fit (Fig 10.4); Adam's monumental design for additions to Barnbougle Castle surrounds the original tower house with a bookmatched Roman scrim that emphasises the stark symmetry of the ancient keep (Fig 10.5). With their crisp proportions and stripped planes, the castles lean away from the picturesque, which relied instead on cobbled-together and decayed vernaculars.[35] Rather, they call to mind a quip by Elizabeth Montagu –

Fig 10.3
Seton Castle, Longniddry, East Lothian, Scotland. A castle-style house of 1789–91.

'queen of the bluestockings' and also Adam's friend and client – who recommended Lord Lyttleton's *History of the Life of King Henry the Second* (1767–71) to Lord Kames by remarking that it was like 'a Gothick building by a Roman architect. The story is Gothick, but expressed with majesty, gravity, and force, without any thing dark, or rude, or perplexed & confused.'[36] These were no romanticised depictions of workaday ruination that projected an unchanging and unchallenged present, as picturesque productions from the period often stand accused of being.[37] Rather, prismatic, severe, out of time and out of place, Adam's castles embody a resisting and uniquely architectural alterity: a heightened sense of artificial order that resonates with John Dalrymple's 'great efforts of art' and speaks to the formulation of an architectural sublime.

Fig 10.4 (above) Dalquharran Castle, South Ayrshire, Scotland; mostly 1780s, extended 1881.

In addition, in their ahistorical recirculation of forms and motifs from architecture's past, Adam's castles are consonant with Hutton's terrestrial cycle, in which the same material was disaggregated and reaggregated into ever-unique configurations. Certainly it seems likely that Robert Adam was familiar with Hutton's ideas. His brother James cited Hutton in his book on agricultural theory, and Robert undertook an extensive topographical sketching tour of Ayrshire with John Clerk of Eldin shortly after the latter covered the same territory on a geological expedition with Hutton. Moreover, Adam frequently resided with his brother-in-law at Eldin and later at his Edinburgh home in Princes Street, and dined at the Oyster Club, which Hutton co-founded, on his summer trips to Scotland.[38]

Above and beyond these biographical links, however, Adam's work should not be viewed as a response to Hutton's ideas but rather as

Fig 10.5 (below) Robert Adam, watercolour elevation of an unexecuted design for additions to Barnbougle Castle, on the Firth of Forth c *1775.*

Fig 10.6
John Clerk of Eldin:
Granitic Veins and Sea Arch, Isle of Man *(top);* Granitic Veining at Cairnsmore of Fleet *(bottom).*

consistent with them. As men associated by place and time and personal connections, they were developing intellectual and imaginative directions that made sense in the context each helped create for the other. In the Scottish Enlightenment, architecture and geology shared not only the same physical landscape but also the same inorganic palette of materials and methodology borrowed from antiquarianism – overlapping concerns and areas of expertise that Hutton certainly took into account in his *Theory*. Alongside instances of architecture as evidence of the brevity and fragility of human endeavours in light of geological time, Hutton repeatedly deployed architectural metaphors to explain the mechanics of the earth as he envisioned them. For example, when trying to give a sense of the discrepancy between the enormous scale and importance of the geological change he described and the minute scale of the observable changes upon which he based his hypothesis, Hutton compared the impact produced by the Egyptian pyramids as completed works with the mundane toil of manoeuvring each block into position: 'We wonder at the whole operation of the pyramid, but not at any one part.'[39] The repetitive processes that produce geological change were equated with the unglamorous labour of building a sublime piece of architecture – a parallel Lord Kames also effected by juxtaposing 'the great pyramid of Egypt' and 'the Alps towering above the clouds' in his discussion of the sublime.[40] In the same vein, Hutton wrote of his discovery that existing rocks are composites of older rocks: 'The thing is evident from inspection, as much as would be the ruins of an ancient city, although there were no record of its history.'[41] Hutton evoked the unmistakable signs of urban settlement, as they had been discovered at Herculaneum, Pompeii and Paestum during his lifetime, to make a case for the similarly distinguishable qualities of conglomerate and metamorphic stone.[42] Throughout his *Theory of the Earth*, pieces of natural evidence are called 'monuments', the planet's strata are 'foundations' and the repair of the eroding world is called 'renovation'.[43] Encounters between architecture and geology were thus characterised by a paradox: 'Time! – Time! – Time!' as the 19th-century volcanologist George Poulett Scrope exclaimed, would reduce all architectural efforts to the basic geological components from which they had been shaped, and yet architectural terms and ideas were appropriated to model its effects.[44]

In a general sense, Hutton's cycle of ruination and reagglomeration emerged at the same time as Adam's uniquely combinatory approach to design. An early example of Adam's technique is the anteroom he designed for Syon House in 1761, just three years after he returned from Italy. As Doreen Yarwood has summarised, the composition contains column shafts supposedly dredged from the Tiber River – allegedly real, ancient elements of Classical antiquity – that Adam embellished with Ionic capitals, an entablature modelled on that of the Greek Erechtheion, and necking decoration inspired by those of Roman baths. The coffered ceiling follows William Kent's design for Houghton Hall, and the gilded trophy panels are based on an engraving of the trophy of Octavianus Augustus by Giovanni Battista Piranesi.[45] It is likely that Piranesi himself, who befriended and mentored the Scottish architect in Rome, encouraged such an approach.[46] Piranesi saw history as a jumble of fragments available for endless combination and recombination – a perspective that resonated with Adam's own innate eclecticism and dovetailed with Hume's characterisation of the past. Certainly the same mix of real and imagined histories is present in Piranesi's archaeological fantasies, some of which seem to evince a stratigraphic sequence of their own.[47]

And Piranesi's compositions became source material for Adam in turn. In a suite of capriccios made in Rome, for example, Adam repeatedly used elements from Piranesi's engravings to populate his own scenes.[48] Like Hutton's cyclical mechanism, which assembled new mineral formations from old ruins, Adam's collage-like approach crafted entirely unique spaces and views from the existing repertoire of architectural elements and images.

In addition to this structural correlation, early geological imagery laid claim to formal analogies with architectural components.[49] The famous example is the engraving of Fingal's Cave based on the illustration by John Cleveley and published in Thomas Pennant's *Tour through Scotland* (1771), which exaggerated the hexagonal regularity of its basalt pillars to stress their resemblance to architectural columns. Clerk of Eldin's geological drawings for Hutton's *Theory of the Earth* function similarly. In one illustration, rectangular fragments of broken rock in the foreground of an exposed section of granitic veining are pictured like cast-aside building blocks, reminding the viewer of the images of architectural ruins in circulation at the time (Fig 10.6, bottom). In another, the metamorphic intrusions that form the scientific subject of Clerk of Eldin's composition are relegated to the background, while an architecture-like sea arch dominates (Fig 10.6, top). It is accurate but insufficient to adduce that the emerging science of geology turned to the representational conventions of architecture because it had yet to develop its own visual language. By privileging architectural *forms* in addition to representational techniques in his renderings of natural subjects, Clerk of Eldin imposed a rigorously artificial framework on the treacherous tumult of the earth.

The inverse of this kind of visual borrowing appears in the cliff-side rendering Adam drew of his design for Culzean Castle (Fig 10.7) – a project that, according to Clerk of Eldin, allowed the architect to 'indulge to the utmost his romantic genius'.[50] The castle is situated on a quintessentially sublime site: a sheer drop down to the Firth of Clyde is broken only by the dark mouths of caves associated with legends of fairies and smugglers. In the romantic imagination, Adam's round tower and battlements accentuate the drama of the landscape, particularly if they are viewed in such a way as to diminish their fundamentally symmetrical arrangement. This is

Fig 10.7
Robert Adam, Cliffside Elevation of Culzean Castle *c 1785.*

what we see in landscape paintings composed after the project was completed. Yet in Adam's own drawing a completely different conception emerges. The cliff seems to bulge and compress underneath the castle, which straightens the landscape's contour lines into its own. Rather than depicting an upward-gazing subject's foreshortened visual experience, Adam turned to the architectural convention of an orthographic projection and presented the imaginary vantage of a gaze capable of hovering in mid-air some distance out over the water. The landscape thus rendered is further architecturalised in two ways: it is stripped of its overhanging and upgrowing vegetation and reduced to stonework as stark as that of the castle above, and its irregular projection towards the viewer to the left of the image is made to cast a clean triangular shadow, as though it were a perfectly rectilinear block. In depicting the landscape and the castle alike as if they were cut from the same cloth, Adam gained for architecture what he deprived from the landscape. The drawing forces an equivalence between the building and the geology that is possible only in the architectural imagination. In his *Theory of the Earth*, Hutton described erosive forces as 'attacks which must, in time, wear away the bulwarks of our soil, and sap the foundations of our dwellings'.[51] In Adam's drawings, by contrast, architecture reaches down to fix the irregular stratifications on which it rests.

A 'human art': Adam's late landscape paintings

Clerk of Eldin's surviving notes for his intended biography of Adam mention the architect's ambition to produce a 'thousand landskips' in his later years. The 400-odd finished watercolours and

Fig 10.8
Robert Adam, Mountainous Landscape with a Castle and Figures, *1784, watercolour. The volcanic cone behind this invented castle echoes its form.*

sketches in Adam's hand that are known to survive today testify to the serious energy he devoted to this project.[52] At the time of Adam's death these paintings were considered as important as his architecture: his obituary in the *Gentleman's Magazine* lauded their 'luxuriance of composition' and 'effect of light and shadow'.[53] By sidestepping the limits of commissioned projects, these works explore architecture's full potential vis-à-vis the newly estranged earth of the Scottish Enlightenment.

At first sight, Adam's landscapes could be interpreted merely as inventive examples of the romantic fantasies just coming into vogue in Britain: with their striking combinations of mountains, rivers, bridges and fortresses, they follow the prescriptions of William Gilpin and echo the fictional compositions of Adam's friend and sketching partner, the topographical artist Paul Sandby.[54] Yet on closer inspection unique traits appear. Fortresses and landforms confront each other or cooperate as if they were equal entities. In some cases, rocky outcrops and castles possess the same apparent heft (Figs 10.8 and 10.9); in others, sombre ridgelines follow the outlines of the architectural compositions they overshadow. In one painting, a rock seems to march towards a hilltop castle like an attacking army. A pair of paintings by Adam and Clerk of Eldin presents a striking inversion (Figs 10.10 and 10.11): in Adam's watercolour a mountainous mass, blurred as if in motion, surges by an inert fortress on an equivalently scaled hill. Clerk of Eldin's take on his friend's composition, 'after R. Adam', takes this scenario to a new extreme, repositioning the castle in front of a rushing landform that rears back as if to engulf the fortress. Taken together with Clerk of Eldin's position as the likely conduit of Hutton's theories and the amount of time these men spent in each

Fig 10.9
Robert Adam, Landscape with a Castle, Lake and Farm Animals *c 1780, pen, ink, watercolour.*

Fig 10.10 (above)
Robert Adam, Castle on Rocks above Lake with Figure and Cart in Foreground *1780s.*

Fig 10.11 (below)
John Clerk of Eldin, Castle in a Landscape after R. Adam 1780s.

other's company during Adam's castle-style period, this startling composition raises the possibility that Clerk of Eldin was a more catalytic figure in Adam's intellectual life than has been previously credited.[55]

Adam had a tendency to include real Scottish landforms in his fictional compositions. A looming ridgeline that overshadows a fortress, mimicking its contours, is a view of the Grampian Mountains near Cluny Castle. The waterfall that drops away from an Italianate tower is recognisable as Cora Linn, a famous feature of the River Clyde. The 'Three Sisters' peaks of Glencoe appear in several paintings (*see* Fig 10.1 and Frontispiece), and the distinctive cone of Ben Lomond dominates the middle distance of a scene with a ruined tower. Fictional elements are also reused: castles from one landscape will appear in another, or invented landforms will be recycled. Treated as a series – and the presence of sequential numbering on the versos of many of these compositions suggests that they were intended to be seen as such – the paintings give the impression of a typology that includes both built and natural forms.

There is, moreover, a laminated quality to many of these works, as though the planes of various two-dimensional tableaux have been stacked together like so many layers of stage scenery. This is a deliberate effect, rather than a lack of ability, as can be seen by comparing a surviving rough sketch from the Blair Adam collection with the completed version at Sir John Soane's Museum. The draft is noticeably seamless compared to the final iteration, which has been adjusted deliberately in order to achieve this kind of effect.[56] In fact, in several of the more highly finished compositions, various perspectival vantages seem to be combined, producing the vertiginous sensation that the landscape before which the viewer is placed is capable of flying apart at any moment (*see* Fig 0.20).[57] In Adam's other drawing projects, this kind of visual fracture is often the signal that some alien element has been allowed to invade the composition. For example, in one fictional Rome-inspired scene, Adam has placed Piranesi's depiction of the Pyramid of Cestius into a perspective to which it does not adhere, literally allowing a piece of his mentor's world to exist in his own composition under the banner of its own representational logic. That this is an intentional aberration on Adam's part is emphasised by the fact that he has left the precise construction lines for his drawing's two-point perspective visible.

The net result of such manoeuvres is to signal that the compositional logic deployed in these watercolours is consistent with Adam's design technique. In other words, Adam was extending his eclectic architectural approach to the landscape as a whole, as if outcrops and rivers, too, were source material for his designs, and even mountains were as available for imaginative recombination as the cornices and capitals he sampled at Syon House. Moreover, in so blatantly articulating the artificial manner of their composition, these sublime scenes invite the power of reason to eclipse the dazzled gaze.

The most commonly repeated of the geological motifs found in Adam's paintings are Arthur's Seat and Salisbury Crags, distinctive landforms that dominate the skyline of modern Edinburgh and would have been an even stronger presence in the 18th century (Figs 10.12 and 10.13).[58] While we know today that Arthur's Seat is an extinct volcano, Hutton vehemently resisted this interpretation. He chemically distinguished between lava thrown up by volcanoes and basalt – the former being free of zeolites and calcareous spar, the latter containing those crystals and representing a subterranean chamber of molten rock that had never erupted at the surface. Since the rock that made up Arthur's Seat conformed to his definition of basalt, Hutton theorised that this outcrop had been slowly exposed as softer rock around it had eroded away – a scenario that rendered it far older than a volcanic interpretation would have and accorded it the status of silent witness to a former state of the earth.[59] Its insistent presence in the background of Adam's fictional compositions both explicitly references the enormous duration of geological time and draws a direct analogy with the fortresses that populate these scenes: in their separate timescales, both are obdurate and anachronistic remnants of a previous order.

Hutton's insistence on the great age of exposed basalt paves the way for a geologically informed interpretation of a remarkable Adam watercolour now in the collection of the RIBA (Fig 10.14). With their vertical scores and block-like forms, the mountains here are depicted as basaltic plugs – explicit references to the extent of geological time. In the foreground, a blackened fortress seems to have grown from its ancient rocky perch. Drum towers and square-cornered ramparts seamlessly extend the landscape's contours; a heavy arch spans a central sheer plunge. Where

Fig 10.12 (above)
Robert Adam, Romantic Castle and Bridge over River with Mountains in Background *1780s.*

Fig 10.13 (below)
John Clerk of Eldin, Arthur's Seat and Salisbury Crags *c 1785, pen, ink, wash.*

two rivers meet, a cataract rushes through the four-arched bridge that connects this grim keep with a white city of spires and domes at the painting's centre. Glittering in a long shaft of sunlight, the city is built on a low spur of the surrounding mountains. Beyond, a hulking basaltic outcrop mirrors the size and weight of the fortress. The two twin masses – one additive, one subtractive – stand like colossal sentinels, guarding the architecture between them from the infinitely slow onrush of the mountains beyond. Their presence will be in vain. Already a rectangular lip of dark landmass opens its mouth over the city; already the mineral landscape takes these pale forms into its jaws like broken teeth. In the context of this bare swirl of natural forces, the formal architectural vocabulary of Classical antiquity is depicted as ephemeral, almost mirage-like. The viewer is compelled to call to mind Hutton's remark about Babylon, also built on a low bend of a river: 'What is become of that city? nothing remains.'[60]

Ultimately, the confrontation to which a painting like this alludes is between more than the earth's forms and humanity's edifices. In the Scottish Enlightenment one of the final props supporting an anthropocentric view of the world slipped – the latest in a series of recognitions that included the discovery of infinity and the Copernican establishment of a heliocentric planetary system. What was at stake was no less than humanity's right to regard itself as the end to which all nature was the means. In this light, paradoxically, Hutton emerges as the more conservative figure. As opposed to Playfair's more scientistic gloss, Hutton's *Theory* is primarily a piece of deistic philosophy that presumes that the self-renovating mechanism of the planet exists to perpetually maintain the earth as an optimal habitat for mankind.[61] By contrast, while Adam's castle-style buildings and landscape paintings seem to register an awareness of the earth's inhuman dimension, they fundamentally propose nothing more than a way to operate in spite of its overtaking flood. Adam's archaeology was always one of fragments rescued willy-nilly from the annihilating crush of time. While both he and Hutton were accustomed to sifting through the incomplete indices of the former states of things, Adam differed from his fellow Edinburgher, and

Fig 10.14
Robert Adam, Romantic Landscape, *1780s.*

indeed from most of his professional colleagues, in his embrace of the past as a hodgepodge from which any meaning gained would always be a contemporary artefact predicated on present tastes. Architecture is a humanistic endeavour. Robert Adam met the Scottish Enlightenment's new and frightening understanding of the earth by extending his disciplinary logic to the landscape.

Acknowledgements

Deep thanks are due to the many scholars who have generously shared their research and ideas with me: the present and former lights at Sir John Soane's Museum – Stephen Astley, Helen Dorey, Susan Palmer and Dr Frances Sands; Iain Gordon Brown, Ian Gow and Professor John Lowrey in Edinburgh; Professor Alistair Rowan in Dublin; Geoffrey Bertram in Somerset; and Dr Adriano Aymonino, Eileen Harris and Colin Thom in London. I must also thank Keith, Elizabeth and Francis Adam of Blair Adam, without whose sustained hospitality and openhandedness this research would have been impossible, and Sir Robert Clerk of Penicuik, for his kind facilitation of access to the fascinating world of John Clerk of Eldin and the collections at Penicuik House. Professor Stuart Monro and Alan McKirdy in Scotland spent hour after hour with me in person and via email, teasing out the real Scottish landforms from the imagined topographies that populate 400-odd Adam landscape watercolours. I am deeply grateful for their boundless forbearance and unmatched expertise. Finally, I must thank my dissertation committee, Professors Antoine Picon, Erika Naginski, Alina Payne and Jimena Canales, at Harvard University's Graduate School of Design, for their essential contributions to and unstinting encouragement of the PhD research from which this piece is drawn.

Notes

1 Thomson, E P 1967 'Time, work-discipline, and industrial capitalism'. *Past & Present* **38** (December 1967), 56–97.
2 Letter to William Strahan, 1770, *in* Greig , J Y T (ed) 1932 *The Letters of David Hume*. Oxford: Clarendon, vol II, 230.
3 For a concise review of Hutton's theory and its context within the Scottish Enlightenment, *see* Dean, D 1992 *James Hutton and the History of Geology*. Ithaca: Cornell University Press, 2–30.
4 Hutton 1795, vol I, 223. Hutton first presented his 'Theory' in 1785 as a lecture to the Royal Society of Edinburgh, who then published it in their *Transactions* in 1788 (vol I, 209–304). An expanded version appeared with a shortened title in two volumes in 1795, which is the edition usually cited in this chapter.
5 Gould, S J 1987 *Time's Arrow, Time's Cycle: Myth and Metaphor in the Discovery of Geological Time*. Cambridge, MA: Harvard University Press, 61–96.
6 Rudwick, M S J 2005 *Bursting the Limits of Time: The Reconstruction of Geohistory in the Age of Revolution*. Chicago: University of Chicago Press, 168–70.
7 Heringman, N 2004 *Romantic Rocks, Aesthetic Geology*. Ithaca: Cornell University Press, 1.
8 Hutton 1795, vol I, 16–17.
9 Gosse E (ed) 1884 *The Works of Thomas Gray in Prose and Verse*. London: Macmillan, vol I, 244; vol II, 36. Quoted and discussed in Dean, D 2007 *Romantic Landscapes: Geology and Its Cultural Influence in Britain, 1765–1835*. Ann Arbor: Scholars' Facsimiles and Reprints, 55–6.
10 Playfair, J 1822 'Biographical account of the late James Hutton, M.D.', repr in *The Works of John Playfair, Esq*. Edinburgh: Archibald Constable, vol IV, 80–1.
11 Hutton 1795, vol II, 469.
12 Longinus, *On the Sublime*, 35. Discussed in Brady, E 2013 *The Sublime in Modern Philosophy: Aesthetics, Ethics, and Nature*. Cambridge: Cambridge University Press, 13.
13 Hume, D 1738 *A Treatise of Human Nature*. Edinburgh, vol II, part III, section VIII, 279.
14 Ibid, 286.
15 Hutton 1795, vol II, 90.
16 Home, H (Lord Kames) 1785 *Elements of Criticism* (reprographic reproduction). London: Routledge/Thoemmes, 1993, vol I, 212. Lord Kames was one of the central figures of the Scottish Enlightenment. He was the recipient in 1763 of the letter from Robert Adam (discussed on pp 17–18), explaining his views on the use of the Classical orders. Kames was also a member of the Edinburgh Council committee that in the 1760s reviewed and advised on James Craig's plans for Edinburgh New Town, as was Robert's brother John Adam. For a discussion of Kames' influence of the Adam brothers, *see* Fleming 1962, 307–11.
17 Dalrymple, Sir J 1823 *An Essay on Landscape Gardening*. Greenwich, 11 (originally published as *Essays on Different Natural Situations of Gardens*, 1774).
18 Ibid, 8.
19 Home, *Elements of Criticism*, vol I, 446–7.
20 Playfair, 'Biographical account', 115. Clerk of Eldin's illustrations were lost after his death and rediscovered in the Clerk family archives at Penicuik House in 1968. *See* Craig *et al* 1978, 6–9.
21 *See*, for example, Arthur T Bolton's appraisal of this period, in Bolton 1922, vol I, 197.
22 Adam and Adam 1773–1822, vol I, preface, i.

23 Robert Adam, quoted in Tait 1993, 3; Sanderson, 1992, 74. After visiting Carlton House, designed by Henry Holland, Horace Walpole famously remarked, 'How sick one shall be, after this chaste palace, of Mr Adam's gingerbread and sippets of embroidery.' *See* Wilton-Ely 2001.
24 Carlyle, A 1860 *Autobiography*, 2 edn. Edinburgh: Blackwood, 361.
25 In 1789 James Adam published two volumes on agricultural theory, *Practical Essays on Agriculture* – *see* pp 23–4 (Adam 1789). His country residence, Shenley Lodge, near Shenley, Hertfordshire, though much altered, still stands, and is now home to the independent Manor Lodge School.
26 For the best account of this fraught period in the Adam family history, *see* Sanderson 1992, 99–107; also Rowan 2007, 36–40.
27 Rowan 1985, 17.
28 Sanderson 1992, 82–3, 102–3, 108–11. By the 1780s James had basically retired from the dwindling practice to focus on his agricultural research. Robert Adam, whose design vision had always been dominant, became solely responsible for the firm's creative direction after that date. Tait 1993, 12.
29 Bolton 1922, vol I, 94.
30 Rowan 1985, 17.
31 Ibid, 18–19; information from Alistair Rowan.
32 Adam and Adam 1773–1822, vol I, iv.
33 Kay 1993.
34 Quoted in Sanderson 1992, 88.
35 Macarthur 2007, 115–18.
36 NRS, GD24/1/573, Elizabeth Montagu to Henry, Lord Kames, 30 July, 1767.
37 Macarthur 2007, 58–64.
38 Adam 1789, vol I, 52; Sanderson 1992, 82, 121, 125.
39 Hutton 1795, vol I, 184.
40 Home, *Elements of Criticism*, vol I, 212.
41 Hutton 1795, vol I, 469–70.
42 Hutton owned a copy of Sir William Hamilton's *Campi Phlegraei* (1776), in which Hamilton gave first-hand accounts of volcanic activity at Mount Vesuvius, with coloured illustrations, and Hutton knew him personally. They agreed that Arthur's Seat was a basaltic extrusion rather than the remnants of a volcano, as we know it to be today. *See* Dean, *James Hutton and the History of Geology*, 13.
43 Hutton 1795, vol I, 19–20, 168, 181, 208, 285, 450.
44 Scrope, G P 1858 *The Geology and Extinct Volcanoes of Central France*, 2 edn. London: John Murray, 209.
45 Yarwood 1970, 56–9.
46 *See* Erika Naginski's forthcoming *Architectures of Retrospection* for more on this connection.
47 *See* his engraving *Remains of the Villa of the Maecenas at Tivoli*, for example (*Vedute di Roma disegnate ed incise da Giambattista Piranesi architetto veneziano*, Rome, 1748–78, part II).
48 *See* the capriccios in Adam's hand in Adam vols 56 and 57 at Sir John Soane's Museum. Thanks to Stephen Astley for identifying the source of these motifs for me.
49 *See*, for example, Rudwick, M J S 1976 'The emergence of a visual language for geological science, 1760–1840'. *History of Science* **14** (3), 149–95. This article was published before Clerk of Eldin's drawings came to light, and while these would have complicated the picture Rudwick paints of a paucity of geological illustrations prior to the 19th century, his overall premise that geological visual materials and modes of communication developed rapidly and idiosyncratically, and are of fundamental importance to historians of science, still holds. Sam Smiles makes the point that architecture and anatomy were the only two such models available for this kind of knowledge-work in Britain at the time. *See* Smiles 2000, 23.
50 NRS, GD18/4981.
51 Hutton 1795, vol I, 185.
52 NRS, GD18/4981. My own interpretation of these paintings is indebted to the pioneering research of A A Tait, who first brought them to scholarly attention. *See* Tait 1993. Towards the end of his life Adam also viewed these landscape works as offering some financial security for his family, and during one particularly bad bout of illness in 1786 gave them to his three sisters in London, lest he die. *See* Sanderson 1992, 103.
53 Astley 2000, 5; *Gentleman's Magazine*, March 1792, 282–3.
54 Adam and Clerk of Eldin were familiar with Gilpin's work and even went so far as to copy his compositions. See BAM, BA151 and BA110. Both men were also influenced in their art by the work of Sandby, who became a friend of the Adam family through his role as a draughtsman to the Board of Ordnance at Fort George, near Inverness, in the early 1750s, when Robert and James Adam were engaged on construction work there for the family practice. *See* Fleming 1962, 84–6, 100–101, 259–60; also Tait 1993, 13–15. The Adam–Clerk–Sandby circle is also discussed in Tait 1980, 110–18.
55 Richard Emerson has revealed the significant extent to which Clerk of Eldin was involved in architectural experimentation, adding weight to this possibility. *See* his 'Robert Adam and John Clerk of Eldin: From primitive hut to temple of religion', *in* Gow and Rowan 1995, 164–79. I am grateful to Ian Gow for first drawing my attention to Richard Emerson's scholarship and to Richard himself for his very helpful remarks and suggestions. Adam and Clerk of Eldin seem to have made a practice both of sketching together and of producing drawings based on each other's compositions or a shared point of reference. In addition to the examples Emerson mentions, the Hunterian Art Gallery at the University of Glasgow owns a *View of Crossraguel Abbey* by Adam after Clerk of Eldin's own etching, *Abbey of Crossraguel* (1762; repr in *A Series of Etchings Chiefly of Views in Scotland*, Edinburgh: Bannatyne Club, 1855). Clerk of Eldin also accompanied Robert Adam on various site visits and to client meetings during Adam's castle period. *See* Sanderson 1992, 116.
56 The draft is from the BAM, BA 128; the final version is SM, 68.1.17.
57 Thanks to Stephen Astley for first pointing out these perspectival anomalies to me. Other particularly vertiginous examples include SM 68.1.7, and D462, D463, D465 and D466 in the National Gallery of Scotland.
58 Other examples included BAM, BA 140, BA 254; and National Gallery of Scotland, D459.
59 Dean, *James Hutton and the History of Geology*, 15.
60 Hutton 1795, vol II, 211.
61 Dean, *James Hutton and the History of Geology*, 26.

11

'The parent style or the original sin': the Adam revival in America

Eileen Harris

Apart from Henry Hope Reed's introduction to the American edition of *The Works in Architecture of Robert and James Adam*, published in New York in 1980, the Adam revival in America has never been systematically explored and brought together.[1] I was only able to scratch the surface in my talk at the 2015 Georgian Group Adam Symposium, and I hope that readers of this printed version will come forth with more examples and more information.

The title of this essay – 'The parent style or the original sin?' – comes from a humorous account of the styles of architecture written by the American architect George P Chappell (d 1933) and illustrated by Rockwell Kent (d 1971), who is best known as a painter and illustrator but was actually trained as an architect at my alma mater, Columbia University. Its appearance in January 1915 in *The Brickbuilder*, an architectural monthly published in Boston, links it to the so-called Adam revival in America.[2] 'The parent style or the original sin' also seems to me to encapsulate the love–hate response to Adam's work from its heyday in the 1760s and 1770s right up to the present. There are admirers and detractors.

Soane praised Adam for replacing the ponderous Palladian style based on Roman temples and public buildings with a light and fanciful antique style of decoration that was better suited to private buildings and to the elegance of modern refinement.[3] But Greek revivalists did not want anything fanciful. They wanted correct Classicism and so dismissed the Adam style as depraved. To Gothic revivalists and admirers of the Italian Renaissance and French baroque and rococo, Adam was too Classical, too light and too delicate – not grand or luxurious enough. Similar ideas were current in America in the late 19th and early 20th centuries when interest in Adam was revived.

Important late 18th-century American architects like Charles Bulfinch (d 1844) in Boston and Washington, DC, and Samuel McIntire (d 1811) in Salem were adapting Adam models in the 1790s. For example, Bulfinch's staircase ceiling in the Perez Morton House in Dorchester, Boston, of 1796 had decorative plasterwork that was unmistakably Adamesque, and the central building of his Tontine Crescent of 1794 (which until 1858 housed the old Boston Library) was modelled on the elevation of the Royal Society of Arts at the Adelphi, published in the Adam brothers' *The Works in Architecture* in 1778. But Bulfinch and McIntire belong to the late Adam or Federal period; they are not Adam revivalists.

In America the Adam revival was a by-product of the resurgence of interest in the Georgian period as a whole, which in turn was part of the Colonial Revival generated by the Philadelphia Centennial Exhibition of 1876, celebrating America's colonial heritage. A pair of Adam-style satinwood and marquetry side tables was shown at Philadelphia by Wright & Mansfield, a high-class and exclusive London firm specialising in Adam and Sheraton revival furniture (the two often went together). It was the purchase by the Victoria and Albert Museum of Wright & Mansfield's prize-winning Adam-style cabinet from the 1867 Paris Exhibition, inlaid with Wedgwood medallions, that launched the revival of the Adam style in Britain.[4]

What influence Wright & Mansfield's Adam-style exhibits had on American designers is not known. My guess is not very much; Adam did not take America by storm. The revival style of decoration, generally known as 'Adams' with an 's', took a while to catch on and was never adopted universally throughout America. Before the First World War the preferred styles were Italian Renaissance and French Louis: Louis Quatorze, Louis Quinze and Louis Seize. Adam was considered the English version of Louis Seize. Creative eclecticism – or what we might

Fig 11.1
Miss Emily Trevor's town house, 15 East 90th Street, New York, designed by Mott Schmidt, 1926. Exterior (top) and dining room (bottom).

call 'more or less styles' – was quite acceptable; reproduction, much less so.

By and large the appeal of the Adam style in late 19th- and early 20th-century America was much the same as it was in Britain. Adam was an elegant, tasteful and affordable alternative to the ostentation of the great houses of the plutocracy – houses like the one built in the mid-1880s for William Henry Vanderbilt on New York's Fifth Avenue between 51st and 52nd Streets (by Charles B Atwood and John B Snook). The Adam-style dining room in Emily Trevor's house at 15 East 90th Street, with its delicate plasterwork ceiling, is a prime example of this shift in taste (Fig 11.1). It was designed in 1926 by Mott Schmidt, the leading exponent of the Adam style in America. Compared with other revival styles, Adam was fairly free from gross eccentricities and easily adapted to modern requirements.

Furnished Adam room-settings shown at the 1876 Centennial Exhibition and also at the enormous 1904 Louisiana Purchase Exposition (or St Louis World's Fair) gave the wider American public a general idea of what the style was like. So too did the miniature Adam rooms later made by Mrs James Ward Thorne and exhibited by her at the World's Fairs in San Francisco in 1939, New York in 1940 and elsewhere, before being given to the Chicago Art Institute, where they remain (Fig 11.2).[5]

Often such revivalist creations were based on the Adams' own work. An 'Adams Room' exhibited at St Louis – one of five furnished rooms by the London firm of George Trollope & Son displaying various eras of British interior design – is a good example. It had a vaulted plasterwork ceiling, frieze, doorways and chimney piece, all said to be 'taken from examples designed by the celebrated architects, Robert and James Adam, in the latter part of the eighteenth century'. Regency striped wall coverings and Sheraton furniture completed the look. Mrs Thorne's miniature Adam dining room was modelled on that at Home House in Portman Square, London, as illustrated in Arthur T Bolton's two-volume monograph *The Architecture of Robert and James Adam*, published by Country Life in 1922. Bolton's study was and still is a must for anyone interested in Adam, and it is not surprising to find the names of leading American Georgian Classicists like Ogden Codman and Mott Schmidt among the subscribers. Inspiring as these three-dimensional exhibition stands and miniature rooms may have been, they were of little practical use to American architects and interior designers.

The London publisher B T Batsford came to the rescue in 1880 with a modestly priced selection of 26 lithographed plates of architecture, decoration and furniture reproduced from the Adam brothers' original *Works in Architecture*, which had become exceedingly rare and expensive. Batsford's reprint was published simultaneously in London, New York and Boston and was followed in 1900 by a further Batsford selection.[6]

In 1915 a portfolio of mantels was published in New York containing 37 drawings in colour reproduced, it was claimed, from recently discovered originals by Robert and James Adam. Its title page was copied from Adam's engraved design for the chimney piece in the Countess of Derby's dressing room at Grosvenor Square, published in *The Works in Architecture*. Most of

Fig 11.2
Mrs James Ward Thorne's miniature English Dining Room of the Georgian Period, 1770–90, *of c 1937.*

the drawings in the portfolio seem to be by Adam, but they were poorly reproduced, and it is doubtful that they aroused much interest. One is of a chimney piece for the great drawing room at Bolton House, for which there is a superb coloured drawing in the Adam collection at the Soane Museum.[7]

By this date a chimney piece in the Adam style was a *sine qua non* of an 'Adams' room and could be bought off the peg from London firms like Litchfield & Co, which in 1910 opened a New York showroom at 200 Fifth Avenue, displaying a large stock of genuine English marble and carved wood mantels. And it wasn't just chimney pieces that could be bought off the peg: adverts of department-store displays of this period show more Adam-inspired ornaments and furniture than most genuine Adam rooms, and one could even buy a whole Adam-style room. For instance, a so-called Angelica Kauffmann Room – a revival room with 'pure Adam panelling', a stucco ceiling and a marble chimney piece, specifically designed to complement original painted panels by Angelica – was offered for sale in 1927 at Arthur S Vernay's antique shop on 45th Street in New York. Angelica Kauffmann and Adam seem to have been a fashionable 'item' of the period, with Angelica receiving all the credit for the work of her husband, Antonio Zucchi, who was Adam's principal decorative painter. According to the antiques dealer Chris Jussel, the Angelica Room was bought in 1930 by the New York business-man and philanthropist Nicholas F Brady and installed at Inisfada, his mansion in Manhasset, Long Island. It was probably removed along with many other fittings when the house became a Jesuit retreat in 1937.

Colour was one of the most important elements of an Adam interior. In 1928 an enterprising American artist, Gerald K Geerlings, published a portfolio of *Color Schemes of Adam Ceilings*, reproducing original drawings at the Soane Museum (Fig 11.3). Having seen Adam's drawings at first hand, Geerlings concluded that Adam-style interiors in America were nothing but colourless 'phantoms of whites and greys'. And this, he claimed, was owing to the fact that there was nothing but black and white engravings and photographs for designers to work from. Worthy as it was, Geerlings' attempt to liven up the limited American palette of Adam green, Wedgwood blue, cream, white and grey had little effect.[8]

In the 1920s, surviving Adam houses, of which there were then still many in England and Scotland, remained by and large private and inaccessible. Not until after the Second World War, when houses were opened to the public and photographed in colour, did the world know what Adam interiors really looked like. And even then, their continuous restoration raised numerous questions.

There is a coloured wall-elevation by Ogden Codman at the Metropolitan Museum in New York, which was probably drawn in the early 1900s when Codman, having previously looked for inspiration only to France and Italy, began occasionally to turn to Adam (Fig 11.4, top). Although panels of multicoloured grotesque ornaments were not exclusive to Adam, and had been used by Codman in 1894 in a French-style room for Cornelius Vanderbilt at The Breakers, his Rhode Island mansion, here their arrangement is unmistakeably English. The cove, in fact, is taken straight from the engraving in *The Works*

Fig 11.3
A lithograph illustration from Gerald Geerlings' Color Schemes of Adam Ceilings, *1928.*

Fig 11.4
A wall elevation by Ogden Codman, early 1900s (top) and detail from Adam's view of the Japanned Room at Buckingham House, from The Works in Architecture *(bottom).*

of Adam's painted ceiling in the Japanned Room at Buckingham House, later Buckingham Palace (Fig 11.4, bottom).

Codman's use of Adam in another undated, uninscribed design – also at the Metropolitan Museum – is much more extensive, and quite easy to identify with the furniture and furnishings designed by Adam in 1774 for Derby House in Grosvenor Square, as illustrated in the rare, hand-coloured copy of *The Works* at the Soane Museum. Derby House was demolished in 1862 but its celebrated commode made by Ince & Mayhew survives at Knowsley, the Earl of Derby's house in Derbyshire.[9]

Though it may be a complete red herring, I can't help wondering whether Codman's two Adam-style rooms were intended for the house at 17 East 51st Street that he designed around 1905 for the stockbroker Frank K Sturgis, president of the New York Stock Exchange and a serious collector of old Wedgwood. The Adam revival and Wedgwood-ware went hand in hand in England – as shown by the Wright & Mansfield cabinet at the V&A – and I suspect this might also have been the case in America. More work is needed on this subject.

An Adam-fronted house like Sturgis House, clearly based on 20 St James's Square, was something of a rarity in Manhattan in the early 1900s, and a rarity in London too. Most of Adam's best-known work had been confined to the interiors of pre-existing houses. The great houses of Fifth Avenue and Park Avenue were in the French or Italian styles – inside and out. The

Adam style, when admitted at all, was relegated to the private apartments on the upper floors. There was a marble 'Adams' bathroom with a vaulted ceiling at the top of the Florentine-style palazzo at 50 East 70th Street, on the south-west corner of Park Avenue, built by the architects Trowbridge and Livingston in 1914 for George Blumenthal, a Wall Street financier, art collector and benefactor of the Metropolitan Museum (Fig 11.5). The bathroom design was closely modelled on Lady Wynn's dressing room or boudoir at 20 St James's Square. (Blumenthal's house was demolished in 1945 for a new apartment block.[10])

The private Adam sitting room at the top of the great house on Fifth Avenue and 67th Street, built in 1907–9 for the railroad millionaire George Jay Gould I by Horace Trombauer, was both a relief from the overpowering French atmosphere of the principal reception rooms on the lower floors and a buffer between Mrs Gould's painted Louis Seize bedroom and her

Fig 11.5
Adam-style bathroom in the home of George and Florence Blumenthal, 50 East 70th Street, New York, photographed in the 1920s.

Fig 11.6
The Adam revival sitting room at George J Gould I's house on Fifth Avenue.

husband's Italian Renaissance room (Fig 11.6). The sitting room's ceiling, with its delicate stucco decorations and inset painted roundels, was presumably based on an Adam model. But the real Adam hallmarks in the room were the large arched mirrors with gilded traceries, taken from the engraving of Adam's second drawing room at Derby House. Derby House also inspired the decorative pilaster strips used to articulate the walls. Somehow Gould House managed to survive until the 1960s, when it was replaced by an apartment block.

There was an extraordinary Adamesque house on East 91st Street, just off Fifth Avenue, which was built sometime between 1916 and 1919 for the industrialist Charles Mather MacNeill to designs by Frederick Sterner, using elements taken from Adam's elevation for the stable block at 20 St James's Square (Fig 11.7). Sterner's addition of an attic storey and balcony above a prominent, overhanging cornice gave the Adamesque house something of an Italian look. C M MacNeill was a pioneering metallurgist who had made a fortune from Colorado's gold and copper mines with his partners Spencer Penrose and C L Tutt senior. MacNeill died in 1923 and was succeeded at 15 East 91st Street by the famous art dealer Joseph Duveen. The house was demolished in 1940, after Duveen's death, and replaced by an apartment block. Sterner is best known for his innovative transformations of Manhattan brownstones, but he also provided early designs for the million-dollar Broadmoor Hotel that MacNeill and Penrose built on Cheyenne Lake in 1916–18 (he was later dismissed and replaced by Warren & Wetmore, architects of the Biltmore Hotel in New York).

Fig 11.7 (right)
The house of C M MacNeill, 15 East 91st Street, New York.

Fig 11.8 (below)
The Astor residence at 130 East 80th Street, in 1943.

Inside MacNeill's house was an Adam-style drawing room with an overmantel mirror that combined Adam's design for the third drawing room at Derby House with the arched top of the mirror from the second drawing room, also used as a source in the Jay Gould house. (Derby House and 20 St James's Square seem to have been the most frequently used sources for American Adam revivalists.) It appears from the reflection seen in the mirror in old photographs that this drawing room also had an oval Adam ceiling.

I came across Sterner's Adam interior for MacNeill in a fascinating trade catalogue issued in 1923 by the Murphy Varnish Company of New Jersey. In addition to essays on various aspects of the Adam style – and other styles too – the catalogue provided a chart of Adam period characteristics that would enable designers to assemble all the appropriate elements, from ceilings to lamps. (There is a photograph of the MacNeill drawing room at the bottom of the chart.[11])

The number of East Coast town houses with Adam façades – though by no means large – is larger than might be expected. Two are worthy of note before we move back indoors to discuss the more widespread use of the Adam revival style for interiors. First is the house at 173 Commonwealth Avenue, Boston, designed, or rather refronted, in 1919 for his own use by the architect J Harleston Parker, of the Boston firm Parker, Thomas & Rice. It was an almost complete rebuilding of an earlier house of the 1870s, though evidently its interiors were not particularly Adamesque. A second example is the elegant front of Vincent Astor's New York house at 130 East 80th Street, designed by Mott Schmidt in 1926 and considered by many to be his best building in the city. Its neoclassical limestone façade is usually regarded as a streamlined version of Adam's Royal Society of Arts, and the similarities are evident (Fig 11.8). But the overall effect of the composition, with the use of slim pilasters rather than columns, and the insertion of an attic floor beneath the pediment, is much more akin to the shallow profiles and proportions of the Adam town houses in Portland Place as they would have looked at this date as a result of Victorian and later alterations (for example, *see* Fig 9.12).

After the Great War the conspicuous display of wealth that had affected earlier houses went out of fashion, and was replaced by Classical simplicity and restraint. Clients with the most money wanted the least show. Added to this was the Anglomania championed by the influential decorator and tastemaker Elsie de Wolfe and the women associated with the Colony Club in New York, all of which focused attention on Adam and brought him out of the private bathroom and into the main reception rooms.

The exterior of the house that Mott Schmidt designed on East 90th Street in 1927 for Emily Trevor is Georgian–Colonial Revival (*see* Fig 11.1, top). Inside it is not exactly Adam in its entirety, but the front hall is unmistakably Adam, as are the drawing room and the dining room (*see* Fig 11.1, bottom). The tasteful but pallid beige, grey and Adam-green colours of these rooms recall Gerald Geerlings' comments on the lack of information in America about the rich variety of colour in Adam's original designs. These interiors still retain much of their original character, but today, when white is virtually the only colour we know, they can look bright, even garish.

Several of the residences designed by Mott Schmidt have staircases with good Adamesque ironwork, such as the country house he created for the investment banker C Douglas Dillon, at Far Hills, New Jersey, dating from 1936, and the lavish duplex apartment he fitted out for John D Rockefeller at 740 Park Avenue, in 1937. For inspiration Schmidt must have turned to Bolton's monograph on the Adam brothers, to which he was a subscriber in 1922. There he would have found photographs of delicate Adam balustrades on the staircases at Osterley, Kenwood and elsewhere. However, I don't know of any ironwork by Adam comparable to the Adam-style grille designed in 1901 by the architects Esenwein and Johnson as part of a new extension wing to the house they built three years earlier for the lawyer Harlow C Curtiss, at 864 Delaware Avenue in Buffalo. Perhaps the vertical arrangement of vases and anthemion on Queen Charlotte's sedan chair was their source of inspiration. It was designed by Adam in 1771 and, having been executed, was published in *The Works*, but is now untraced.

The vocabulary of Classical ornaments used by Adam and his contemporaries lent itself to virtually limitless combinations in a wide variety of materials, including wallpaper. Interesting and rather late examples are the Adamesque wallpapers used to decorate the interiors of West-Chester Hall, a Tudor Revival addition to the campus of the Buffalo Seminary, built in 1929 to designs by the local practice of Bley and Lyman (Fig 11.9). Though its exterior harmonises with the Collegiate Gothic of the original seminary building of 1909, its delicate interiors are mostly in the Federal or Adam revival style.

But the real hotspots of the Adam style in America are always said to be found in luxury hotels and Broadway theatres: two phenomena of the post-First World War period. So far I have not found much evidence of this, possibly because my definition of 'Adam style' is quite precise, while the norm is – or was – to apply Adam's name to everything neoclassical and decorative, everything with a few swags and anthemia, be it neo-Renaissance or Pompeian. Still, I suspect more Adamesque theatres and hotels will soon emerge in unsuspected places in America.

Fig 11.9
Adam-style wallpaper in the 1920s addition to the Buffalo Seminary at West-Chester Hall.

The Lucas Theatre in Savannah, Georgia, is a good example of an American Adam-style theatre (Fig 11.10). It was opened in 1921 by the pioneer film distributor Arthur Lucas as the first movie palace in Savannah, to designs by the Virginian architect Claude K Howell. Its auditorium, which reopened as a live theatre in 2001 after a major restoration overseen by Geoffrey Steward, has a central dome reminiscent perhaps of London's Theatre Royal, Drury Lane, as altered and decorated by Robert Adam for David Garrick in 1774. Howell is best known for his Colonial Revival mansions in the Monument Avenue district of Richmond (designed with his partner F W Scarborough) and also for several variety theatres across the southern states and movie theatres like the National in Virginia (1923), which also has Adam-style interior features. There was even an Adam-style cigar and theatre ticket sales counter at the Hotel Cadillac, at Broadway and 43rd Street, in the early 1900s. But the star of Adamesque theatres has to be New York's Ambassador, at 219 West 49th Street, designed in 1921 by the prolific Broadway theatre architect and designer Herbert J Krapp for the Schubert brothers. Unusually laid out with a diagonal orientation, to make the most of a constricted site, the theatre was decorated inside with Krapp's favourite Adam revival motifs and decorative panels (Fig 11.11).

There was, of course, more scope for Adam revival design in the public function rooms of luxury hotels – the lobbies, dining rooms and ballrooms – than in theatres. And it is worth remembering that we are looking at the roaring twenties, the Jazz Age, when modern styles were just as glamorous and as highly valued as historical ones. The Atlanta Biltmore Hotel, the brainchild of the Coca-Cola heir William Candler, opened in 1924 with the grandest display of Adam interiors anywhere in America – a positive Adam fanfare. The designers were the New York firm of architects Schultze and Weaver, favourites of the owner of the Biltmore chain, John McEtee Bowman, and renowned for their luxury hotels, including the Waldorf Astoria in New York and The Breakers in Palm Beach. A T Bolton's monograph on the Adams had just come out when Schultze and Weaver decided to 'go Adam'. They were among the first American architects to use the style.

The Atlanta Biltmore, known as the South's '*Supreme*' Hotel, closed in 1982 and was left to decay. In 1999 the ground-floor Adam rooms were restored by Geoffrey Steward's International

Fine Art Conservation Studios, and the upper 10 storeys were converted to offices. The hotel's original dining room, now the Imperial Ballroom of the restored hotel and apartments, has a ceiling design obviously derived from Adam's ceiling in the Long Gallery at Syon, with its intricate pattern of interlocking octagons. However, a comparison of the two shows just how much more sophisticated and varied Adam's ceiling at Syon was. The so-called Georgian Ballroom on the opposite side of the lobby at the Atlanta Biltmore is reminiscent of the music room at Home House on Portman Square (Fig 11.12). Tall, vaulted aisles run down both sides of the ballroom, their ceilings perhaps inspired by engravings of Adam's vaulted drawing room at Derby House, Grosvenor Square, a room widely regarded as one of his greatest achievements.

As well as the creation of these new interiors and buildings, this early 20th-century revival of interest in the 'Adam style' in America also included the rescue and reuse of authentic

Fig 11.10 (opposite) Details of the Adam revival auditorium at the Lucas Theatre.

Fig 11.11 (left and below) Details of the Ambassador Theatre, New York.

Fig 11.12
The Georgian Ballroom at the Atlanta Biltmore Hotel.

Fig 11.13
Altered façade at 481 Delaware Avenue, Buffalo, built in 1894 as the home of Dr Bernard Bartow (right). Detail of Robert Adam's design for a mirror for the first drawing room at Derby House, Grosvenor Square, from The Works in Architecture of Robert and James Adam *(far right).*

interiors taken from houses in Britain. The earliest examples of genuine Adam work imported to America are in the Center for Inter-American Relations, a neo-Georgian house on the corner of Park Avenue and 68th Street, designed by McKim, Mead and White for the financier and philanthropist Percy Rivington Pyne in 1909. As luck would have it, the completion of the shell of the Pyne House in 1911 coincided exactly with the demolition of the Duke of Bolton's house and its Adam interiors on London's Russell Square for a new hotel. Pyne bought the ceiling of the Duchess of Bolton's dressing room for his drawing room, and the marble chimney piece in the Duke of Bolton's drawing room for his front hall. In order to make Adam's square dressing room into a rectangular drawing room, fairly convincing Adam-style compartments were added over the chimney piece and at the opposite end of the room. Adam's deep cove was not used, nor was the bow. A comparison of Adam's exceptional coloured design in the Soane Museum with the painted ceiling in New York is further proof of the anaemic American interpretation of Adam interiors referred to by Geerlings. Percy Pyne's pioneering Adam imports in 1911 were followed in the 1930s by the purchases of the Lansdowne House dining room (by the Metropolitan Museum in New York) and drawing room (by the Philadelphia Museum of Art), the Croome Court Tapestry Room (by the Met) and the furniture from the gallery at Croome (by Philadelphia).[12] (This early 20th-century transatlantic trade in period rooms is rightfully my husband John Harris's subject, not mine, and is covered in his book *Moving Rooms*.[13])

I would like to finish with what I think is the most imaginative and humorous example of the use of the Adam style in America: the shopfront of a bakery at 481 Delaware Avenue, Buffalo, a historic terraced house built in 1894 by the architects Marling and Johnson for the distinguished orthopaedic surgeon Dr Bernard Bartow (d 1920). Although the designer of the shopfront and its precise date are not known, its source is easy to identify. It is Adam's chimney-glass in the first drawing room at Derby House, Grosvenor Square, and it is a particularly effective and dramatic use of an Adam interior motif on the elevation of a Colonial Revival house (Fig 11.13).

Notes

1 Reed, H H 1980 *The Works in Architecture of Robert and James Adam*, with a new introduction by Henry Hope Reed. New York: Dover.

2 *The Brickbuilder*, January 1915, 23.

3 Bolton 1929, 172 (lecture 11, 16 March 1815).

4 For Wright & Mansfield's work and the English contribution to the Philadelphia Exhibition, *see* Harris, E 1996 'Adams in the family: Wright and Mansfield at Haddo, Guisachan, Brook House and Grosvenor Square'. *Furniture History* **32**, 141–58, and Donnelly, M 2001 'British furniture at the Philadelphia Centennial Exhibition, 1876'. *Furniture History* **37**, 91–120; *also* Symonds, R W and Whineray, B B 1962 *Victorian Furniture*. London: Country Life, 45.

5 For Mrs Thorne's miniature rooms, *see* Boyer, H and Weingartner, F 2005 *Miniature Rooms: The Thorne Rooms at the Art Institute of Chicago* (2 rev edn). Chicago: Art Institute of Chicago.

6 *The Architecture, Decoration and Furniture of R. & J. Adam*. London: Batsford, 1880; *The Decorative Work of Robert and James Adam, Being a Reproduction of the Plates illustrating ... their Works in Architecture Published 1778–1812*. London: Batsford, 1900.

7 *A Book of Mantels: Thirty-seven Drawings in Color Reproduced from recently discovered originals by R. & J. Adam*. New York: Architectural Book, 1915; SM, Adam vol 22/276–7.

8 Geerlings, G K 1928 *Color Schemes of Adam Ceilings*. New York: Charles Scribner's Sons.

9 For the Derby House commode, *see* Roberts, H 1985 'The Derby House commode'. *Burlington Magazine*, May 1985, 275–82.

10 A comprehensive photographic survey of the house and its contents was made in the 1920s and survives in the Thomas J Watson Library, Metropolitan Museum of Art, New York (http://library.metmuseum.org/record=b1683771~S1) [accessed 18 April 2018]).

11 Murphy Varnish Company 1923 *The Book of Decoration*. Newark, New Jersey: Murphy Varnish Company.

12 For the Bolton House interior and its installation in the Pyne House, *see* Harris 1995; for the Lansdowne House drawing room, *see* Fiske Kimball, F 1943 'Lansdowne House redivivus'. *Philadelphia Museum Bulletin* **39** (no 199, November 1943), np.

13 Harris, J 2007 *Moving Rooms: The Trade in Architectural Salvages*. London and New Haven: Yale University Press.

Bibliography

Adam, J 1789 *Practical Essays on Agriculture*, 2 vols. London: T Cadell

[Adam, J] 1831 'Journal of a tour in Italy: By Robert Adam, Esq'. *Library of the Fine Arts or Repertory of Painting, Sculpture, Architecture and Engraving* **2** (9), October–November 1831, 165–78, 235–45

Adam, R 1764 *Ruins of the Palace of the Emperor Diocletian at Spalatro in Dalmatia*. London: The Author

Adam, R and Adam, J 1773–1822 *The Works in Architecture of Robert and James Adam, Esquires*, 3 vols. London: The Authors

Adam, W 1812 *Vitruvius Scoticus: Plans, Elevations and Sections of Public Buildings, Noblemen's and Gentlemen's Houses in Scotland*. Edinburgh and London: Adam Black and J J Robertson; T Underwood and J Taylor

Allan, D G C 1974 *The Houses of the Royal Society of Arts: A History and a Guide*, 2 edn. London: Royal Society of Arts

Ansiaux, S 1952 'Les dessins d'Italie de Laurent-Benoit Dewez'. *Bulletin de l'Institut Historique Belge de Rome*, 1–15

Ashby, T 1914 'Drawings of ancient paintings in English collections'. *Papers of the British School at Rome* **7** (1), 1–62

Astley, S 2000 *Robert Adam's Castles* London: Soane Gallery

Aymonino, A 2010a 'Decorum and celebration of the family line: Robert Adam's monuments to the 1st Duchess of Northumberland'. *Burlington Magazine* **152**, 288–96

Aymonino, A 2010b 'Syon House e l'uso delle fonti antiquarie nella decorazione inglese del Settecento', *in* Brook, C and Curzi, V (eds) *Roma e l'antico: Visione e realtà nel Settecento* (exhibition catalogue). Turin: Skira, 207–12

Aymonino, A forthcoming *Georgian Taste: The 1st Duke and Duchess of Northumberland as Patrons of Robert Adam*. New Haven and London: Yale University Press

Aymonino, A, with Gwynn, L and Modolo, M 2013 *Paper Palaces: The Topham Collection as a Source for British Neo-Classicism*. Eton: Eton College Press

Bagneris, M L 2010 *Agostino Brunias: Capturing the Carribean* [sic] *(c 1770–1800)*. London: Robilant + Voena

Beard, G W 1953 'Robert Adam at Croome Court'. *The Connoisseur* **132**, 73–6

Beard, G 1967 'The Rose Family of Plasterers'. *Apollo* **85**, 266–77

Beard, G 1975 *Decorative Plasterwork in Great Britain*. London: Phaidon

Beard, G 1978 *The Work of Robert Adam*. Edinburgh: J Bartholomew

Beard, G W 1981 *Craftsmen and Interior Decoration in England, 1660–1820*. Edinburgh: J Bartholomew

Beard, G 1992 'Robert Adam's craftsmen', *in* Worsely, G (ed) *Adam in Context: Georgian Group Symposium*. London: Georgian Group, 24–7

Belmarić, J and Šverko, A (eds) 2017 *Robert Adam and Diocletian's Palace in Split*. Zagreb: Školska knjiga

Bignamini, I, Hornsby, C, Giovanpaola, I D and Yarker, J 2010 *Digging and Dealing in Eighteenth-Century Rome*, 2 vols. New Haven and London: Yale University Press

Bolton, A T 1917 'Robert Adam, R.S., F.S.A., architect to King George III and to Queen Charlotte, as a bibliographer, publisher and designer of libraries: A paper read before the Bibliographical Society, January 15, 1917'. *Transactions of the Bibliographical Society* **14** (1915–17), 243–310

Bolton, A T 1922 *The Architecture of Robert and James Adam*, 2 vols. London: Country Life and George Newnes

Bolton, A T (ed) 1929 *Lectures on Architecture by Sir John Soane*. London: Soane Museum

Bristow, I 1986 'The room in the context of Robert Adam's work'. *Philadelphia Museum of Art Bulletin* **82** (351/2, Summer 1986), 12–19

Bristow, I 1996 *Architectural Colour in British Interiors, 1615–1840*. New Haven and London: Yale University Press

Brown, I G 1992a *Monumental Reputation: Robert Adam and the Emperor's Palace* (exhibition catalogue). Edinburgh: National Library of Scotland

Brown, I G 1992b 'The picturesque vision: Fact and fancy in the capriccio plates of Robert Adam's Spalatro'. *Apollo* **136** (366), 76–82

Brown, I G 1998 'Spalatro *redivivus:* An Italian reworking of Robert Adam's plates of Diocletian's Palace'. *Apollo* **146** (431), 32–6

Bryant, J 2006 '"The purest taste": James "Athenian" Stuart's work in villas and country houses', *in* Weber Soros, S (ed) *James 'Athenian' Stuart, 1713–1788: The Rediscovery of Antiquity* (exhibition catalogue). New Haven and London: Yale University Press, 265–315

Calderón, L 2008 'Headfort House: Architecture, history and conservation'. Unpublished MA thesis, University College Dublin

Cameron, C 1772 *Baths of the Romans*. London: The Author

Cameron, N M 1993 'Adam and "Gothick" at Yester Chapel', *in* Lowrey, J (ed) *Architectural Heritage IV: Robert Adam*. Edinburgh: Architectural Heritage Society of Scotland, 39–44

Cassidy, B 2011 *The Life & Letters of Gavin Hamilton: Artist & Art Dealer in Eighteenth-Century Rome*, 2 vols. London: Harvey Miller

Chevtchenko, V, Cotté, S and Pinault Sørensen, M (eds) 1995 *Charles-Louis Clérisseau (1721–1820): Dessins du musée de l'Ermitage Saint-Pétersbourg* (exhibition catalogue). Paris: Réunion des Musées Nationaux

Colvin, H 1986 'The beginnings of the architectural profession in Scotland'. *Architectural History* **29**, 168–82

Colvin, H 1998 'Croome Church and its architect'. *Georgian Group Journal* **8**, 28–32

Colvin, H 2008 *A Biographical Dictionary of British Architects, 1600–1840*, 4 edn. New Haven and London: Yale University Press

Connor Bulman, L M 2001 '"All the profusion of Eaton & Santo Bartoli": The first collections of ancient painting in Britain', *in* Jones, L R and Matthew L C (eds) *Coming About ... a Festschrift for John Shearman*. Cambridge, MA: Harvard University Art Museums, 343–48

Cornforth, J 1973 'The genesis of Headfort'. *Country Life* **153**, 947–8
Cornforth, J 2004 *Early Georgian Interiors*. New Haven and London: Yale University Press for Paul Mellon Centre for Studies in British Art
Cosh, M 2003 *Edinburgh: The Golden Age*. Edinburgh: John Donald
Craig, G Y, McIntyre, D B and Waterson, C D 1978 *James Hutton's Theory of the Earth: The Lost Drawings*. Edinburgh: Scottish Academic
Croft-Murray, E 1962–70 *Decorative Painting in England, 1537–1837*, 2 vols. Worthing: Littlehampton
Curzi, V 2010 'La tradizione del classico come legittimazione culturale e politica: Modelli della pittura romana antica in Inghilterra e in Russia', in Brook, C and Curzi, V (eds) *Roma e l'antico: Visione e realtà nel Settecento* (exhibition catalogue). Turin: Skira, 197–206
Dodsley, J and Dodsley, R 1761 *London and Its Environs*, 6 vols. London: The Authors
Drury, P J 1984 'Joseph Rose senior's site workshop at Audley End, Essex: Aspects of the development of decorative plasterwork technology in Britain during the eighteenth century'. *Antiquaries Journal* **64** (1), 62–83
Dunbar, J G 1972 'The building of Yester House, 1670–1878'. *Transactions of the East Lothian Antiquarian and Field Naturalists' Society* **13**, 31
Evans, R (ed) 1997 *Translations from Drawing to Building and Other Essays*. London: Architectural Association
Fleming, J 1955 'Robert Adam, the grand tourist'. *Cornhill Magazine* **168** (1004), 116–37
Fleming, J 1956 'Allan Ramsay and Robert Adam in Italy'. *The Connoisseur* **137** (March 1956), 79–84
Fleming, J 1958a 'Adam Gothic'. *The Connoisseur* **142** (1958), 75–9
Fleming, J 1958b 'Cardinal Albani's drawings at Windsor: Their purchase by James Adam for George III'. *The Connoisseur* **142** (1958), 164–9
Fleming, J 1958c 'An Adam miscellany: The journey to Spalatro'. *Architectural Review* **123** (733), 103–7
Fleming, J 1959 'Messrs Robert & James Adam: Art dealers (I)'. *The Connoisseur* **137** (November 1959), 168–71
Fleming, J 1962 *Robert Adam and His Circle in Edinburgh & Rome*. London: John Murray
Fleming, J 1968a 'Robert Adam's castle style – I'. *Country Life* **143**, 1356–9
Fleming, J 1968b 'Seton Castle's debt to ancient Rome: Robert Adam's castle style – II'. *Country Life* **143**, 1444
Fleming, J 1968c 'A "retrospective view" by John Clerk of Eldin, with some comments on Adam's castle style', in Summerson, J (ed) *Concerning Architecture: Essays on Architectural Writers and Writing presented to Nikolaus Pevsner*. London: Allen Lane, Penguin, 75–84
Fleming, J 1984 'James Adam in Naples', *in* Chaney, E and Ritchie, N (eds) *Oxford, China and Italy: Writings in Honour of Sir Harold Acton*. London: Thames & Hudson, 169–76
Fowler, J and Cornforth, J 1978 *English Decoration in the 18th Century*, 2 edn. London: Barrie & Jenkins
Fraser, A G 1989 *The Building of Old College: Adam, Playfair & the University of Edinburgh*. Edinburgh: Edinburgh University Press
Gerson, M B 1981 'A glossary of Robert Adam's neo-classical ornament'. *Architectural History* **24**, 59–82
Gifford, J 1989 *William Adam, 1689–1748: A Life and Times of Scotland's Universal Architect*. Edinburgh: Mainstream
Gifford, J, McWilliam, C and Walker, D 1984 *The Buildings of Scotland: Edinburgh*. London: Penguin
Gifford, J and Walker, F A 2002 *The Buildings of Scotland: Stirling and Central Scotland*. New Haven: Yale University Press
Goodsir, S 2010, 'George Steuart and Robert Adam: A professional relationship revealed'. *Georgian Group Journal* **18**, 91–104
Gow, I 1990 'William Adam: A planner of genius', *in* Howard, D (ed) *Architectural Heritage I: William Adam*. Edinburgh: Architectural Heritage Society of Scotland, 62–73
Gow, I and Rowan, A (eds) 1995 *Scottish Country Houses, 1600–1914*. Edinburgh: Edinburgh University Press
Green, S 2014 *Dumfries House: An Architectural Story*. Edinburgh: Royal Commission on the Ancient and Historical Monuments of Scotland
Guilding, R 2014 *Owning the Past: Why the English Collected Antique Sculpture, 1640–1840*. New Haven and London: Yale University Press for the Paul Mellon Centre for Studies in British Art
Hanson, B 2003 *Constructing Authority: Architects and the 'Building World' from Chambers to Ruskin*. Cambridge: Cambridge University Press
Hardy, J 1978 'Robert Adam and the furnishing of Kedleston Hall'. *Connoisseur* **198** (July 1978), 197–208
Harris, E 1963 *The Furniture of Robert Adam*. London: Alec Tiranti
Harris, E 1977 'Batty Langley: A tutor to Freemasons (1696–1751)'. *Burlington Magazine* **119**, 327–35
Harris, E 1992 'The Lansdowne House drawing room: Reconstructing Adam – properly'. *Apollo* **136** (366, August 1992), 83–6
Harris, E 1993 'The parent style or the original sin? Adam revived', *in* Lowrey, J (ed) *Architectural Heritage IV: Robert Adam*. Edinburgh: Architectural Heritage Society of Scotland, 45–57
Harris, E 1995 'Robert Adam on Park Avenue: The interiors for Bolton House'. *Burlington Magazine* **137** (1103, February 1995), 68–75
Harris, E 2001 *The Genius of Robert Adam: His Interiors*. New Haven and London: Yale University Press
Harris, E 2007 *The Country Houses of Robert Adam from the Archives of Country Life*. London: Aurum
Harris, E and Savage, N 1990 *British Architectural Books and Writers, 1556–85*. Cambridge: Cambridge University Press
Harris, J 1958 'The transparent portico'. *Architectural Review* **123** (733), 108–9
Harris, J 1970 *Sir William Chambers, Knight of the Polar Star*. London: Zwemmer
Harris, J 1971 *A Catalogue of British Drawings for Architecture, Decoration, Sculpture and Landscape Gardening in American Collections, 1550–1900*. Upper Saddle River, NJ: Gregg
Harris, J 1973 *Headfort House and Robert Adam: Drawings from the Collection of Mr & Mrs Paul Mellon*. London: RIBA
Harris, J and Snodin, M (eds) 1996 *Sir William Chambers, Architect to George III*. New Haven and London: Yale University Press
Harris, L 1987 *Robert Adam and Kedleston: The Making of a Neo-Classical Masterpiece*. London: The National Trust
Haskell, F and Penny, N 1981 *Taste and the Antique: The Lure of Classical Sculpture, 1500–1900*. New Haven and London: Yale University Press
Hewlings, R 2006 'The London houses', in Weber Soros, S (ed) *James 'Athenian' Stuart, 1713–1788: The Rediscovery of Antiquity* (exhibition catalogue). New Haven and London: Yale University Press, 195–263
Howard, D (ed) 1990 *Architectural Heritage I: William Adam*. Edinburgh: Architectural Heritage Society of Scotland
Hussey, C 1936 'Headfort, Co Meath, the seat of the Marquess of Headfort – I, II & III'. *Country Life* **79**, 300–5, 326–31, 352–8

Huth, H 1962 'Agostino Brunias, Romano': Robert Adam's "Bred Painter". *The Connoisseur* **151** (December 1962), 265–9

Hutton, J 1795 *Theory of the Earth, with Proofs and Illustrations*. London and Edinburgh: Messrs Cadell junior and Messrs Creech

Ingamells, J (ed) 1997 *A Dictionary of British and Irish Travellers in Italy, 1701–1800*. New Haven and London: Yale University Press

Ireland, R 2008 'Headfort House: Decorative investigation'. Unpublished report, 3 vols

Ireland, R and Gundry, D 2011 'Headfort: A Robert Adam designed house interior rediscovered'. *Journal of Architectural Conservation* **17**, 59–80

Jacobus, L 1998 'On "whether a man could see before him and behind him both at once": The role of drawing in the design of interior space in England, *c* 1600–1800'. *Architectural History* **31**, 148–65

Jenkins, S 2012 'The decoration of the library at Kenwood'. *English Heritage Historical Review* **7**, 66–77

Kay, W R M 1990 'What's his line? Would the real William Adam please stand up? Some recent research discoveries', *in* Howard, D (ed) *Architectural Heritage I: William Adam*. Edinburgh: Architectural Heritage Society of Scotland, 49–61

Kay, W 1993 'Robert Adam: Some responses to a Scottish background', *in* Lowrey, J (ed) *Architectural Heritage IV: Robert Adam*. Edinburgh: Architectural Heritage Society of Scotland, 23–38

Kedleston 1769 *Catalogue of the Pictures, Statues, etc. with Some Accounts of the Architecture at Kedleston*. Kedleston: privately printed [updated in 1770, 1778 and 1796]

Kelsall, F 1984 'Liardet versus Adam'. *Architectural History* **27**, 118–26

King, D 1991 *The Complete Works of Robert and James Adam*. Oxford: Butterworth Architecture

King, D 2001 *The Complete Works of Robert and James Adam and Unbuilt Adam*. Oxford: Architectural Press

Kondo, A 2012 *Robert and James Adam, Architects of the Age of Enlightenment*. London: Pickering & Chatto

Kriz, K D 2008 *Slavery, Sugar, and the Culture of Refinement: Picturing the British West Indies, 1700–1840*. New Haven and London: Yale University Press

Lees-Milne, J 1947 *The Age of Adam*. London: Batsford

Lehmann, K 1961 'Piranesi as interpreter of Roman architecture', in *Piranesi* (exhibition catalogue). Northampton, MA: Smith College Museum of Art, 88–98

Lewis, L 1961 *Connoisseurs and Secret Agents in Eighteenth Century Rome*. London: Chatto & Windus

Lewis, W S (ed) 1937–83 *The Yale Edition of Horace Walpole's Correspondence*. New Haven: Yale University Press

Lindfield, P N 2014 'Serious Gothic and "doing the ancient buildings": Batty Langley's *Ancient Architecture* and *Principal Geometric Elevations*'. *Architectural History* **57**, 150–5

Lindfield, P N 2016 *Georgian Gothic: Medievalist Architecture, Furniture and Interiors, 1730–1840*. Woodbridge: Boydell & Brewer

Lowrey, J (ed) 1993 *Architectural Heritage IV: Robert Adam*. Edinburgh: Architectural Heritage Society of Scotland

Macarthur, J 2007 *The Picturesque: Architecture, Disgust and other Irregularities*. London and New York: Routledge

McCarthy, M 1987 *The Origins of the Gothic Revival*. New Haven and London: Yale University Press

Macaulay, J 1975 *The Gothic Revival, 1745–1845*. Glasgow and London: Blackie

Macaulay, J 1987 *The Classical Country House in Scotland, 1660–1800*. London and Boston: Faber & Faber

McCormick, T J 1978 'Piranesi and Clérisseau's vision of Classical antiquity', in Brunel, G (ed) *Piranèse et les Français: Colloque tenu à la Villa Médicis, 12–14 mai 1976*. Rome: Académie de France à Rome, Elefante, 308–10

McCormick, T J 1990 *Charles-Louis Clérisseau and the Genesis of Neo-classicism*. New York, Cambridge and London: Architectural History Foundation: MIT Press

MacInnes, R 1993 'Robert Adam's public buildings', *in* Lowrey, J (ed) *Architectural Heritage IV: Robert Adam*. Edinburgh: Architectural Heritage Society of Scotland, 10–22

MacInnes, R 1997 'Robert Adam's bridges', in Mays, D (ed) *The Architecture of Scottish Cities: Essays in Honour of David Walker*. East Linton, Scotland: Tuckwell, 78–86

Mayer, L 2011 'Landscape as legacy: Elizabeth Percy, 1st Duchess of Northumberland, and the Gothick garden buildings of Alnwick, Northumberland'. *Garden History* **39**, 34–9

Michel, O 1978 'Recherches sur Jean-Baptiste Lallemand à Rome', in Brunel, G (ed) *Piranèse et les Français: Colloque tenu à la Villa Médicis, 12–14 mai 1976*. Rome: Académie de France à Rome, Elefante, 327–44

Middleton, R, Beasley, G and Savage, N 1998 *The Mark J Millard Architectural Collection*. Vol II. *British Books, Seventeenth through Nineteenth Centuries*. Washington: National Gallery of Art

Miele, C 1998 '"The first architect in the world" in Brighton: Robert Adam, Marlborough House and Mrs Fitzherbert'. *Sussex Archaeological Collections* **136**, 149–75

Mowat, I 1979 'An 18th century private library: The books of William and John Adam at Blair Adam'. *The Library Review* **28** (Spring 1979), 8–13

Mowat, I R M 2001 'John Adam: The forgotten brother'. *Scottish Genealogist* **48** (2), 62–71

Murray, A 2012 'Joseph Rose and Company'. *Georgian Group Journal* **20**, 103–18

Musson, J and Barker, P 2017 *Robert Adam: Country House Design, Decoration and the Art of Elegance*. New York: Rizzoli International Publications Inc

Neale, J P 1818–23 *Views of the Seats of Noblemen and Gentlemen*. London: W H Reid

Nenadic, S 2012 'Architect builders in London and Edinburgh, *c* 1750–1800, and the market for expertise'. *Historical Journal* **55**, 597–617

Parissien, S 1992 *Adam Style*. London: Phaidon

Parker, J 1968 'Patrons of Robert Adam at the Metropolitan Museum'. *Metropolitan Museum of Art Journal* **1**, 109–24

Pilkington, P 2015 'Kerry House, Portland Place: A proposed Adam town house for the third Earl of Kerry'. *Georgian Group Journal* **23**, 147–66

Pinto, J A 2012 *Speaking Ruins: Piranesi, Architects and Antiquity in Eighteenth-Century Rome*. Ann Arbor, MI: University of Michigan Press

Pinto, J A 2017 '"The beautiful spirit of antiquity": Robert Adam and Diocletian's Palace', *in* Belmarić, J and Šverko, A (eds) *Robert Adam and Diocletian's Palace in Split*. Zagreb: Školska knjiga, 357–77

Polignac, F de and Raspi Serra, J (eds) 1998 *La fascination de l'antique, 1700–1770: Rome découverte, Rome inventée* (exhibition catalogue). Paris: Somogy éditions d'art

Reeve, M M and Lindfield, P N 2015 '"A child of strawberry": Thomas Barrett and Lee Priory, Kent'. *Burlington Magazine* **157**, 836–42

Rice, C 2007 *The Emergence of the Interior: Architecture, Modernity, Domesticity*. London and New York: Routledge

Robinson, J M 2012 *James Wyatt (1746–1813): Architect to George III*. New Haven and London: Yale University Press

Roscoe, I, Hardy, E and Sullivan, M G 2009 *A Biographical Dictionary of Sculptors in Britain, 1660–1851*. New Haven and London: Yale University Press

Rowan, A 1967 'Paxton House, Berwickshire, I, II & III'. *Country Life* **142** (17, 24 and 31 August 1967), 364–7, 422–5, 470–3

Rowan, A 1974 'After the Adelphi: Forgotten years in the Adam brothers' practice (I. William Adam and Company; II. The Adam castle style; III. Ideal villas as projected and built)'. *Journal of the Royal Society of Arts* **122**, 659–710

Rowan, A J 1975 'Batty Langley's Gothic', *in* Robertson, G and Henderson, G (eds) *Studies in Memory of David Talbot Rice*. Edinburgh: Edinburgh University Press, 197–215

Rowan, A 1983 'Sunnyside and Rosebank: Suburban villas by the Adam brothers'. *AA Files* **4** (July 1983), 29–39

Rowan, A 1984 'The building of Hopetoun'. *Architectural History* **27**, 183–209

Rowan, A J 1985 *Designs for Castles and Country Villas by Robert and James Adam*. Oxford: Phaidon

Rowan, A 1988 *Robert Adam: Catalogue of Drawings in the Victoria and Albert Museum*. London: Victoria & Albert Museum

Rowan, A 1990 'William Adam's library', *in* Howard, D (ed) *Architectural Heritage I: William Adam*. Edinburgh: Architectural Heritage Society of Scotland, 8–25

Rowan, A 1993 'The Adam brothers and contemporary office practice', in Worsley, G (ed) *Adam in Context (Papers given at the Georgian Group Symposium 1992)*. London: The Georgian Group, 42–9

Rowan, A 1997 'Kinross and Edinburgh: Some ideas of Robert Adam on the proper improvement of towns', *in* Mays, D (ed) *The Architecture of Scottish Cities: Essays in Honour of David Walker*. East Linton, Scotland: Tuckwell, 70–77

Rowan, A 1998 'The Duke of Northumberland's garden house at Hulne Priory'. *Architectural History* **41**, 265–73

Rowan, A 2003 *'Bob the Roman': Heroic Antiquity & the Architecture of Robert Adam* (exhibition catalogue). London: Sir John Soane's Museum

Rowan, A 2004 'Robert Adam's ideas for the North Bridge in Edinburgh'. *Architectural Heritage* **15** (1), 34–9

Rowan, A 2007 *Vaulting Ambition. The Adam Brothers: Contractors to the Metropolis in the Reign of George III* (exhibition catalogue). London: Sir John Soane's Museum

Ruddock, E C 1974–6 'The building of the North Bridge, Edinburgh, 1763–1775'. *Transactions of the Newcomen Society* **47**, 9–33

Rykwert, J and Rykwert, A 1984 *The Brothers Adam: The Men and Their Style*. London: William Collins

Salter, A 2013 *Four Emperors and an Architect: How Robert Adam Rediscovered the Tetrarchy*. Alresford: Lexicon Publishing

Sanderson, M H B 1982 'Robert Adam's last visit to Scotland, 1791'. *Architectural History* **25**, 34–46

Sanderson, M H B 1992 *Robert Adam and Scotland: Portrait of an Architect*. Edinburgh: HMSO

Sanderson, M H B 1993 'Trivial pursuit? Portrait of the artist in letters and diaries', *in* Lowrey, J (ed) *Architectural Heritage IV: Robert Adam*. Edinburgh: Architectural Heritage Society of Scotland, 66–81

Sands, F 2011 'The art of collaboration: Antonio Zucchi at Nostell Priory'. *Georgian Group Journal* **19**, 106–19

Sands, F 2016 *Robert Adam's London*. Oxford: Archaeopress Archaeology

Sands, F 2017 'Reconstructed and in ruins: The influence of Diocletian's Palace within the drawings of Robert Adam', *in* Belmarić, J and Šverko, A (eds) *Robert Adam and Diocletian's Palace in Split*. Zagreb: Školska knjiga, 379–99

Scott, J 2004 *The Pleasures of Antiquity: British Collectors of Greece and Rome*. New Haven and London: Yale University Press

Sicca, C M 1986 'On William Kent's Roman sources'. *Architectural History* **29**, 134–57

Simpson, J 1980. *Vitruvius Scoticus: William Adam*, facsim edn (from the copy in the University of Glasgow Library, with an Introduction by J Simpson). Edinburgh: Paul Harris

Smiles, S 2000 *Eye Witness: Artists and Visual Documentation in Britain, 1770–1830*. Aldershot, Hants: Ashgate

Stillman, D 1966. *The Decorative Work of Robert Adam*. London: Alec Tiranti

Stillman, D 1967 'Robert Adam and Piranesi', *in* Fraser, D, Hibbard, H and Lewine, M J *Essays in the History of Architecture presented to Rudolf Wittkower*. London: Phaidon, 197–206

Stillman, D 1973 'British Architects and Italian Architectural Competitions, 1758–1780'. *Journal of the Society of Architectural Historians* **32** (1 March 1973), 43–66

Summerson, J 1962 *Georgian London: An Architectural Study*, rev edn. London: Penguin

Swarbrick, J 1915 *Robert Adam and His Brothers: Their Lives, Work & Influence on English Architecture, Decoration and Furniture*. London: B T Batsford

Tait, A A 1971 'The picturesque drawings of Robert Adam'. *Master Drawings* **9**, 164–8

Tait, A A 1972 *Robert Adam and Scotland: The Picturesque Drawings* (exhibition catalogue). Edinburgh: Scottish Arts Council

Tait, A A 1978a 'Robert Adam and John Clerk of Eldin'. *Master Drawings* **16**, 53–7, 109–11

Tait, A A 1978b 'The sale of Robert Adam's drawings'. *Burlington Magazine* **120**, 451–4

Tait, A A 1980 *The Landscape Garden in Scotland, 1735–1835*. Edinburgh: Edinburgh University Press

Tait, A A 1981 'Robert Adam's picturesque architecture'. *Burlington Magazine* **123**, 421–4

Tait, A A 1984 'Reading the ruins: Robert Adam and Piranesi in Rome'. *Architectural History* **27**, 524–33

Tait, A A 1993 *Robert Adam: Drawings and Imagination*. Cambridge: Cambridge University Press

Tait, A A 2004 'Adam, Robert (1728–1792)'. *Oxford Dictionary of National Biography*. Oxford University Press, www.oxforddnb.com/view/article/105 [accessed 7 June 2016]

Tait, A A 2008 *The Adam Brothers in Rome: Drawings from the Grand Tour*. London: Sir John Soane's Museum

Thom, C 2015 'Robert Adam's first Marylebone house: The story of General Robert Clerk, the Countess of Warwick and their mansion in Mansfield Street'. *Georgian Group Journal* **23**, 125–46

Thom, C 2017 '"Spalatro on Thames": How Diocletian's Palace inspired Robert Adam's most audacious development – the Adelphi', *in* Belmarić, J and Šverko, A (eds) *Robert Adam and Diocletian's Palace in Split*. Zagreb: Školska knjiga, 401–27

Watkin, D (ed) 1972 *Sale Catalogues of Libraries of Eminent Persons*. IV. *Architects*. London: Mansell with Sotheby

Weber, S (ed) 2013 *William Kent: Designing Georgian Britain* (exhibition catalogue). New Haven and London: Yale University Press

Williams, G 2016 'The hidden hand of genius: Robert Adam and the Pulteney Estate in Shropshire'. *Georgian Group Journal* **24**, 65–80

Wilson-North, R and Porter, S 1997 'Witham, Somerset: From Carthusian monastery to country house to Gothic folly'. *Architectural History* **40**, 81–98

Wilton-Ely, J 1978 *The Mind and Art of Giovanni Battista Piranesi*. London: Thames & Hudson

Wilton-Ely, J 1993 *Piranesi as Architect and Designer*. New Haven, New York and London: Yale University Press/Pierpont Morgan Library

Wilton-Ely, J 1995 'Piranesi, Clérisseau and the Adam brothers'. *Bulletin de l'Association des Historiens de l'Art Italien* **2**, 15–24

Wilton-Ely, J 2001 '"Gingerbread and sippets of Embroidery": Horace Walpole and Robert Adam'. *Eighteenth-Century Life* **25**, 149–51

Wilton-Ely, J 2006 '"Amazing and ingenious fancies": Piranesi and the Adam brothers', *in* Bevilacqua, M, Hyde Minor, H and Barry, F (eds) *The Serpent and the Stylus: Essays on G B Piranesi*. Memoirs of the American Academy in Rome. Supplementary vol 4. Ann Arbor, MI: University of Michigan Press for the American Academy in Rome, 213–37

Wilton-Ely, J 2011 'Style and serendipity: Adam, Walpole and Strawberry Hill'. *British Art Journal* **11**, 3–14

Worsley, G (ed) 1993 *Adam in Context (Papers given at the Georgian Group Symposium 1992)*. London: The Georgian Group

Yarwood, D 1970 *Robert Adam*. London: Dent

Manuscripts and archive sources

(and abbreviations used in the notes and references)

BAM Blair Adam House, by Kelty, Fife, Scotland, Blair Adam Muniments

BL British Library

BM British Museum

LMA London Metropolitan Archives

NLI National Library of Ireland, Headfort papers:
MS 25,307(17), 'Book of returns from May 1774 to May 1775'
MS 25,307(18), 'Book of returns from May 1776 to May 1777'
MS 25,307(22), 'Book of returns from May 1780 to May 1781'
MS 25,307(26), 'Book of returns from May 1784 to April 1785'

NRS National Records of Scotland, Clerk of Penicuik papers (GD18)

RIBA Royal Institute of British Architects

SM Sir John Soane's Museum

TNA The National Archives

UL University of London, Senate House, Headfort papers, 'Book entering receipts for bills paid'

Index

Page numbers in *italics* refer to figures.

A

B

C

D

E

F

P

Q

R

S

T

U

V

W

Y

Z

Illustration credits

Historic England would like to thank the listed copyright holders for giving permission to reproduce their images within this book. Images reproduced courtesy of the Sir John Soane's Museum can be found on their website: https://www.soane.org.

Frontispiece © Keith Adam of Blair Adam, Blair Adam Muniments 61

Introduction

Fig 0.1 © National Portrait Gallery, NPG 2953

Fig 0.2 © Keith Adam of Blair Adam, photo by John McKenzie

Fig 0.3 Laing Art Gallery, Newcastle-upon-Tyne, UK © Tyne & Wear Archives & Museums/Bridgeman Images, TW426951

Fig 0.4 Photo and © Nick Haynes

Fig 0.5 © The Bute Archive at Mount Stuart, SC1135260

Fig 0.6 © Historic England Archive, BB95/11533

Fig 0.7 Courtesy Yale Center for British Art, Paul Mellon Collection, B1977.14.18468

Fig 0.8 Photo and © Nick Haynes

Fig 0.9 © Sir John Soane's Museum, Adam vol 11/002

Fig 0.10 © Historic England Archive, DP033522

Fig 0.11 © National Trust, NT 132456

Fig 0.12 © Sir John Soane's Museum, Adam vol 56/092

Fig 0.13 Photo and © Nick Haynes

Fig 0.14 © National Trust, NT 176494

Fig 0.15 © Sir John Soane's Museum, Adam vol 40/003

Fig 0.16 Royal Collection Trust © Her Majesty Queen Elizabeth II 2019, RCIN 917643

Fig 0.17 From *The Works in Architecture of Robert and James Adam*, vol II, No II, Plate V

Fig 0.18 (top) © Sir John Soane's Museum, Adam vol 34/033

Fig 0.18 (below) © Sir John Soane's Museum, Adam vol 34/005

Fig 0.19 (left) © Sir John Soane's Museum, Adam vol 46/064

Fig 0.19 (right) © Sir John Soane's Museum, Adam vol 46/065

Fig 0.20 © Sir John Soane's Museum, Adam Drawings SM 68/1/16

Fig 0.21 Photo and © Nick Haynes

Fig 0.22 Photo and © Nick Haynes

Chapter 1

Fig 1.1 © British Library Board. All Rights Reserved/Bridgeman Images, BL5242087, pl 17

Fig 1.2 Photo and © Nick Haynes

Fig 1.3 Photo and © Nick Haynes

Fig 1.4 Photo and © Nick Haynes

Fig 1.5 © Crown Copyright Historic Environment Scotland (HES), DP083154

Fig 1.6 © Crown Copyright Historic Environment Scotland (HES), DP033666

Fig 1.7 © Keith Adam of Blair Adam, photo by John McKenzie

Fig 1.8 Clerk of Penicuik Muniments © Sir Robert Clerk of Penicuik, photo by John McKenzie

Fig 1.9 Clerk of Penicuik Muniments © Sir Robert Clerk of Penicuik, photo by John McKenzie

Fig 1.10 © British Library Board. All Rights Reserved/Bridgeman Images, BL5242085, pl 123

Fig 1.11 Photo and © Alistair Rowan

Fig 1.12 Photo and © Nick Haynes

Fig 1.13 Clerk of Penicuik Muniments © Sir Robert Clerk of Penicuik, photo by John McKenzie

Fig 1.14 © Irish Architectural Archive

Fig 1.15 © Crown Copyright Historic Environment Scotland (HES), SC765862

Fig 1.16 Photo and © Alistair Rowan

Chapter 2

Fig 2.1 © The Pierpont Morgan Library & Museum 1977.43. Purchased on the fellows fund, with special assistance of Mr Roland Burdon-Muller, Mrs W Rodman Fay, Mrs Enid A Haupt, Mrs Gerard B Lambert. Mr Robert B O'Connor, and Mr John S Thacher.

Fig 2.2 From Caveceppi's, Raccolta d'antiche statue, vol I, 1768. Taken from Public Domain, Scan 27: https://arachne.uni-koeln.de/arachne/index.php?view[layout]=buchseite_item&search[constraints][buchseite][PS_BuchseiteID]=215829&view[section]=uebersicht&view[pag

Fig 2.3 (far left) William Randolph Hearst Collection (49.23.12). Taken from Public Domain: https://collections.lacma.org/node/229788 (ma-78467)

Fig 2.3 (left) From *Cavaceppi's Raccolta d'antiche statue*, vol I, 1768, plate 21; Public Domain, Scan 69: https://arachne.unikoeln.de/arachne/index.php?view[layout]=buchseite_item&search[constraints][buchseite][PS_BuchseiteID]=215871&view[section]=uebersicht&vie

Fig 2.4 (far left) © National Trust, CMS_NOS0456

Fig 2.4 (left) From *Cavaceppi's Raccolta d'antiche statue*, vol 1, 1768, plate 16; Public Domain, Scan 59: http://arachne.uni-koeln.de/arachne/index.php?view[layout]=buchseite_item&search[constraints][buchseite][searchSeriennummer]=215861&view[section]=uebersicht

Fig 2.5 (below) © Sir John Soane's Museum, Adam vol 26/131

Fig 2.5 (bottom) From *Cavaceppi's Raccolta d'antiche statue*, vol III, 1772, plate 39; Public Domain, Scan 107: https://arachne.uni-koeln.de/arachne/index.php?view[layout]=buchseite_item&search[constraints][buchseite][PS_BuchseiteID]=217820&view[section]=uebersic

Fig 2.6 (top) © Sir John Soane's Museum, Adam vol 26/081

Fig 2.6 (above) From Cavaceppi's Raccolta d'antiche statue, vol III, 1772, plate 34; Public Domain, Scan 97: https://arachne.uni-koeln.de/arachne/index.php?view[layout]=buchseite_item&search[constraints][buchseite][PS_BuchseiteID]=217810&view[section]=uebersich

Fig 2.7 © Historic England Archive, K960113

Fig 2.8 © Historic England Archive, DP103377

Fig 2.9 © Country Life Picture Library, CL5439513

Fig 2.10 Gift of John D. Rockefeller Jr., 1932 , Accession Number: 32.11.4 , The Metropolitan Museum of Art, New York. Taken from Public Domain: https://www.metmuseum.org/art/collection/search/253373

Chapter 3

Fig 3.1 Courtesy of Adam Williams Ltd, New York

Fig 3.2 From *Compendio delle Vite de' Pittori Veneziani istorici*, Plate XXI © The Trustees of the British Museum, Dept of Prints & Drawings, Museum No 1938,0226.1.1-53

Fig 3.3 (left) From *Varie Fabriche Antiche, e Moderne della Citta di Verona* © The Trustees of the British Museum, Dept of Prints & Drawings, Museum No 1952,0206.1.4, plate 22

Fig 3.3 (right) From *Varie Fabriche Antiche, e Moderne della Citta di Verona* © The Trustees of the British Museum, Dept of Prints & Drawings, Museum No 1952,0206.1.4, plate 22

Fig 3.4 Photo and © akg-images/Cameraphoto, AKG291091

Fig 3.5 © Historic England Archive, K030572

Fig 3.6 Private Collection. Taken from Public Domain: http://www.the-athenaeum.org/art/detail.php?ID=118176

Fig 3.7 © The Trustees of the British Museum, Dept of Prints & Drawings, Museum No Ff,3.218

Fig 3.8 © The Robert Lehman Collection, The Metropolitan Museum of Art, New York, 1975.1.555. Public Domain: https://www.metmuseum.org/art/collection/search/459951

Fig 3.9 (far left) © Sir John Soane's Museum, Adam vol 125/104

Fig 3.9 (left) © Sir John Soane's Museum, Adam vol 26/070

Fig 3.10 Courtesy of Adam Williams Ltd, New York

Fig 3.11 © The Trustees of the British Museum, Dept of Prints & Drawings, U.923.55

Fig 3.12 Courtesy of Adam Williams Ltd, New York

Fig 3.13 From Baron von Stosch's Gemmae antiquae, Plate XXII

Fig 3.14 © Historic England Archive, DP103211

Fig 3.15 Courtesy of Adam Williams Ltd, New York

Fig 3.16 © Sir John Soane's Museum, Adam vol 7/069

Chapter 4

Fig 4.1 From John Woolfe and James Gandon, *Vitruvius Britannicus*, IV, London, 1767, pl 46

Fig 4.2 © National Trust, Scarsdale Collection, Kedleston Hall, CMS_109133

Fig 4.3 © National Trust, Scarsdale Collection, Kedleston Hall, CMS_109132

Fig 4.4 © National Trust, Scarsdale Collection, Kedleston Hall, CMS_109134

Fig 4.5 © Victoria and Albert Museum, London, W.36.1 – 1975 to W.40.1 – 1975

Fig 4.5 © Victoria and Albert Museum, London, W.36.1 – 1975 to W.40.1 – 1976

Fig 4.5 © Victoria and Albert Museum, London, W.36.1 – 1975 to W.40.1 – 1977

Fig 4.5 © Victoria and Albert Museum, London, W.36.1 – 1975 to W.40.1 – 1978

Fig 4.5 © Victoria and Albert Museum, London, W.36.1 – 1975 to W.40.1 – 1979

Fig 4.6 From *Admiranda Romanarum antiquitatum*, Warburg Institute. Taken from Public Domain: https://digi.ub.uni-heidelberg.de/diglit/bartoli1693/0038/image

Fig 4.6 From *Admiranda Romanarum antiquitatum*, Warburg Institute. Taken from Public Domain: https://digi.ub.uni-heidelberg.de/diglit/bartoli1693/0039/image

Fig 4.6 From *Admiranda Romanarum antiquitatum*, Warburg Institute. Taken from Public Domain: https://digi.ub.uni-heidelberg.de/diglit/bartoli1693/0040/image

Fig 4.6 From *Admiranda Romanarum antiquitatum*, Warburg Institute. Taken from Public Domain https://digi.ub.uni-heidelberg.de/diglit/bartoli1693/0042/image

Fig 4.7 © National Trust, Scarsdale Collection, Kedleston Hall, CMS_109135

Fig 4.8 From Bernard de Montfaucon, *L' antiquité expliquée et représentée en figures*, 15 vols, Paris, Florentin Delaulne [et al], 1719–24, supplement III (1724), plate 59

Fig 4.9 © Eton College Library, Topham Collection, Bn.6.34

Fig 4.10 © Eton College Library, Topham Collection, TB 15; Ex Mahogany case

Fig 4.11 © RIBA Drawing & Archives Collections, vol 54, fol 24

Fig 4.12 © Eton College Library, Topham Collection, TB 13; Ex Mahogany case

Fig 4.13 © RIBA Drawing & Archive Collections, vol 54, fol 20

Fig 4.14 © Sir John Soane's Museum, London, Adam vol 54/01/048

Fig 4.15 © Sir John Soane's Museum, Adam vol 27/074–075

Fig 4.15 © Sir John Soane's Museum, Adam vol 27/074–075

Fig 4.16 (left) Getty Research Institute, Los Angeles, 2002.M.22*, nos 17–18. Taken from Public Domain, Leaf 19: https://rosettaapp.getty.edu/delivery/DeliveryManagerServlet?dps_pid=IE378583

Fig 4.16 (far left) Getty Research Institute, Los Angeles, 2002.M.22*, nos 17–18. Taken from Public Domain, Leaf 20: https://rosettaapp.getty.edu/delivery/DeliveryManagerServlet?dps_pid=IE378583

Fig 4.17 © Eton College Library, Topham Collection, Bn.7.39

Fig 4.18 © Eton College Library, Topham Collection, Bn.7.95

Fig 4.19 © Eton College Library, Topham Collection, Bn.5.26

Fig 4.2O © RIBA Drawing & Archive Collections, vol 54, fol 14

Chapter 5

Fig 5.1 From *The Works in Architecture of Robert and James Adam*, vol II, 1778, Plate VII

Fig 5.2 © English Heritiage Trust, EH14484

Fig 5.3 © Kent Wang, self photographed. Taken from Creative Commons.org/licenses/by-sa/2.0: https://www.flickr.com/photos/kentwang/33741948060/

Fig 5.4 From John Swarbrick *Robert Adam & His Brothers*, 1915, 146

Fig 5.5 From *The Works in Architecture of Robert and James Adam*, vol I, 1773, Plate V

Fig 5.6 (below) © Stourhead House @ National Trust/David Cousins, CMS_STO05364

Fig 5.6 (below right) Public domain sourced and © Rijksmuseum; https://www.rijksmuseum.nl/en/search?q=rp-p-ob-10.350&ii=0&p=1

Fig 5.7 © National Trust, NT1284879

Fig 5.8 © Country Life Picture Library, CL519272

Fig 5.9 From *The works in architecture of Robert and James Adam*, vol III (1822), Plate XX

Fig 5.1O (below) Reproduced by kind permission of the Earl and Countess of Derby © The Right Hon. Earl of Derby/Bridgman Images, KNW204107

Fig 5.1O (left) Reproduced by kind permission of the Earl and Countess of Derby © The Right Hon. Earl of Derby/Bridgman Images, KNW204108

Fig 5.11 © Historic England Archive, FF98/00336

Fig 5.12 © Fitzwilliam Museum, Cambridge, 3607

Fig 5.13 © Kent Wang, self photographed. Taken from Creative Commons.org/licenses/by-sa/2.0. Taken from https://www.flickr.com/photos/kentwang/33741948060/

Fig 5.14 Reproduced courtesy of Miranda Hausberg, photo by Melanie Radawiec

Chapter 6

Fig 6.1 Photo and © Richard Ireland

Fig 6.2 Photo and © Richard Ireland

Fig 6.2 Photo and © Richard Ireland

Fig 6.3 Photo and © Conor Lucey

Fig 6.4 Courtesy Yale Center for British Art, Paul Mellon Collection B1975.2.791

Fig 6.5 © Sir John Soane's Museum: Adam Drawings, vol 50/057

Fig 6.6 Courtesy Yale Center for British Art, Paul Mellon Collection B1975.2.791

Fig 6.7 Courtesy Yale Center for British Art, Paul Mellon Collection, B1975.2.790

Fig 6.8 Courtesy Yale Center for British Art, Paul Mellon Collection, B1975.2.787

Fig 6.9 © Sir John Soane's Museum, Adam Drawings, vol 12/097

Fig 6.1O Courtesy Yale Center for British Art, Paul Mellon Collection B1975.2.803

Fig 6.11 Courtesy Yale Center for British Art, Paul Mellon Collection B1975.2.235

Fig 6.12 Courtesy Yale Center for British Art, Paul Mellon Collection, B1975.2.795

Chapter 7

Fig 7.1 (above right) © RIBA Drawing & Archive Collections, RIBA 98522, folio 17

Fig 7.1 (above) © RIBA Collections, Drawing Collections, RIBA 98521, folio 19

Fig 7.2 Courtesy of The Lewis Walpole Library, Yale University

Fig 7.3 © Sir John Soane's Museum, Adam Drawings, vol 56/036

Fig 7.4 Courtesy of the Yale Center for British Art, Paul Mellon Collection, L211.33, Folio A

Fig 7.5 © Sir John Soane's Museum, Adam Drawings, vol 55/018

Fig 7.6 © Sir John Soane's Museum, Adam Drawings, vol 50/016

Fig 7.7 Courtesy of the Yale Center for British Art, Paul Mellon Collection; plate LI, Folio A N 63 Copy 1

Fig 7.8 © Sir John Soane's Museum, Adam Drawings, vol 50/015

Fig 7.9 © National Trust Images/James Dobson, NT 1302528

Fig 7.1O (top) © Sir John Soane's Museum, Adam Drawings, vol 50/019

Fig 7.10 © Sir John Soane's Museum, Adam Drawings, vol 50/018

Fig 7.11 © Sir John Soane's Museum, Adam Drawings, vol 22/228

Fig 7.11 © Sir John Soane's Museum, Adam Drawings, vol 22/227

Fig 7.11 © Sir John Soane's Museum, Adam Drawings, vol 18/601

Fig 7.11 © Sir John Soane's Museum: Adam Drawings, vol 22/229

Fig 7.12 Courtesy of the Lewis Walpole Library, Yale University

Fig 7.13 Courtesy of the Lewis Walpole Library, Yale University

Fig 7.14 © Sir John Soane's Museum, Adam Drawings, vol 24/214

Fig 7.15 © Sir John Soane's Museum, Adam Drawings, vol 22/050

Fig 7.16 © Sir John Soane's Museum, Adam Drawings, vol 17/202

Fig 7.17 © Sir John Soane's Museum, Adam Drawings, vol 11/030

Fig 7.18 (right) © Sir John Soane's Museum, Adam Drawings, vol 19/156

Fig 7.18 (far right) Images of England © Mr Peter Oakley Clarke, IoE236396

Chapter 8

Fig 8.1 © Sir John Soane's Museum, Adam vol 30/074

Fig 8.2a Plans devised by David King

Fig 8.2b Plans devised by David King

Fig 8.2c Plans devised by David King

Fig 8.2d Plans devised by David King

Fig 8.3 © Sir John Soane's Museum, Adam vol 33/101

Fig 8.4 (top) © Sir John Soane's Museum, Adam vol 48/075

Fig 8.4 (bottom) © Sir John Soane's Museum, Adam vol 48/079

Fig 8.5 © Sir John Soane's Museum, Adam vol 48/083

Fig 8.6 From Neale's *Views of the Seats of Noblemen and Gentlemen*, 1818–23, Vol VI, Plate 25

Fig 8.7 (bottom) © Sir John Soane's Museum, Adam vol 34/064

Fig 8.7 (right) © Sir John Soane's Museum, Adam vol 34/066

Fig 8.8 (top) © Sir John Soane's Museum, Adam vol 46/152

Fig 8.8 (centre) © Sir John Soane's Museum, Adam vol 1/223

Fig 8.8 (bottom) © Sir John Soane's Museum, Adam vol 46/153

Fig 8.9a Plans devised by David King

Fig 8.9b Plans devised by David King

Fig 8.9c Plans devised by David King

Fig 8.9d Plans devised by David King

Fig 8.9e Plans devised by David King

Fig 8.9f Plans devised by David King

Fig 8.10 © Sir John Soane's Museum, Adam vol 31/033

Fig 8.11 (top) © Sir John Soane's Museum, Adam Drawings, vol 31/037

Fig 8.11 (bottom) © Sir John Soane's Museum: Adam Drawings, vol 31/035

Chapter 9

Fig 9.1 Drawing by Helen Jones

Fig 9.2 © The Trustees of the British Museum, Dept of Prints & Drawings, Crace Collection, vol XXIX, no 58, museum no 1880,1113.4519

Fig 9.3 Sir John Soane's Museum, Adam Drawings, vol 48/095

Fig 9.4 © Sir John Soane's Museum, Adam Drawings, vol 4/157

Fig 9.5 © Sir John Soane's Museum, Adam Drawings, vol 6/136

Fig 9.6 © Sir John Soane's Museum, Adam Drawings, vol 36/036

Fig 9.7 © Sir John Soane's Museum, Adam Drawings, vol 32/016

Fig 9.8 © Historic England Archive, DP185299

Fig 9.9 Taken from Public Domain licenced for resuse under Creative Commons Licence CC0 1.0: https://commons.wikimedia.org/wiki/File:Detail Constantine%27s Arch Round relief Rome 1.jpg

Fig 9.10 From Bernard de Montfaucon, L' antiquité expliquée et représentée en figures, vol V, 1719, Paris, Florentin Delaulne [et al], 1719–24 supplement III (1724), plate 59

Fig 9.10 From Bernard de Montfaucon, L' antiquité expliquée et représentée en figures, vol V, 1719, Paris, Florentin Delaulne [et al], 1719–24 supplement III (1724), plate 59

Fig 9.11 From *Vasi, candelabra, cippi*, vol II, Plate 79, 1778

Fig 9.12 © Historic England Archive, DP185269

Fig 9.13 Photo and © Colin Thom

Fig 9.14 © Historic England Archive, DP232928

Fig 9.15 (below) © Historic England Archive, DP185253

Fig 9.15 (bottom) © Historic England Archive, DP185304

Fig 9.16 (top) © Sir John Soane's Museum: Adam Drawings, vol 13/137

Fig 9.16 (bottom) © Sir John Soane's Museum: Adam Drawings, vol 13/139

Fig 9.17 © Historic England Archive, DP232932

Fig 9.18 © Sir John Soane's Museum, Adam Drawings, vol 53/060

Chapter 10

Fig 10.1 © National Gallery of Scotland, D444

Fig 10.2 Blair Adam Muniments, 418 © Keith Adam of Blair Adam

Fig 10.3 Reproduced with kind permission of Neale Smith

Fig 10.4 © Crown Copyright Historic Environment Scotland (HES), DP 062703

Fig 10.5 © National Gallery of Scotland, D467

Fig 10.6 (top) Clerk of Penicuik Muniments © Sir Robert Clerk of Penicuik

Fig 10.6 (bottom) Clerk of Penicuik Muniments © Sir Robert Clerk of Penicuik

Fig 10.7 © Sir John Soane's Museum, Adam Drawings, vol 37/001

Fig 10.8 © National Gallery of Scotland, D445

Fig 10.9 © National Gallery of Scotland, D454

Fig 10.10 Blair Adam Muniments 702 © Keith Adam of Blair Adam

Fig 10.11 © In a Private Collection, National Gallery of Scotland, H2430

Fig 10.12 © Keith Adam of Blair Adam, Blair Adam Muniments

Fig 10.13 Clerk of Penicuik Muniments © Sir Robert Clerk of Penicuik

Fig 10.14 © RIBA Drawing & Archive Collections, RIBA 34637

Chapter 11

Fig 11.1 (top) Trevor House, New York NY Mott B Schmidt © Peter Mauss/Esto. All rights reserved, 87M22 5015905599

Fig 11.1 (bottom) Trevor House, New York NY Mott B Schmidt © Peter Mauss/Esto. All rights reserved, 87M22 5520000001

Fig 11.2 Gift of Mrs. James Ward Thorne, 1941.1195. Chicago (IL), Art Institute of Chicago. © 2019. The Art Institute of Chicago / Art Resource, NY/ Scala, Florence, Ref A545097

Fig 11.3 Out of copyright

Fig 11.4 The Gift of Estate of Ogden Codman, 1951, Accession Number: 51.644.80(11) New York, Metropolitan Museum of Art. © 2019. Image copyright The Metropolitan Museum of Art/Art Resource/Scala, Florence

Fig 11.4 From *The Works in Architecture of Robert and James Adam*, Vol I, No V, Plate VII

Fig 11.5 The Thomas J. Watson Library, Metropolitan Museum of Art, New York, b1683771_076. Taken from Public Domain: http://libmma.contentdm.oclc.org/cdm/compoundobject/collection/p16028coll4/id/24752/rec/5

Fig 11.6 © Pach Brothers Glass Plate Negatives Collection, Monmouth County Historical Association, Freehold, NJ

Fig 11.7 © Photography Collection, Miriam and Ira D. Wallach Division of Art, Prints and Photographs, The New York Public Library, Astor, Lenox and Tilden Foundations

Fig 11.8 © Wurts Bros. (New York, N.Y.) / Museum of the City of New York. X2010.7.1.8419, MNY241974

Fig 11.9 © 2004 Chuck LaChiusa *see* http://www.buffaloah.com/a/bidwell/205/wc/

Fig 11.9 © 2004 Chuck LaChiusa *see* http://www.buffaloah.com/a/bidwell/205/wc/

Fig 11.10 Reproduced with kind permission of Lott + Barbar, Inc, restoration architects of the Lucas Theater and the photographers

Fig 11.10 Reproduced with kind permission of Lott + Barbar Inc, restoration architects of the Lucas Theater and the photographers

Fig 11.11 (left) Photo Whitney Cox © The Schubert Organization, NYC

Fig 11.11 (below) Photo Whitney Cox © The Schubert Organization, NYC

Fig 11.12 Photo and © Once Like a Spark Photography

Fig 11.13 (right) Reproduced with kind permission of Newell Nussbaumer/ Buffalo Rising

Fig 11.13 (far right) From *The Works in Architecture of Robert and James Adam*, vol II, No I, Plate III